System Reliability Analysis

The text covers both basic and advanced techniques based on state performance systems and binary systems. The chapters will highlight reliability prediction, series-parallel, and complex modeling. It presents a dynamic reliability analysis of safety-critical systems using Petri nets, and dynamic resource allocation modeling of software with patching. The text illustrates a semi-Markov analysis of systems with a Weibull interface.

This book

- discusses in a comprehensive manner the reliability-centered maintenance modeling of electric vehicle systems;
- covers the reliability modeling of multi-state systems under the product development stage, and the reliability assessment of a multi-state degraded system;
- examines the role of nature-inspired techniques in the reliability optimization of systems;
- explores the practical challenges and solutions for RAMS management of train control systems; and
- showcases the methodology for the assessment of multi-state system reliability of traction electric drives, including overload modes.

It is primarily written for graduate students and academic researchers in the fields of industrial engineering, systems engineering, manufacturing engineering, production engineering, mechanical engineering, and mathematics.

Advanced Research in Reliability and System Assurance Engineering

Series Editor: Mangey Ram, Professor, Graphic Era University, Uttarakhand, India

Applied Systems Analysis
Science and Art of Solving Real-Life Problems
F. P. Tarasenko

Stochastic Models in Reliability Engineering
Lirong Cui, Ilia Frenkel, and Anatoly Lisnianski

Predictive Analytics
Modeling and Optimization
Vijay Kumar and Mangey Ram

Design of Mechanical Systems Based on Statistics
A Guide to Improving Product Reliability
Seong-woo Woo

Social Networks
Modeling and Analysis
Niyati Aggrawal and Adarsh Anand

Operations Research
Methods, Techniques, and Advancements
Edited by Amit Kumar and Mangey Ram

Statistical Modeling of Reliability Structures and Industrial Processes
Edited by Ioannis S. Triantafyllou and Mangey Ram

Industrial Reliability and Safety Engineering
Applications and Practices
Edited by Dilbagh Panchal, Mangey Ram, Prasenjit Chatterjee, and Anish Kumar Sachdeva

Reliability and Maintenance Modeling with Optimization
Advances and Applications
Edited by Mitsutaka Kimura, Satoshi Mizutani, Mitsuhiro Imaizumi, and Kodo Ito

Intelligent Prognostics for Engineering Systems with Machine Learning Techniques
Edited by Gunjan Soni, Om Prakash Yadav, Gaurav Kumar Badhotiya, and Mangey Ram

For more information about this series, please visit: https://www.routledge.com/Advanced-Research-in-Reliability-and-System-Assurance-Engineering/book-series/CRCARRSAE

System Reliability Analysis

Transition from Binary to Multi-state Models

Edited by
Akshay Kumar, Mangey Ram,
Rajesh S. Prabhu Gaonkar,
and Yury Klochkov

CRC Press is an imprint of the
Taylor & Francis Group, an informa business

Designed cover image: Shutterstock

First edition published 2025
by CRC Press
2385 NW Executive Center Drive, Suite 320, Boca Raton FL 33431

and by CRC Press
4 Park Square, Milton Park, Abingdon, Oxon, OX14 4RN

CRC Press is an imprint of Taylor & Francis Group, LLC

ISBN: 978-1-032-55456-3 (hbk)
ISBN: 978-1-032-90116-9 (pbk)
ISBN: 978-1-003-54621-4 (ebk)

DOI: 10.1201/9781003546214

Typeset in Sabon
by KnowledgeWorks Global Ltd.

Contents

Preface

In recent years reliability analysis and its measure have wide area in sciences and technology. There are various methods for calculating reliability in the case of binary and multi-state systems such as engineering, industry, and information technology. There are two states in binary system: totally working or totally failed but in the context of multi-state system reliability depends on performance of elements. This book will be dedicated to discussing various binary systems and multi-state systems and their applications in different branches in engineering and sciences. Every chapter is covered with reliability analysis modeling through more specific techniques. This book can be used by research scholars and scientists in the field of science and engineering. The book not only aims to introduce to students to reliability analysis of binary and multi-state systems, applied mathematics and applications, etc. but it also aims to give the students a panoramic view of applications of applied mathematics in field of science and technology.

Feature

- Jump Diffusion Reliability Modeling Based on Different Scenarios for OSS Multi Upgradation.
- Models and Methods of Realization of Optimal Reliability of Water Supply System at Condensing Thermal Power Plant.
- Semi-Markov Analysis of Systems with Weibull Interface.
- Reliability Assessment of a Multistate Degraded System.
- System Reliability Analysis: Transition from Binary to Multi-state Models.
- Topology Optimization of Binary-State Network with use of Cumulative Updating of Network Reliability Bounds.
- Quantitative Analysis of Reliability and Cost Factors in Pulse Radar Systems.
- Methodology for the Assessment on Multi-State System Reliability of Traction Electric Drives Including Overload Modes.
- Reliability and Sensitivity Analysis of a Network Structure.

- Generic Model and Reliability Measures of Multi-State System Analysis with a Case Study from Industry.
- Reliability-Centered Maintenance (RCM) for Multistate Systems.
- Reliability Analysis of Multiple Cylinder System with MADM Techniques in Diesel Engine, Employed in 3 Out of 5 Subsystems.

About the editors

Akshay Kumar received the B.Sc. and M.Sc. degree in science from Chaudhary Charan Singh University, Meerut, India, in 2010 and 2012, and the Ph.D. degree major in mathematics and minor in computer science from G. B. Pant University of Agriculture and Technology, Pantnagar, India, in 2017. He is currently working as Assistant Professor in the Department of Mathematics, Graphic Era Hill University, Dehradun in India. He is currently teaching LLP and Probability and Statistics in B.Sc.(H) and taught Group Theory, Numerical Methods, Business Statistics, Discrete Mathematics, Engineering Mathematics-III and in PG level taught OR, Linear Algebra and Discrete Structure. He is a regular reviewer for more than 43 international journals. He has published 80 plus research papers in *Communications in Statistics – Simulation and Computation*, *Emerald*, *Inderscience*, *International Journal of Mathematical, Engineering and Management Sciences*, *IGI Global Publisher*, *MDPI*, *Cambridge Scientific Publishers*, *Springer Nature*, *Elsevier Multimedia Tools and Applications*, and many other national and international journals of repute and presented his works at national and international conferences. Also, he has published three books (edited) with international publishers. His fields of research are reliability theory, fuzzy reliability, signature reliability, and applied mathematics. He has been awarded recently "Outstanding Researcher Award" in 2019–2020 at Graphic Era Hill University, Dehradun, India and in 15th and 16th Uttarakhand State Science and Technology Congress 2020–2022 for his significant contribution in academics and research at Graphic Era Deemed to be University, Dehradun, India.

Mangey Ram received the Ph.D. degree major in mathematics and minor in computer science from G. B. Pant University of Agriculture and Technology, Pantnagar, India in 2008. He has been Faculty Member for around 15 years and has taught several core courses in pure and applied mathematics at undergraduate, postgraduate, and doctorate levels. He is currently *Research Professor* at Graphic Era (Deemed to be University), Dehradun, India. Before joining the Graphic Era, he was Deputy Manager (Probationary Officer) with Syndicate Bank for a short period. He is Editor-in-Chief of the *International Journal of Mathematical, Engineering* *and Management Sciences*, *Journal of Reliability and Statistical Studies*, *Journal of Graphic Era University*; Series Editor of six Book Series with reputed publishers and Guest Editor and Associate Editor with various journals. He has published 400 plus publications (journal articles/books/book chapters/conference articles) in many national and international journals and conferences. Also, he has published more than 60 books (authored/edited) with international publishers. His fields of research are reliability theory and applied mathematics. Dr. Ram is Senior Member of the IEEE, Senior Life Member of the Operational Research Society of India, the Society for Reliability Engineering, Quality and Operations Management in India, Indian Society of Industrial and Applied Mathematics. He has been a member of the organizing committee of a number of international and national conferences, seminars, and workshops. He has been conferred with the "*Young Scientist Award*" by the Uttarakhand State Council for Science and Technology, Dehradun, in 2009. He has been awarded the *Best Faculty Award* in 2011; the "Research Excellence Award" in 2015; "*Outstanding Researcher Award*" in 2018 for his significant contribution to academics and research at Graphic Era Deemed to be University, Dehradun, India. Also, he has received the "*Excellence in Research of the Year-2021 Award*" by the Honorable Chief Minister of Uttarakhand State, India, and the "Emerging Mathematician of Uttarakhand" state award by the Director, Uttarakhand Higher Education. Recently, he received the "*Distinguished Service Award-2023*" for the subject and nation development by Vijñāna Parishad of India.

Rajesh S. Prabhu Gaonkar is presently Professor on Deputation at the School of Mechanical Sciences, Indian Institute of Technology Goa (IIT Goa). He is B.E. (Mechanical Eng., 1993), M.E. (Industrial Eng., 1997), P.G.D.O.M. (I.G.N.O.U., 2000), Ph.D. (I.I.T. Bombay, 2007), and Research Fellow at National University of Singapore (2010–2011). He is Former Professor and Head of the Mechanical Engineering Department at Goa College of Engineering and Former Dean of Faculty of Engineering at the Goa University. He is Associate Editor of the International Journal of System Assurance Engineering and Management and Editorial Board Member of the *International Journal of Mathematical, Engineering and Management Sciences*. He is on reviewer panel of more than 20 international journals. Prof. Rajesh S. Prabhu Gaonkar is first recipient (awardee) of the Goa State Award for Meritorious Teacher in Technical Education for the year 2015. This is awarded in public recognition of valuable services to the community as a teacher of outstanding merit. He is recipient of prestigious Fellowship of the Indian Institution of Industrial Engineering (IIIE) for his immense contribution in the field of industrial engineering in the state of Goa and recipient of Dr. J. M. Mahajan Award for the year 2016–2017 by the Indian Institution of Industrial Engineering for contributing substantially to the spread/propagation of Industrial Engineering in Education i.e. in institutes and in university. He is also Fellow of the Institution of Engineers (India). Prof. Rajesh S. Prabhu Gaonkar was awarded "Diploma of Excellence" for the paper presentation at DQM International Reliability Summer School by D.Q.M. Research Center in Serbia in 2005. He has also bagged best paper award at four international conferences. He has published over 90 research papers in various national and international conferences and journals, co-authored a book on "Fuzzy Reliability – Concepts and Applications" (2007), and co-authored a monograph "Six-Sigma: A Key to Enterprise Excellence" (2008). He was involved in organization of about 30 international conferences in various capacities. He has total 30 years of teaching and research experience. He has four book chapters, three Indian Patents and two Australian Innovation Patents to his credit. He has successfully guided four Ph.D. students as on date. Prof. Rajesh S. Prabhu Gaonkar is Founder Chairman of Mechanical Engineering Students Association (MESA) formed to organize technical events in Mechanical Engineering and related fields at Goa College of Engineering and Founder Chairman of the Indian Institution of Industrial Engineering (IIIE) Goa Chapter. He conceived and implemented MESSERGY (Mechanical Engineering Students' Social and Education Responsibility and Green Initiates) voluntary sustainable project for Mechanical Engineering Students of Goa College of Engineering. His teaching areas include Industrial Engineering, Operations Research, Advanced Optimization, Reliability Engineering/Reliability-based Design, CAD/CAM, Maintenance Engineering and Management, Operations Management, System Modeling and Simulation. His research areas include Fuzzy Set Applications in Reliability and Maintenance Engineering, and Multi-Attribute Decision Making.

Yury Klochkov holds the post of Rector of Tyumen Industrial University. He has the degree of mechanical engineering from the Samara State Transport University in 2001, postgraduate studies majoring in 05.02.23 standardization and quality management from Samara State Aerospace University (National Research University) (SSAU) in 2006 and doctorate degree majoring in 05.02.23 standardization and quality management from Samara State Aerospace University (National Research University) (SSAU) 2012. He has achieved many awards, certificates, and acknowledgments in 2007 the Potanin Foundation for young teachers grant holder, in 2014 the certificate of little Russian Quality Leader, Amity University Professor Emeritus certificate in 2016. In 2018 Dr. Yury Klochkov was the St. Petersburg Government prize winner for excellence in higher education and secondary professional education and in 2019 he got the acknowledgment for dedicated work and high standard of professionalism by Peter the Great St. Petersburg Polytechnic University. He is also Guest Editor of many special issues of magazines like *Key Engineering Materials, Scientific.Net, Resources, MDPI, International Journal of System Assurance Engineering and Management, Springer, International Journal of Reliability, Quality* *and Safety Engineering*, *World Scientific*, and *International Journal of Mathematical, Engineering and Management Sciences* (*IJMEMS*). He is Visiting Professor in Amity University and University of Kragujevac conferences. His total number of published research articles is 193. His H-index in Scopus with 58 publications is 20 and in WoS with 45 publications is 15. His total number of teaching and learning publications is 16. He also was certified by the Academic Methodological Association and the Ministry of Education and Science of the Russian Federation of the most significant teaching and learning publications.

Contributors

A. Migov Denis
Institute of Computational Mathematics and Mathematical Geophysics SB RAS, Novosibirsk, Russia

A. Nechunaeva Kseniya
Institute of Computational Mathematics and Mathematical Geophysics SB RAS, Novosibirsk, Russia

Ali Quadri Sarfraz
Mechanical Engineering Department, MGM University, India

B Pawar Shrikrishna
Mechanical Engineering Department, MGM University, India

B. Yamuna
Department of Mathematics, Dayananda Sagar College of Engineering, Bangalore, Karnataka, India

B. Rane Santosh
Department of Mechanical Engineering, Sardar Patel College of Engineering, Govt. Aided Autonomous Institute affiliated to University of Mumbai Bhavan's Campus, India

Bisht Soni
Department of Mathematics, Eternal University, Baru Sahib, Himachal Pradesh, India

Bolvashenkov Igor
Professorship of Energy Conversion Technology, Technical University of Munich (TUM), Germany

Goyal Nupur
Department of Mathematics, Graphic Era Deemed to be University, India

Gupta Radha
Department of Mathematics, Dayananda Sagar College of Engineering, Bangalore, Karnataka, India

H. N. Suresh
Former head, Department of Automobile Engineering, Dayananda Sagar College of Engineering, Bangalore, India

Jadhav Varsha
Department of AI & DS, VIIT Pune, India

Janičić Milovanović Valentina
Independent Researcher, Banja Luka, Bosnia & Herzegovina

Kharola Shristi
Department of Mathematics, Graphic Era Deemed to be University, India

Kumar Akshay
Department of Mathematics, Graphic Era Hill University, India

Kumar Ashish
Department of Mathematics, Graphic Era Hill University, India

Kumar Singh Lalit
Artificial Intelligence Applications, NPCIL, Department of Atomic Energy Govt. of India

L. O'Neal Dennis
Baylor University

L. Shet Verenkar Vishwanath
Goa Shipyard Ltd., Goa, India

Lj. Branković Dejan
Department of Hydro and Thermal Engineering, University of Banja Luka, Faculty of Mechanical Engineering Banja Luka, Banja Luka, Bosnia & Herzegovina

V. Mariappan
Mechanical Engineering Department, Agnel Institute of Technology and Design, Assagao, Goa, India

Mebrahtu Gezae
Ethiopian Technical University

Meena Jaishree
Amity University, India

Mohammed Hassen Yimer
Ethiopian Technical University, Ethiopia

N. K. Geetha
Department of Mathematics, Dayananda Sagar College of Engineering, Bangalore, India

N. Milovanović Zdravko
Department of Hydro and Thermal Engineering, University of Banja Luka, Faculty of Mechanical Engineering Banja Luka, Banja Luka, Bosnia & Herzegovina

P. Pai Sunay
Institute of Maritime Studies, Vasco – Da-Gama, Goa, India

Pandey Riya
Department of Mathematics, Graphic Era Hill University, India

R. Dolas Dhananjay
Mechanical Engineering Department, MGM University, India

R. Potdar Prathamesh
Parul Institute of Technology, Parul University, India

Ram Mangey
Department of Mathematics & Computer Sciences, Graphic Era Deemed to be University, India

Rana Sarita
Department of Mathematics, Graphic Era Hill University, India

S. Prabhu Gaonkar Rajesh
School of Mechanical Sciences, Indian Institute of Technology Goa (IIT Goa) Farmagudi, Ponda, India

S. Rodionov Alexey
Institute of Computational Mathematics and Mathematical Geophysics SB RAS, Novosibirsk, Russia

Sakhardande Milind
Mechanical Engineering Department, Goa College of Engineering, Goa, India

Shivani
Department of Mathematics, Graphic Era Deemed to be University, India

Singh Bhandari Ashok
Department of Mathematics, School of Basic & Applied Sciences, Shri Guru Ram Rai University, Dehradun, India

Singh Pankaj
Department of Mathematics, Graphic Era Hill University, India

Singh Pooja
SIES Graduate School of Technology, India

A. Srividya
Retired Professor (Civil Engineering), IIT Bombay, India

Tamura Yoshinobu
Graduate School of Sciences and Technology for Innovation, Yamaguchi University, Japan

Vinod Gopika
Bhabha Atomic Research Centre (BARC), India

Woo Seongwoo
Faculty of Mech. Eng., Ethiopian Technical University, Addis Ababa, Ethiopia

Yamada Shigeru
Department of Social Management Engineering, Graduate School of Engineering, Tottori University, Koyama-cho, Tottori-shi, Tottori, Japan

Chapter 1

Optimization by jump-diffusion process modelling based on different scenarios for open-source software multi upgradation

Yoshinobu Tamura and Shigeru Yamada

1.1 INTRODUCTION

A lot of software reliability growth models have been structured by several researchers. In particular, many software reliability growth models based on the non-homogeneous Poisson process are practically used in the actual software system testing phase of software development. As the characteristic models, there are several stochastic differential equation models derived from the non-homogeneous Poisson process models [1–3].

Moreover, the optimum software version-upgradation problems based on the software reliability growth models are well known as the estimation of optimum release time. Then, there are the optimum version-upgradation problems based on the stochastic differential equation models. These optimum version-upgradation problems assume that the cost parameters are fixed by using the rates based on the standard values.

In the optimum version-upgradation problem, we focus on that the cost parameters are fixed by using the rates based on the standard values. Many open-source software [4] have been used in cloud computing [5–7] and the edge computing [8, 9]. Then, we assume the environment of edge computing. Also, there are several stochastic differential equation models [10–13]. We have proposed jump-diffusion process models in the past [14–16]. We propose the optimum version-upgradation problems derived from the proposed jump-diffusion process models in the past. In this chapter, we apply the cost functions with noises to the cost parameters.

1.2 EFFORT MODELLING BASED ON THE WIENER AND JUMP-DIFFUSION PROCESSES

We have proposed the jump-diffusion process model in the past [17]. We show the jump-diffusion process model as follows:

$$M_j(t) = \mathrm{O}\left[1 - \frac{1+p}{1+p\cdot\exp(-\delta t)}\exp\{-\delta t - c_w(t)\sigma\omega(t)\} - \sum_{k=1}^{K} c_j^k(t) \sum_{i=1}^{\rho_t(\lambda_k)} \log U_i^k\right]. \tag{1.1}$$

DOI: 10.1201/9781003546214-1

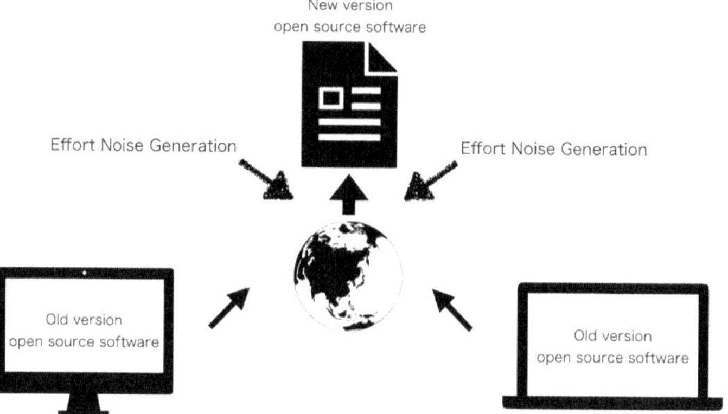

Figure 1.1 Noisy cases for the man-days effort estimation considering the OSS multi upgradation.

The above-mentioned model is based on the cyclic jump-diffusion process considering the different man-days effort consumption scenarios for open-source software (OSS) multi upgradation, where $t^k\left(k=1,2,\cdots,K\right)$ is kth specific time for the multi version-upgradation of edge OSS maintenance, and K is the number of multi version-up of edge OSS maintenance. Moreover, $c_i^k\left(t\right)$ is the cyclic functions obtained from the Fourier series expansion after kth jump. For example, the man-days effort of the development has the noisy case, when newly version open-source software is made by using many man-days effort such as Figure 1.1.

1.3 EFFORT OPTIMIZATION MODELS BASED ON THE JUMP-DIFFUSION PROCESS

Considering the multi version-upgradation, we apply the noisy cases to the man-days effort optimization models. In particular, we assume that the man-days effort parameters depend on the noisy cases.

This chapter discusses the man-days effort optimization with noisy case based on the existing optimal software version-up problems. Conventionally, the man-days effort parameters have been assumed as the given parameters based on the existing optimal software version-upgradation problem. We propose the man-days effort parameters with noisy cases in this chapter. Based on the existing optimal version-upgradation problems, we propose the optimal open-source software maintenance problem with noisy parameters considering the software man-days effort. Then, we assume the following man-days effort parameters:

r_1: the fixing man-days effort per fault under the edge open-source software maintenance,

r_2: the man-days effort per unit time under the edge open-source software maintenance,

r_3: the edge open-source software maintenance man-days effort per required man-days effort after the maintenance.

Then, the edge open-source software man-days effort in the maintenance phase can be formulated as follows:

$$E_1(t) = (r_1 + W(t))M_j(t) + (r_2 + W(t))t \tag{1.2}$$

where $W(t)$ is the Wiener process at time t. Also, the man-days effort of edge open-source software after the maintenance is represented as follows:

$$E_2(t) = (r_3 + W(t))M_j\{O - M_j(t)\}. \tag{1.3}$$

Consequently, from Eqs. (1.2) and (1.3), the total man-days effort of edge open-source software is given by

$$E(t) = E_1(t) + E_2(t). \tag{1.4}$$

The optimum maintenance time t^* of edge open-source software based on the man-days effort is obtained by minimizing $E(t)$ in Eq. (1.4).

Considering the noisy cases, we apply the Wiener process to the man-days effort parameters.

1.4 NUMERICAL EXAMPLES

This chapter discusses the numerical illustrations of the proposed edge open-source software optimum maintenance problem by using the actual man-days effort data of edge open-source software maintenance. The OpenStack Project [18] includes several edge components. Firstly, we show the conventional total optimal maintenance man-days effort without the noisy case in Figure 1.2.

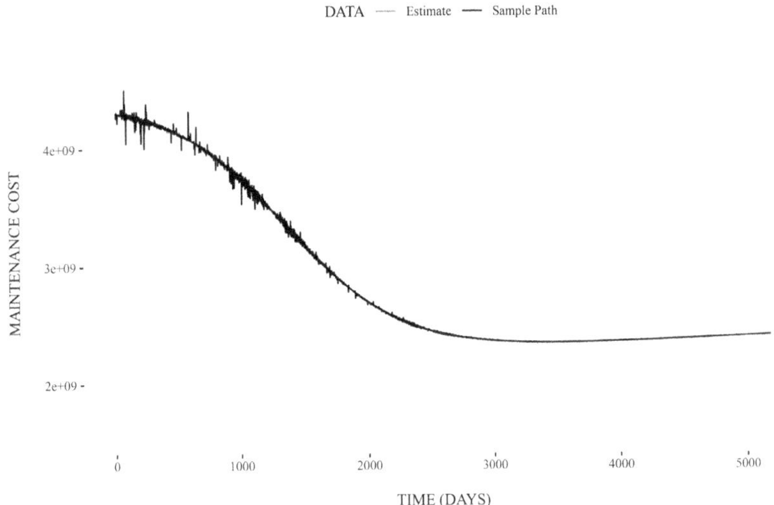

Figure 1.2 The estimated total man-days effort.

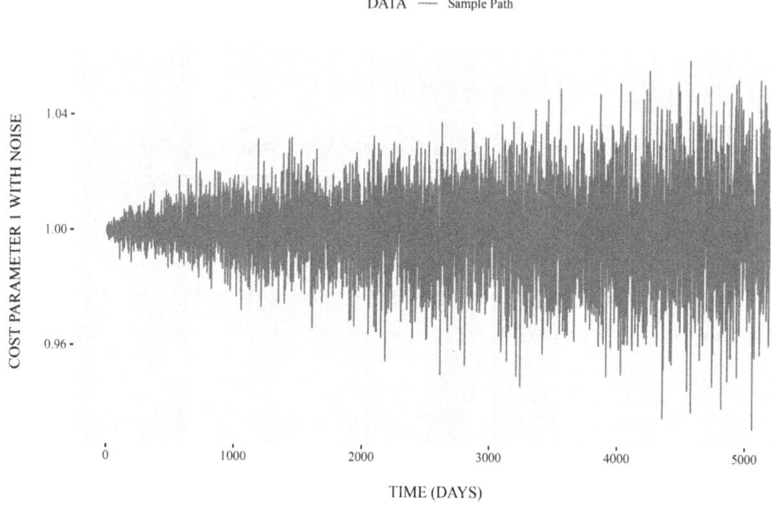

Figure 1.3 The cost parameter r_1 with Wiener process noise.

Considering the noisy cases, Figures 1.3–1.5 show the cost parameter r_1, r_2, and r_3 with Wiener process noise, respectively. Then, we show the estimation results based on the noises of the Wiener process in Eq. (1.4). Figures 1.6–1.8 show the estimated total man-days effort considering the noisy case in parameter r_1, r_2, and r_3, respectively.

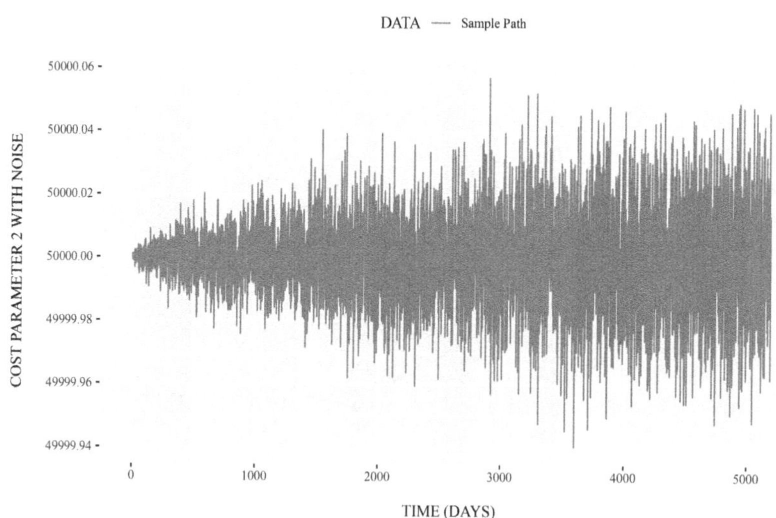

Figure 1.4 The cost parameter r_2 with Wiener process noise.

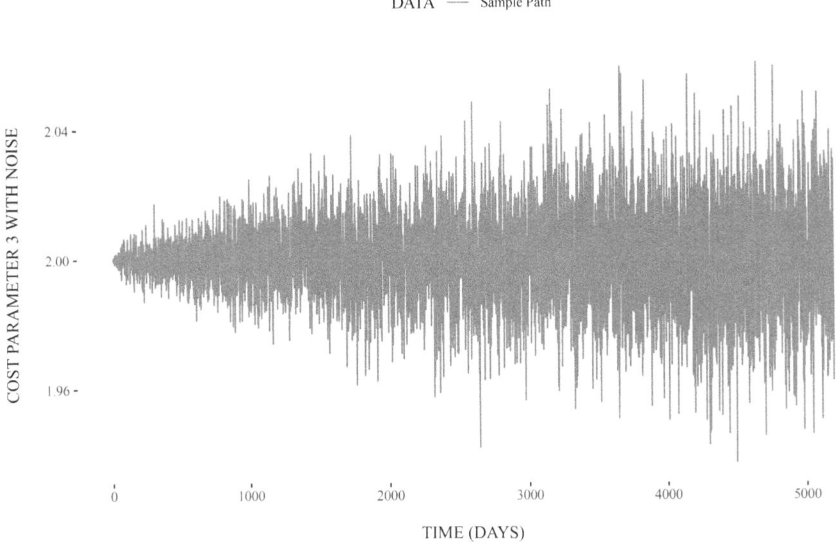

Figure 1.5 The cost parameter r_3 with Wiener process noise.

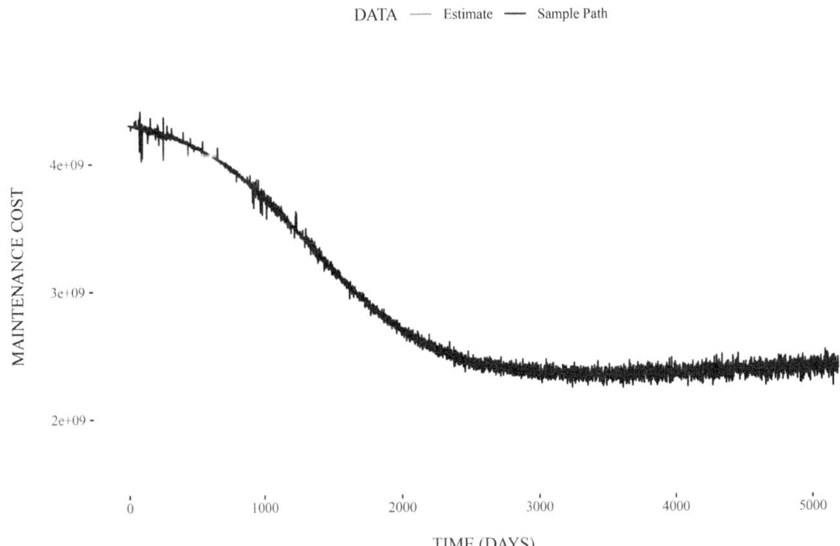

Figure 1.6 The estimated total man-days effort considering the noisy case in parameter r_1.

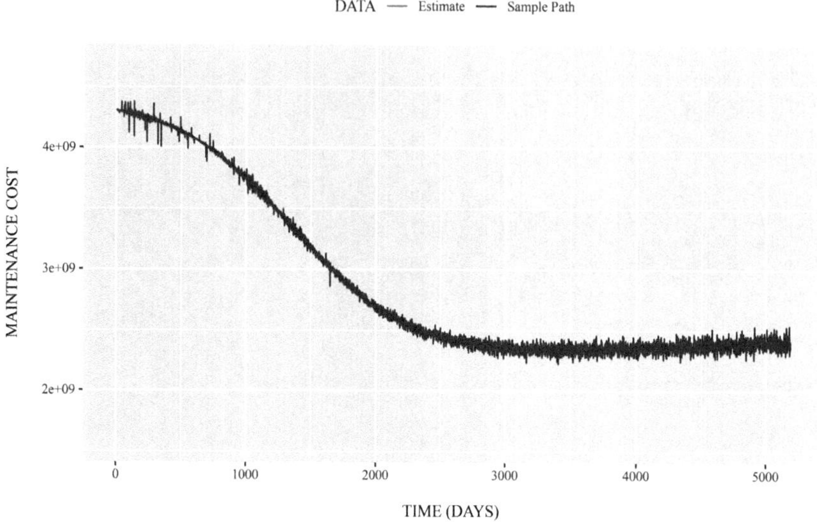

Figure 1.7 The estimated total man-days effort considering the noisy case in parameter r_2.

The noise of total maintenance man-days effort becomes small according to the maintenance time procedures go on. From the above-mentioned results, we have found that the noise of the proposed total maintenance man-days effort becomes large according to the maintenance time procedures go on.

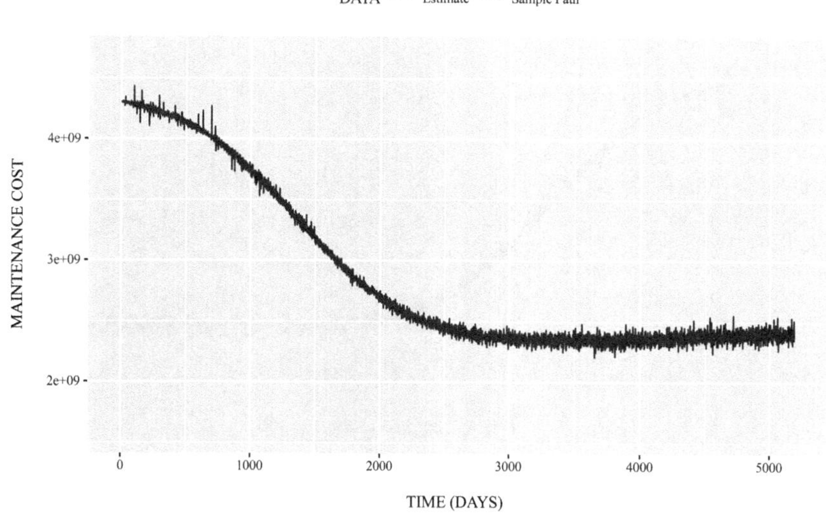

Figure 1.8 The estimated total man-days effort considering the noisy case in parameter r_3.

1.5 CONCLUSION

We have discussed the edge open-source software maintenance control with the man-days effort. Then, we have focused on the version-upgradation, the open-source software man-days effort data, and the cyclic man-days effort of network environment. In the past, we have discussed the cyclic multiple jump-diffusion process model in order to comprehend the characteristics of software man-days effort data, network environment, and open-source software, respectively [17]. The characteristics of the proposed man-days effort estimation method are as follows:

- The man-days effort parameters have the noisy case.
- The noisy cases are assumed as the Wiener processes.
- The each noise is the independent with each other.

The method of this chapter may be useful as the estimation of man-days effort optimization under the edge open-source software computing management.

ACKNOWLEDGEMENT

This work was supported in part by the JSPS KAKENHI Grant No. 23K11066 in Japan.

REFERENCES

1. M.R. Lyu, ed., *Handbook of Software Reliability Engineering*. IEEE Computer Society Press, Los Alamitos, CA, 1996.
2. S. Yamada, *Software Reliability Modeling: Fundamentals and Applications*. Springer-Verlag, Tokyo/Heidelberg, 2014
3. P.K. Kapur, H. Pham, A. Gupta, and P.C. Jha, *Software Reliability Assessment with OR Applications*, Springer-Verlag, London, 2011.
4. S. Yamada and Y. Tamura, *OSS Reliability Measurement and Assessment*. Springer International Publishing Switzerland, 2016.
5. I.M. Ibrahim, M.G.M. Mostafa, S.H.N. El-Din, R. Elgohary, and H. Faheem, "A robust generic multi-authority attributes management system for cloud storage services," *IEEE Transactions on Cloud Computing*, 2018, doi: 10.1109/TCC.2018.2867871.
6. A.A. Al-Said and P. Andras, "Scalability analysis comparisons of cloud-based software services," *Journal of Cloud Computing: Advances, Systems and Applications*, 2019, doi: 10.1186/s13677-019-0134-y.
7. M.O. Ozcan, F. Odaci, and I. Ari, "Remote debugging for containerized applications in edge computing environments," *Proceedings of 2019 IEEE International Conference on Edge Computing (EDGE)*, 2019, pp. 30–32, doi: 10.1109/EDGE.2019.00021.
8. P. Hu and W. Chen, "Software-defined edge computing (SDEC): Principles, open system architecture and challenges," *2019 IEEE SmartWorld, Ubiquitous*

Intelligence & Computing, Advanced & Trusted Computing, Scalable Computing & Communications, Cloud & Big Data Computing, Internet of People and Smart City Innovation (SmartWorld/SCALCOM/UIC/ATC/CBDCom/IOP/SCI), 2019, pp. 8–16, doi: 10.1109/SmartWorld-UIC-ATC-SCALCOM-IOP-SCI.2019.00047.

9. Y. Alsenani, G. Crosby, and T. Velasco, "SaRa: A stochastic model to estimate reliability of edge resources in volunteer cloud," *Proceedings of 2018 IEEE International Conference on Edge Computing (EDGE)*, 2018, pp. 121–124, doi: 10.1109/EDGE.2018.00024.
10. L. Arnold, *Stochastic Differential Equations-Theory and Applications*. John Wiley & Sons, 1974.
11. S. Yamada, M. Kimura, H. Tanaka, and S. Osaki, "Software reliability measurement and assessment with stochastic differential equations," *IEICE Transactions on Fundamentals*, Vol. E77-A, No. 1, pp. 109–116, 1994.
12. R.C. Merton, "Option pricing when underlying stock returns are discontinuous," *Journal of Financial Economics*, Vol. 3, No. 1–2, pp. 125–144, 1976.
13. P. Honor'e, "Pitfalls in Estimating Jump-Diffusion Models," Working Paper Series 18, University of Aarhus, School of Business, 1998.
14. Y. Tamura and S. Yamada, "Reliability analysis based on a jump diffusion model with two Wiener processes for cloud computing with big data," *Entropy*, Vol. 17, No. 7, pp. 4533–4546, 2015.
15. Y. Tamura, H. Sone, and S. Yamada, "Productivity assessment based on jump diffusion model considering the effort management for OSS project," *International Journal of Reliability, Quality and Safety Engineering*, Vol. 26, No. 5, World Scientific, pp. 1950022-1–1950022-22, 2019.
16. Y. Tamura and S. Yamada, "Maintenance effort management based on double jump diffusion model for OSS project," *Annals of Operations Research*, Springer US, pp. 1–16, 2019, doi: 10.1007/s10479-019-03170-w.
17. Y. Tamura, A. Anand, P.K. Kapur, and S. Yamada, Cyclic jump diffusion process modeling based on different effort consumption scenarios for OSS multi up-gradation. *Proceedings of the IEEE International Conference on Industrial Engineering and Engineering Management*, Malaysia, December 7–10, 2022, (Reliability and Maintenance Engineering 1).
18. The OpenStack project, OpenStack. 2024 [Online]. Available: http://www.openstack.org/

Chapter 2

Models and methods of realization of optimal reliability of water supply system at condensing thermal power plant

Zdravko Milovanović, Dejan Branković, and Valentina Janičić-Milovanović

2.1 INTRODUCTION

Expensive complex thermal energy systems, whose downtime causes large direct production (absence of production) and indirect non-production (purchase of electricity in the system as an alternative, failure of the transmission or distribution network within the power system as a result of falling from the state "in operation" to the state "in failure" of thermal energy facility, etc.) losses, require (apart from corrective) also the planned preventive maintenance activities. When it comes to complexity, it is reflected in a large number of components, subsystems, assemblies, subsystems, and system elements but also in their diversity (nominal power, different modes of operation, different roles within a higher hierarchical system, etc.). In addition, the complexity of the technical system according to the number of components that make it up differs from the complexity according to the type of components. The question of the feasibility of calculations as an adequate decomposition of multiple complex systems, especially in conditions of a small number of failures, becomes very, very significant. Each part of the system, which is of a different type, requires different parameters of the probability distribution of failure occurrence (if the model includes failure occurrences on these parts) or requires a different theoretical distribution of failure occurrence or even a whole new model. By additionally considering the requirement that the data be linked in time in the shortest possible intervals for the sake of calculation precision, new obstacles may appear for the correct definition of the reliability assessment model, which would adequately enable the execution of adequate calculations in a reasonable time frame during the working life of a complex technical system.

Special attention is paid to large periodical repairs – overhauls that are usually carried out once a year (ongoing overhaul lasting 30–45 days financed from regular operations) on a slightly smaller scale and once every four or five years (capital overhaul lasting 60–90 days, financed from additional credit debt or accumulated funds on the basis of depreciation). The appearance of repeated preventive maintenance activities of thermal energy systems can result in a loss of insight into the current state of the technical system and its reliability, and reliability assessment models must take this

DOI: 10.1201/9781003546214-2

into account. An additional difficulty is the failure to achieve the optimal level of reliability of the system during its operational life, as a result of repairs on only one part of the system (auxiliary technical systems, such as the water supply system at the thermal power plant). Those that are in reliability models most often include, in addition to the occurrence of changes in the level of reliability (increase, stagnation, decrease), the changes in the risk of failure, and/or the function that describes it as well. Also, issues related to the interaction between system components and defining relationships in terms of the consequences that the failure of one part of the system can cause on another part of the system in terms of reliability changes are also of particular importance. The number of available models that consider imperfect preventive maintenance actions, in the case when maintenance activities do not raise the level of reliability to 100%, is relatively small. Their integration with block diagrams applied to complex technical systems is even less. As an example, a model that connects the reliability of complex systems with a block diagram is often cited as the SSA (Split System Approach) model. Research related to interactions between parts of technical systems and their failures or reliability prediction using very little or no failure data is also very scarce and their application is very rare.

Reliability analysis based on established models for assessing the state and reliability of technical systems within a complex thermal power plant complex should contribute to the efficient and simpler determination of parameters for making relative decisions for the reliability of complex technical systems. Based on the parameters obtained in this way, the necessary activities related to timing are defined, making decisions about maintenance actions based on the required level of reliability (optimal moments of time for replacing/repairing parts of the technical system before its failure or the need to act correctively). Consequently, these models, by implementing adequate maintenance actions, serve to ensure the level of reliability of complex plants in thermal power plants (TPP). Whether it is new equipment or it has already been in operation, the maintenance service, based on the recommendations of the manufacturer and its own experience, must form an initial database, with basic technical characteristics and nominal (budget) conditions under which the equipment (machine, component, device, plant) is functioning. On the other hand, based on the planned operating modes of the plant as a whole, it is necessary to define the requirements that the equipment must fulfill. Technical systems, such as the water supply system at TPP, represent an organized set of elements, united by a common goal function – supplying the power plant with water. At the beginning of the design of the power plant, it is necessary to draw up the *water balance*, as one of the basic elements on the basis of which the determination is made for the selection of the macro location and finally the most optimal micro location of the facilities.

In modern water supply systems, it is necessary to use pumps, which are used to transport neutral or chemically aggressive fluids, clean or dirty, cold or hot, toxic and environmentally toxic fluids. The working fluid can be lubricant, emulsion, thermal oil, water (hot/warm/cold), acid and base solutions, and other chemicals. The pump is the main element in every industrial process. An improperly selected pump can be the cause of numerous problems in the operation of the system and the cause of production stoppages, as well as the cause of large losses. Centrifugal pumps are used more often than other types in the process industry, thermo technics, thermal power, and hydropower. They have a wide range of applications and are available in a wide range of dimensions and capacities. Their advantages are continuous flow, relatively simple construction, low purchase price, small dimensions, low maintenance costs and quiet operation. They are used to working in unfavorable working conditions; where they work for a long time, multiple shifts, and seasonally. Damage and failure of vital elements of machine systems can cause the failure of the entire machine. If repairs and replacements are not done in time, much greater damage to other machine parts can occur. Although, as a rule, the price of the bearing itself is relatively small, any damage to the bearing that reduces the functional correctness of the system or causes failure can cause large material costs.

2.2 OVERVIEW OF PREVIOUS RESEARCH

Numerous literatures have been published on the topic of maintenance technology and reliability models of technical systems. Starting from the first strategic decisions related to corrective maintenance of technical systems during the exploitation (period 1940–1960), then through preventive maintenance (period 1960–1975) up to the strategy of predictive maintenance, i.e., maintenance according to the condition, which belongs to the third generation of maintenance that dates back to the mid-1970s of the last century. Maintenance activities are based on the state of the system or its parts, whereby the maintenance goals become a higher level of plant reliability and availability, a higher level of operational safety, better product quality, longer equipment life, etc. Proactive maintenance represents the next generation of strategic maintenance characteristic of the end of the last century, with the defined goal of completely avoiding the occurrence of failure or reducing its consequences in the event of its occurrence. Self-maintenance or cloning for the purpose of renewing technical systems or "maintenance-free" maintenance (the use of disposable systems or devices whose favorable price enables the acquisition and replacement with new ones, with the recycling of discarded systems) are directions for strategic maintenance decisions for the period up to 2050.

Research activities in the field of maintenance and reliability of complex technical systems, as well as the considered water supply system as an auxiliary technological system within the thermal power plant, are grouped into several areas:

- Research activities on defining the reliability of complex machine systems (Table 2.1),
- Overview of reliability research on water supply systems at TPP (Table 2.2),

Table 2.1 Research activities on defining the reliability of complex machine systems

Authors	*Research area*	*A brief summary of the research*
Radionov, Evdokimov, Petukhova, Shokhina and Yabbarova (2016)	Application of vibrodiagnostics in condition-based maintenance	An example of the transition from the usual corrective maintenance method to the state-based maintenance system by introducing a vibrodiagnostic monitoring of the state of vibrations is given.
Mikić (2016)	Application of methods of analysis and synthesis, modeling using transformation and probability matrices	Confirmation that the final research results depend on the applied research methodology, on the analysis of existing methods for the development of reliability models of technical systems, on the comparative analysis of different theoretical models of the reliability of technical systems, as well as the development of reliability models of technical systems that best describe the experimental data.
Branković (2018)	Analysis of the effectiveness of a real production system	The dissertation proves the positive effect of increasing the reliability of the industrial plant for the production of hygienic paper after the installation and operation of the system for technical diagnostics of the critical positions of the technical system of the paper machine.
Milovanović and Branković (2021)	Maintainability of Industrial Systems	The monograph partly talks about the safety of the technical system, which can be viewed from two aspects: safety in relation to humans and the conditions in which the technical device is used, and safety of the equipment itself against permanent damage.
Milovanović et al. (2021)	Qualitative analysis	The analysis and monitoring of reliability level, as a specific project task, are conducted through all stages of the life cycle of a steam turbine plant. Analysis of the reliability of components or the entire technical system requires the application of models, which can be graphical or mathematical.
Branković, Milovanović and Papić (2021)	Analysis of the technical system reliability	Ensuring the conditions for a continuous process of production of a complex paper machine system enables the fulfillment of production plans and the survival of the complete business system in the conditions of market competition.

(Continued)

Table 2.1 (Continued)

Authors	*Research area*	*A brief summary of the research*
Milovanović et al. (2020)	Methods for Prognosis and Optimization of Energy Plants Efficiency	Within this chapter, appropriate methods will be provided for prognosis and optimizing the effectiveness based on the quality of design, production and testing, assembly and trial release, exploitation, development of procedures for prognosis of the complex systems behavior based on the characteristics of certain constituent elements of the system and the possible impact of human factors and the environment itself on the system.
Paunjorić (2016)	The influence of maintenance methods on the reliability of complex technical systems in surface mines	Analysis of maintenance methods and their impact on the reliability of complex machines in surface mines.
Milovanović, Branković and Janičić-Milovanović (2023)	Efficiency of condensing thermal power plant as a complex system	The objectives of reliability prediction, i.e., the process of determining numerical values for the ability of the structure to meet the required reliability requirements during certain stages of the life cycle, are feasibility assessment, comparison of possible solutions, identification of possible problems, supply and maintenance planning, datagaps, harmonization in cases of interdependence parameters, allocation of reliability, and measurement of progress in achieving the set reliability.
Branković, Milovanović and Janičić-Milovanović (2023)	Maintenance and safety of industrial systems	Achieving the projected degree of effectiveness of industrial plants is possible if the normal and safe functioning of all its components is ensured. The previously presented algorithm of the criticality assessments is a practical tool for defining the impact that parts of industrial systems have on safety, quality of operation, and safety for workers. In relation to the organization and management of maintenance processes, the part related to the degree of reliability and the impact of costs incurred in maintenance processes is particularly important.

Table 2.2 Overview of reliability research on water supply systems at thermal power plants

Authors	*Research area*	*A brief summary of the research*
Shina and Junb (2015)	Condition-based maintenance strategy	Condition-based maintenance strategy is currently the best in the field of preventive maintenance because maintenance decisions are made based on the current state of the system.
Asadzadehd, Salehib and Firoozic (2015)	Combined Markov simulation model	A Markov simulation model was used to evaluate the effectiveness of the model based on state diagnostics.
Liu, Liu, Cai, Zhang and Zheng (2015)	New opportunities in reliability prediction accuracy	Within this paper, the authors also used the Bayesian approach to describe failures with a common cause.
Belitser, Serra and Zanten (2015)	A combination of the Bayesian method with the Poisson process	In this paper, a combination of the Bayesian method with the Poisson process is given.
Lee and Sohn (2015)	Markov chain Monte Carlo (MCMC) simulation	There is a combination with as many as three models each, such as Bayesian, Markov chains, and the Monte Carlo method. Given route usage patterns by subway passengers based only on travel time data within a Bayesian framework using reversible jump Markov chain Monte Carlo simulation (MCMC).
Liu and Li (2016)	Complex-valued Bayesian parameter estimation via Markov chain Monte Carlo	The basis of the research is complexly evaluated Bayesian estimation of parameters via Markov chain Monte Carlo.
He, Gong, Xie, Zhang, Zhang and Hong (2016)	Hybrid models for system failure prediction	The perspective of the application of hybrid models in terms of scientific research (universality of solutions and accompanying problems at the level of applicability).
Yi, Bao, Jiang and Xue (2015)	Cascading failures and common cause failures	Research based on the modeling of cascading failures with the occurrence of errors related to trust in social networks.
Henneaux (2015)	The probability of cascading failures of technical systems	Analysis of the probability of cascading failures of overloaded lines.
Wu and Wu (2015)	Failures with a common cause of occurrence	An Extended Object-Oriented Petri Net Model for Mission Reliability Simulation of Repairable preventive maintenance system with Common Cause Failures.

- Overview of reliability research on pumps as a basic element of the water supply system for the power plant with the necessary water (Table 2.3),
- Review of research on other influential elements of the water supply system of the power plant with necessary water (pipelines with necessary fittings and measuring devices), Table 2.4.

Table 2.3 Overview of reliability research on pumps as a basic element of the water supply system for the power plant with the necessary water

Authors	*Research area*	*A brief summary of the research*
Ašonja (2012)	Measurement of diagnostic parameters of bearings on a test bench in the laboratory	The laboratory test table and machines for dynamic and static testing are primarily intended for testing the reliability of bearings, as well as testing the coupling joints in cars as well as other mechanical power transmissions.
Ašonja et al. (2012)	Measurement of diagnostic parameters of bearings on a test bench in the laboratory	By appropriate adaptation of the external elements, it is possible to place any power transmission on the test bench and perform the desired tests (diagnostic, tribological, degree of utilization, service life, etc.). On the already existing model of the test table, under certain conditions, ball bearings were tested, on which diagnostic measurements were made.
Mikić (2016)	Automated approach to diagnostics of mechanical technical systems and removal of influences accompanying natural processes	The scientific contribution of the dissertation is reflected in the establishment of cooperation that is of general interest between scientific research activities dealing with modeling, diagnostics, reliability, mechanics, mathematics, computing, engineering, telecommunications and social sciences. This dissertation should contribute to preventive, technical, diagnostic approaches to new technology, the use of diagnostic devices.
Ugechi, Ogbonnaya, Lilly, Ogaji and Probert (2009)	Application of the diagnostic model of vibration analysis for the case of centrifugal pumps of power from 15 to 300 kW	The authors provide the results of the analysis, which, among other things, showed a suitable value of the vibration level of 0.9–2.7 mm/s for the safe operation of the plant, while measures are necessary for vibration values of 2.8–7.0 mm/s continuous measurement of the state of the system as it moves into a state of increased risk. The described diagnostic model uses computer support within the model of artificial intelligence and neural networks and is applicable for other rotary elements that are within the indicated power ranges.

(Continued)

Table 2.3 (Continued)

Authors	*Research area*	*A brief summary of the research*
Milošević (2015)	Analysis of models for ensuring the reliability of complex plants in thermal power plants	Using exploitation research carried out in the company "Termoelektrane i kopovi Kostolac" d.o.o., Kostolac and the company "Termoelektrane Nikola Tesla" d.o.o., Obrenovac in the period from 2011 to 2014, the author in his doctoral dissertation provides an analysis of the model for ensuring the reliability of complex plants in thermal power plants.

Table 2.4 Review of research on other influential elements of the water supply system of the power plant with necessary water (pipelines with necessary fittings and measuring devices)

Authors	*Research area*	*A brief summary of the research*
Miličić and Milovanović (2010)	Energy machines: diagnosis and maintenance of steam turbines in thermal power plants	Monography called "Energy-generating machines – Steam turbines" aims at including all of the phases of life cycle as far as steam turbines are concerned. Besides this, problems regarding design and construction of modern energy-generating steam turbines have been looked at as a part of a higher hierarchy system, from the aspect of their purpose and conditions for their safe and secure exploitation, together with accomplishment of the required level, protection of personnel, as well as an environment where exploitation is performed.
Milovanović (2011b)	Designing, exploitation, and monitoring with diagnostics during exploitation and maintenance of thermal energy plants and equipment	A special part of Chapter 4 refers to the definition of the testing methodology and determination of the remaining service life of individual elements within the thermal power plant (steam boiler and high-pressure pipelines, turbine plant and other equipment). Repair activities at thermal power plants are also shown, with a special focus on the steam boiler and turbogenerator plant. The basics related to the part of the extended working life, which includes the implementation of activities related to the reconstruction, modernization, and revitalization of certain parts of thermal power plants (criteria and methodology, an inspection of damage according to CEN CWA 15740:2008, technical diagnostics, warranty, and normative tests, etc.) are also given.

(Continued)

Table 2.4 (Continued)

Authors	*Research area*	*A brief summary of the research*
Milovanović and Miličić (2012)	Operation and maintenance of steam turbines for cogeneration systems	Details on transient operating modes and the causes of turbine operating mode variability are given in Chapter 7, which deals with load control systems. Variable operating modes result in changing the values of other operating parameters (change in speed, reactivity, steam flow, stage efficiency, etc.), and turbine operation in such conditions represents the variable operating mode of a steam turbine. The complexity of the mentioned facilities also does not exclude the possibility of their failure and disruption of their operating modes, which can cause danger to the turbines. The result of the use of computer technology is the improvement of technical diagnostics of individual elements of the system or the turbine as a whole (especially important when introducing maintenance according to condition), better overall economy, and greater availability and safety in the exploitation of the turbine. The development of computer technology also enabled the accompanying development of expert systems for various purposes, with special participation in energy and process plants.
Concawe (2010)	Indicators that affect the leakage of pipelines of different purposes	Based on statistical data on pipeline leaks and the analysis of available literature, the following indicators that affect leaks can be defined: damage caused by the influence of a third party, corrosion, pipeline design, and construction, improper handling, additional factors that affect leaks.
Šimunović (2012)	Microbiological corrosion of welded joints of stainless steel in water	Microbiological corrosion as damage to steel or iron that is a direct or indirect result of the metabolic activity of microorganisms. It also appears in the literature as bacteriologically induced corrosion. In the doctoral thesis, among other things, cases of sudden and unexpected microbiologically induced damage to various plants are considered.

2.2.1 Research activities on defining the reliability of complex mechanical systems

The problem of studying the reliability of technical mechanical systems, as a complex multidisciplinary activity, requires fundamental long-term and extensive research, as well as accompanying experimental activities with the aim of determining and forecasting the state of the technical system as a whole or its most critical components (Mikić 2016). The general guidelines on which the research activities of the reliability of machine systems were carried out are presented through the following scientific research (Adamović and Ašonja 2014):

- Theoretical research was carried out in the sense of defining the appropriate forms of functions and laws of reliability theory, modified and adapted to the specific characteristics of the technical systems whose reliability is being investigated (both from the aspect of the construction of the observed system and in relation to the type and character of the malfunctions and the possibility of their removal),
- Theoretical research was carried out in the sense of defining appropriate forms of functions and laws of reliability theory in order to determine criteria for more precisely defining the actual impact of individual elements on the reliability of the system as a whole (determining critical units or elements in the system),
- Laboratory tests to quickly determine the reliability characteristics of the observed elements of machine systems in optimal laboratory conditions (analysis of existing methods and the introduction of appropriate improvements to them in order to more fully approach real conditions similar to those occurring in exploitation),
- Exploitative studies of operating parameters (operating modes of operation) such as number of revolutions, limit number of revolutions, temperature, lubrication, and external load have a great influence on the working characteristics of the bearing (they are carried out with the aim of understanding the basic laws of the occurrence of malfunctions and getting to know the actual operating modes, i.e., the actual laws of changing workloads, working conditions in exploitation and other relevant factors),
- Statistical processing of the research results was carried out according to the methods of reliability theory, i.e., statistical mathematics, in order to quantitatively determine the reliability of the tested systems, machine assemblies (reliability based on analogs and similarity theory), and
- Finding appropriate mathematical models (statistical structures), i.e., the occurrence of malfunctions that lead to certain failures of technical systems or their components.

In Table 2.1, an overview of relevant publications is given, with a presentation of the brief contents that they treat.

2.2.2 Review of reliability research on water supply systems at thermal power plants

The analysis of the impact of reliability and the consequences of failure, as well as other factors that affect the maintenance and production of the thermal power plant as an integral part of the power system, is necessary in order to achieve the goal of reducing maintenance costs and optimizing the maintenance strategy. Development of new technical solutions of certain basic (elements located in the main power facility: steam generator/boiler, turbine, and electric generator with associated equipment) and auxiliary systems (system for preparation and delivery of fuel, a system for shipping non-products – slag and ash from the power plant, system for supply and chemical preparation of water, condensation plant, a system for collection and purification of waste water, etc.) at TPP leads to the fact that condition and failure diagnostics are of increasing importance in maintenance and reliability research (maintenance according to condition and/or proactive maintenance). The application of continuous diagnostics is based on the use of available techniques such as thermal imaging, ultrasound, vibrodiagnostics, and acoustics. Failure cause and effect analysis is used to discover the cause of failure and the consequences for other subsystems that are directly dependent on the observed element or subsystem of the thermal power plant. If a system is multifunctional, different block diagrams can be created. This method is widely applicable and can be combined with many other techniques, e.g., with the FTA (Fault Tree Analysis) method. However, it can be said that the failure tree is more suitable for searching for the cause of failure and the reliability block diagram is more suitable for quantitative reliability analysis. When used for qualitative analysis, a reliability block diagram can be used to identify whether a system is in a functional state or not, under certain conditions. The structure function is binary. Using the basic principles of probability, several models have been developed for predicting the reliability of repairable systems. Models based on probabilistic principles have been developed to determine the most suitable preventive maintenance time according to a reliability function or a distribution function. The most commonly used distribution function is the Weibull distribution, and the normal and exponential distributions are also used as special cases of the Weibull distribution. It should be noted that significant efforts have recently been made to improve them through various combinations with other methods, which aim to increase their applicability. Probability principles are mostly used in the case when it comes to researching preventive and/or proactive time-based maintenance models for complex systems where failure data is rare or not precisely managed.

So far, the modeling of various failure mechanisms of complex technical systems, as well as their reliability, have shown that there are failures that cannot be subsumed under either independent or specified groups of failures in which there is interaction. Apart from the fact that failures can interact unidirectionally, possible options are for component failures to mutually interact, either in a positive or negative direction, as well as from one component (subassembly or assembly) to multiple components and vice versa. As a priority for realistic modeling of the reliability of such technical systems, the assessment of the probability of failure of components that are exposed to failure interaction is imposed. Effective models and techniques used for this quantitative analysis of the probability of failure universally developed do not yet exist or are inapplicable to certain technical systems. In the case of more modern technical systems, where state diagnostics can be used to assess the reliability of the system, the need for failure modeling ceases. On the other hand, an excessive number of inspections can cause large losses when the system has to be stopped as planned or due to the cost of inspections, which significantly affects the overall effectiveness of the system. Reliability models that were developed in the previous period mostly rely on the theory of probability, where they usually reflect very little experience from practice or realized experimental research. On the other hand, reliability models that include empiricism and experiment, despite the aspiration to get closer to real conditions, often have serious shortcomings. Thus, models that calculate system reliability after preventive activities during exploitation and maintenance become inadequate and inapplicable. Reliability assessment models of complex repairable systems often look at the system integrally, giving a simplified and imprecise assessment of system reliability (they often do not decompose the system as a whole and consider the reliability of individual system components). Models with continuous interactions between the components of complex technical systems have not yet been developed to the required extent in which the interaction between system component failures would be adequately modeled (existing models related to dependent failures consider mostly one-way failure effects). Therefore, there are still no adequate models that have been developed to assess system reliability based on a small number of failures or in the case where they do not exist. The theoretical aspect of modeling compared to the experimental approach of analysis can give good results only if it recognizes the advantages of the other and removes its own shortcomings as much as possible. Individually, without previously given directions of action, no approach gives good results and does not cover the entire working life of the considered technical system. The very starting point of reliability modeling in complex thermal energy systems is based on taking available knowledge based on an interdisciplinary approach, with their summarization into practically applicable modular units, governed by modes of operation and regulation within the power higher hierarchical system and requirements with a special emphasis on the applicability and expediency of the model. The accuracy of the reliability model still provides great opportunities for progress, which is of particular importance when it comes to reliability prediction as a basis for planning and making decisions about optimal maintenance actions according to the given (most often economic and safety) criteria. Reliability models related to complex TPP have not yet been developed as sufficiently accurate, and applicable and do not include important real factors, so it is necessary to continue work in these directions.

In Table 2.2, an overview of the relevant publications is given, with a presentation of the brief contents that they treat.

2.2.3 Review of reliability research on pumps as a basic element of the system for supplying the power plant with the necessary water

The system for supplying the power plant with the necessary water (pumping plant) is a technological system consisting of the basic components: a pump (most often centrifugal), a driving machine for the pump (most often an electric motor), a distribution plant with a transformer, suction and discharge pipelines with associated fittings (manometer, vacuum meter, valves, shutters, etc.), suction basket, filling pipeline (depending on the type of pump), connecting elements, etc. Figure 2.1 shows a pump plant with a horizontal centrifugal pump.

The basic function of this system is the provision of a given flow of water (in the general case of fluid) over time (the aim function of the pumping plant). The pumping plant system can be decomposed into three functional subsystems connected by functions representing their input and output quantities: the water pumping subsystem, the propulsion subsystem, and the power subsystem (Figure 2.2). Functional subsystems are connected by functions

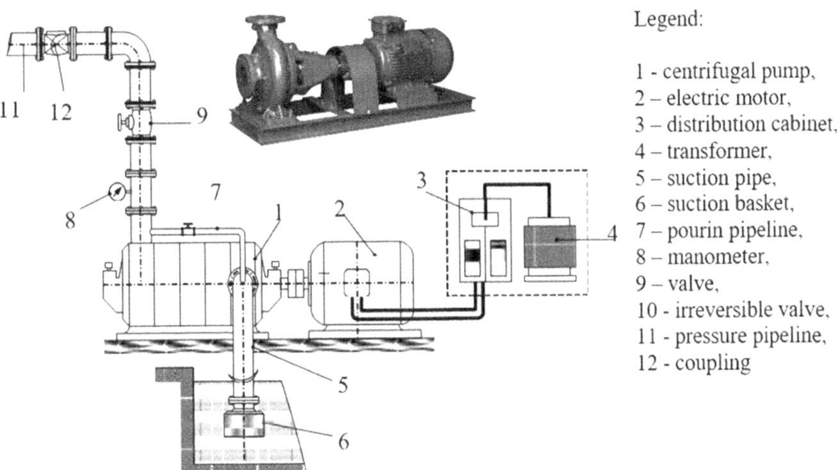

Figure 2.1 Pumping plant with a centrifugal horizontal pump.

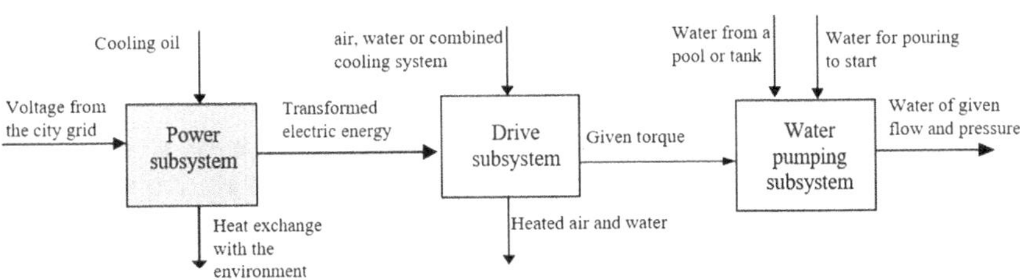

Figure 2.2 Functional block diagram of the water supply system of the pumping plant.

representing their input and output quantities (Petrović, Radičević, Vukičević and Bjelić 2009).

In order to achieve the pumping of water of the given flow and pressure, it is necessary to ensure the supply of water from the tank (pool) and the nominal torque on the shaft of the most commonly used horizontal centrifugal pump, which requires a full suction pipeline for the start because the pump can only deliver sufficient kinetic energy to the liquid to start and later continuously pump the working fluid. Most often, this function of filling with water for the start pump is performed using a filling pipe. The second functional subsystem is the drive, whose basic function is to generate the nominal torque required for the pump drive on the output shaft of the electric motor (three-phase cage asynchronous motors or SUS motors are most often used). In order for these motors to successfully perform their function, it is necessary to ensure their cooling, usually with air through a fan. Also, the conditions for unhindered air (or water) drainage used in the cooling function must be met. Pumping plants of this type are often of high power, so a satisfactory power supply must be provided, that is, a substation with a transformer that generates electricity of a certain strength and voltage. Each of the components in the system performs a specific function that needs to be preserved over time, in order for the entire system to function properly.

In addition to being used in water supply systems, centrifugal pumps are also widely used for transporting neutral or chemically aggressive fluids, clean or dirty, cold or hot, toxic and environmentally toxic fluids, etc. Changes in the mechanical structures in the deposit and the difficult operating conditions of the system in exploitation lead to a significant decrease in the reliability of this technical system. In addition, these defects also affect insufficient operational safety in the event of an increase in temperature, vibrations, and noise. Therefore, there is an understandable tendency to constantly monitor the state of the system and improve its characteristics such as flow, pressure head and utilization, and the like. In Table 2.3, an overview of relevant publications is given, with a presentation of the brief contents that they treat.

2.2.4 Review of research on other elements of the water supply system for the power plant with the necessary water (pipelines with the necessary fittings and measuring devices)

2.2.4.1 Pipelines as an element of the water supply system

The pipeline consists of a series of interconnected pipes with accompanying elements (compensators, appropriate fittings, measuring and control equipment), which ensures impermeability. Pipelines transport liquid, gaseous, pasty, and granular hard (solid) matters (substances). A pipeline network is a group of pipelines connected into a functional unit. Networks are widely represented: drinking water supply, waste water drainage (sewage), natural gas supply, central heating, etc. Pipeline construction is approached very cautiously, according to strictly defined regulations and standards. Their exploitation must be adapted so that, from the ecological aspect, their negative impact on the environment is minimal. When planning and designing trunk pipelines, many factors should be discussed and considered: the nature and volume of the liquid to be transported, the length of the pipeline, the type of terrain, environmental restrictions, proximity to settlements and roads, etc. In order to achieve the optimal construction of the pipe system, complex engineering studies are necessary to define the basic parameters: pipeline diameter, used material, pipeline routes, and required capacities of pumps that compress the transported fluid.

The flow of the transported medium (fluid) is established through the pipeline: by suction into the area of lower pressure (pump), due to the different heights of the entrance and exit from the pipeline, as well as by pushing with pumps or blowers. In the first two cases, the required flows and/or pressures are established by the suitable supply of energy (electrical). In addition to mechanical (strength) and thermal (heat exchange) requirements, as a rule, requirements related to the interaction of the pipeline with the transported medium and the environment (corrosion, corrosion products) should also be met. Devices (machines) of the production technique include pipelines for water supply/drainage (cooling), lubricants (lubrication), and their dispersions (fluids for lubrication/cooling in metal processing by separating chips). In the energy industry, energy is transported along with the medium (heated water, water vapor) through pipelines, and in the process technology, various materials necessary for the process to be supplied/removed – physical (e.g., distillation) and chemical (chemical reactors). With the use of a carrier medium (air, water), pasty substances (concrete, sludge) and granular solids (cereals, sand, cement) can be transported through pipelines. In the case of regulation systems (pneumatic with air and hydraulic with oil), pressure is transmitted from the unit to the actuators (executive bodies) through pipelines. Pipelines transport large quantities of fluid in a continuous flow. In order to achieve the optimal construction of the pipeline system, complex engineering studies are necessary to define the basic parameters: pipeline diameter, material used, pipeline route, and required capacities of pumps that compress the transported fluid.

The pipeline, as a unique structural unit, is achieved by interconnecting each individual pipe. Continuation of pipes (seamed and seamless) is carried out by applying: welding technology, as non-separable connections, as well as using flange connections and hermetic joints (for small diameter pipes), as separable connections. Cast pipes are produced (cast) together with flanges. Pipelines can be simple (rubber hose) and complex structures. Pipelines of complex construction consist of: pipe as a basic element, elements for connection and continuation, elements for sealing, elements for changing the direction (elbows) and compensating expansions, elements for regulating and closing the flow, safety elements, instruments of supporting elements (e.g., for condensate separation and pressure regulation) and pipe support elements. In order to meet the needs of changing the direction of the media flow, branching of the flow, changing the section, and closing the parts of the flow, there are special ready-made parts that enable the satisfaction of these needs. Pipe joints enable connecting a series of pipes into a pipeline, connecting pipes to devices, connecting to fittings, changing the diameter and direction of the pipeline, and closing the pipeline.

In the Republic of Srpska, as well as in Bosnia and Herzegovina as a whole, standards and related documents issued by the Institute for Standardization are marked with a symbol that begins with the abbreviation BAS. This chapter presents the calculation of the required wall thickness of straight pipes and pipe elbows according to the standard BAS EN 13480-3: 2017 – Industrial metal pipelines – Part 3: Design and calculation.

Pipe joints are assemblies that serve to interconnect pipes, compensators, and fittings into one continuous hermetic pipeline, through which the working fluid flows under a certain pressure. The mutual connection of pipes depends on the material of which the pipes are made, the technical requirements regarding the strength of the composition, and the severability of the connection. The connection between the pipe and the fitting should be easily separable so that the fitting can be separated from the pipeline in the event of failure, replaced, and repaired. According to their purpose, they can be inseparable (welded, soldered, glued) or separable. There are various constructions of detachable pipe joints, but the most important among them are flange joints. Other types are of greater importance only in special cases. Detachable connections include connecting pipes using: headers, pipe thread flanges, and hose fittings, using the so-called swivel or articulated connections. Inseparable connections can be made by: riveting, brazing, and welding. The pipeline, as a unique structural unit, is achieved by interconnecting each individual pipe. Continuation of pipes (seamed and seamless) is carried out by applying: welding technology, as non-separable connections, as well as using flange connections and hermetic joints (for small diameter pipes), as separable connections. Cast pipes are produced (cast) together with flanges.

By using free flanges, considerable savings can be achieved in special alloy steels or other expensive materials. In addition, the temperature field in free flanges is much more homogeneous than in rigidly connected flanges, and their expansion in the radial direction is unhindered. The ultimate consequence is that the thermal stresses in free flanges are lower than in rigidly connected ones. Connecting pipes using headers is used mainly for the continuation of cast steel, plastic, and ceramic pipes. Heads are used for lower pressures and where there is a risk of corrosion and therefore flanges cannot be used because the screws would fail in a short time. The place where the headers are joined is filled with melted bitumen (for ceramic pipes) or melted lead (for cast steel pipes). Connection using flanges is most often used for the continuation of steel pipes and connection with fittings. The connection using flanges is easily accessible, resistant to high pressures, hermetic, but expensive. The flange can be factory-made together with the pipe, rolled, welded, riveted, broken, or threaded connected to the pipe. Rolling, due to the more difficult execution, is used less often, and soldering and riveting are excluded for higher pressures. The connection using flanges is not suitable for aggressive environments, where the screws corrode quickly and are difficult to remove. Flanges are not suitable for use in confined spaces, where removal and adjustment are difficult. Free flanges are especially used for high pressures and temperatures because the pipeline can freely expand and contract independently of the flange, which significantly relieves the influence of stress due to heat. Connection using a pipe thread is used especially for connecting fittings and pipelines, as well as pipes of smaller diameter placed in inaccessible places. The ends of the pipes that are being assembled are supplied with pipe clamps, which are connected via a pipe nut. Hermeticity connection is achieved by direct fitting of the conical coil, or by means of soaked tow or a similar means. Hermeticity can be achieved by using special sealants. The hose extension is used exclusively for the extension of flexible lines, such as rubber and plastic hoses. If the connection needs to be assembled and disassembled quickly, then the so-called claw coupling. If the connection is not separated often, then metal extensions are used. Inseparable connections can be realized by: riveting (rarely used, mainly for pipes with a large diameter, e.g., pipes for conveying water to the turbine circuit), hard soldering (used for connecting pipes made of non-ferrous metals, for this purpose a pipe connector is used which is hard-soldered for pipes that continue), and by welding (this method is most often applied because it is fast, cheap and the joint is leak-proof and of sufficient strength).

2.2.4.2 *Pipeline armature as an element of the thermal power plant water supply system*

Armature means all devices whose main purpose is to control the flow and transport of the working material, as well as the protection of all systems in which the transport, flow, processing, and storage of the working material is carried out, which can be a fluid, a mixture of various fluid phases and a mixture of fluids and particles of solid materials. As a rule, fittings are always made of corrosion-resistant materials: copper and zinc alloy (cast or pressed brass), copper, zinc and tin alloy (red cast iron), stainless steel, and polymer materials.

The correct choice of fittings is an important and responsible component of system design because the occurrence of failure during the designed working life can cause great material damage and endanger the lives of personnel and the environment. For these reasons, the technical-technological data necessary for the correct selection of fittings are the purpose of the fittings and the way of management (regulation), properties of the working material, working pressure and temperature, installation and dimensional requirements, nominal diameter, type of connection and restrictions on dimensions and weight, and additional requirements related to safety and reliability in work and working life. When choosing reinforcement, one should strive to apply general-purpose reinforcement in accordance with the principles of general unification and standardization of products. Special-purpose fittings are used exclusively when general-purpose fittings cannot meet all technical and technological requirements, which ensure the permanent and reliable functioning of the system as a whole.

2.2.4.3 *Supports and carriers of pipelines and plants*

Pipeline supports are elements through which loads from the weight of the pipeline, reinforcement, insulation, transported (working) fluid, possible loads from snow and wind, as well as forces and moments caused by interference with the thermal expansion of the pipeline, effects of internal pressure, changes the amount of movement and friction of the fluid against the pipeline. According to the EN 13480-3 standard, support elements are defined as devices that connect the pipeline to the surrounding structure, where the supports should fulfill the following tasks: to carry the mass of the pipeline, including the equipment installed on it, to control and manage the movements of the pipeline and to guide and transfer loads from the pipe to the surrounding structure. According to the cause of their origin, the forces acting on the pipeline are divided into three groups:

a. I group: forces that arise as a result of the mechanical action of the transported (working) fluid on the pipeline (due to changes in the amount of movement, pressure, and friction),
b. II group: forces arising as a result of the thermal effect of the environment and/or the transported (working) fluid on the pipeline (forces due to interference with temperature expansions),
c. III group: forces resulting from external mechanical influences and weight loads (weight of pipelines, fittings, insulation, transported (working) fluid or fluid for testing pipelines and/or snow, as well as horizontal wind forces).

Depending on the restrictions they impose on the pipeline, the supports are divided into stationary and mobile supports. When it is necessary to accept weight loads with simultaneous vertical mobility, vertically movable supports will be applied, which can be of variable or constant load capacity. If the weight load is directly compensated by the force in the spring, the bearing capacity of the support changes depending on the stiffness of the spring, and proportionally by the vertical displacement of the support. These so-called ordinary spring supports can be: movable spring support with lateral guidance, planar spring support, stop spring point, and hanging spring support. According to the type of movement, movable supports can be divided into: sliding supports, in which the moving elements slide on stationary ones, rolling supports, in which the movement of the moving elements is realized as a result of elastic deformation (most often shearing) of a special elastic element, and pivoting supports, in which the desired movement made possible by turning (tilting) the supporting element: braces (hangers) or supports (legs). In the case of the last two types (elastic and especially pivoting) supports, unwanted vertical movements occur, which are proportional to horizontal movements. Constructive execution of supports depends on several factors:

- from the type of support (function intended for it),
- from the type of material, nominal diameter, and own stiffness of the cross-section of the pipeline,
- from the magnitude and characteristics of the load (forces and moments) that must be transferred to the supporting structure,
- from spatial (dimension) limitations regarding the location of the support,
- from the operating temperature and internal pressure of the pipeline,
- from the permitted heating or cooling of the roof structure through the support (which acts as a thermal bridge),
- from working conditions (increased corrosive effects of the environment, access for maintenance and tightening, control and/or lubrication, etc.),
- restrictions regarding assembly conditions (e.g., prohibited welding during assembly and/or disassembly).

In Table 2.4, an overview of the relevant publications is given, with a presentation of the brief contents that they treat.

2.3 BASIC THEORETICAL CONSIDERATIONS

2.3.1 Theoretical foundations of the reliability of complex technical systems

A decision support system is defined as a set of mathematical models and analyses that use available data and create information and knowledge from them that are useful in the process of making more or less important decisions at various hierarchical levels. Those decisions can have a long-term or short-term effect, and may (or may not) involve personnel. Most decisions are made using a simple and intuitive method, based on the subjective predispositions of the decision maker (experience, knowledge, skills) and available information. As the decision-making processes in energy and process technology are often complex and dynamic, therefore not applicable for assessments and decisions made in an intuitive way, a stricter approach based on analytical methods and mathematical models is used.

The main role of a decision support system is to provide tools and methods that enable effective and timely decisions to be made. At the same time, the application of strict analytical methods enables decision makers to rely on more reliable information and knowledge. The result is the making of better decisions and action plans that enable more effective achievement of goals. As analytical methods require an explicit description of the criteria for evaluating alternatives and mechanisms that regulate the problem, an understanding of the basic and the logic of the decision-making process. When decision makers are faced with problems or possible directions for the development of certain events, it is necessary to find answers to the questions raised through activities related to the realization of an appropriate analysis. Usually, several options are examined and compared, in order to choose the best decision given the conditions and constraints considered. By relying on a decision support system, decision makers can expect additional improvements in the overall quality of the decision-making process. With the help of mathematical models and algorithms, through the analysis of a large number of alternative actions, better conclusions and more efficient and timely decisions are reached. Increasing the efficiency of the decision-making process is the main advantage derived from the adoption of a decision support system.

Data, even if collected and stored in a systematic way, cannot be used directly for decision-making, but must be processed with appropriate extraction tools and analytical methods, capable of transforming them into information and knowledge, which could be used for decision-making. Decision support is based on the choice of a classifier whose mode of operation is described by the most important parameters (Marsland 2014): the classification accuracy is shown by the percentage of correctly classified instances, although some errors may occur (error control is very important) and the speed of the classifier may in some cases be extremely important (a 90% accuracy classifier may be a better choice than a classifier that achieves 95% accuracy if it is much faster). In a rapidly changing environment, the ability to quickly learn classification rules is very important. The importance of data preparation, which includes the steps of data cleaning and reduction of the number of attributes, as a bottleneck of the entire process of discovering knowledge in data is considered the most important step in the process of discovering knowledge in data, and a good selection of attributes can significantly speed up the entire process of classification. Classification accuracy (the ability of the model to correctly determine the class belonging to new data) and the speed of attribute selection are used as evaluation criteria. To measure accuracy, the confusion matrix (for two classes given in Table 2.5) is used, which is a useful tool for analyzing the extent to which the results obtained by sorting the samples differ from the actual values (Očević 2015). If there are m classes, the confusion matrix is a table of size at least $m \times m$. The classifier gives good accuracy if most of the samples are on the diagonal and the values outside the diagonal are close to zero (0).

Maintaining the reliability of complex technical systems such as a condensing thermal power plant at a given level requires continuous monitoring and corrective activities during exploitation in the function of a given operating

Table 2.5 Confusion matrix and measures used in confusion matrices

Confusion matrix		*Assumed classification*	
		Negative	*Positive*
Actual classification	Negative	TN – number of correctly predicted negative outcomes	FP – number of incorrectly predicted positive outcomes
	Positive	FN – number of wrongly predicted negative outcomes	TP – number of correctly predicted positive outcomes
Measures used in the confusion matrix			
Term	*Definition*		*Formula*
Accuracy	Ratio of samples whose class is correctly predicted and the total number of samples		(TN + TP)/(TN + FP + FN + TP)
Recall	Measure of correctly predicted positive samples (true positive rate)		TP/(FN + TP)
False positive rate	Ratio of samples that are wrongly classified into the positive class and the total number of negative samples		FP/(FN + TP)
True negative rate or specificity	Correctly predicted negative samples		TN/(TN + FP)
False negative rate	Ratio of samples that were wrongly classified in the negative class to the total number of positive samples		FN/(FN + TP)
Precision	The ratio of correctly predicted positive samples to the total number of samples for which the positive class is predicted		TP/(FP + TP)

mode. In order to achieve their operational reliability in the process of system operation (operation without failure), it is necessary to provide the default nominal conditions in a certain period of time with the agreed norms of maintenance measures and accompanying overhaul activities (current and capital overhaul). The definition of reliability based only on the data on the parameters of the system elements and the mean time between failures (MTBFs) is not sufficient. It is completed by the mentioned exploitation factors. They enable obtaining a more complete characteristic of the reliability of the system: the qualification of the servicing personnel, the quality and quantity of the works carried out for servicing and maintaining its reliability, the availability of spare parts, the use of measurement and equipment, its timely overhaul and repair, the availability of technical description, instructions for the exploitation of the system and measuring accessories, load conditions and transport systems and recommendations on how to assemble and dismantle them.

System efficiency parameters, such as its investment value (price), mass, storage conditions, possible modernization of components and miniaturization of the structure, and improvement of failure detection methods, are taken in the case of reliability assessment of machine systems most often partially, which significantly affects the obtained results and assessment accuracy. In order to examine the influence of such a large number of factors that allow obtaining information for predicting the level of reliability and evaluating the effectiveness of the system in exploitation, system evaluation can only be done using a computer. This approach to analysis provides successful measures, as well as technical implementation of measures to increase the level of reliability of machine systems. Evaluation of the effectiveness of measures to increase the level of reliability of mechanical technical systems in the conditions of exploitation can be done on the basis of forecasting, simulation on a prototype, or results obtained on the basis of analogs, as well as static and dynamic modeling methods.

2.3.1.1 Quantitative reliability indicators and mathematical models of reliability

One of the basic properties of the reliability of a complex technical system such as a condensing thermal power plant is its failure-free operation, which is quantitatively evaluated by the probability of failure-free operation of the system as a whole during the interval of use, that is, the duration of exploitation. The most important indicators of the reliability of non-renewable facilities include, in addition to the probability of failure-free operation, the failure density function, the intensity of failure, and the mean time to failure. Observed through the time picture of the process of exploitation, temporary repair, or renewal of the thermal power plant, it is short compared to the time of its failure-free operation, which means that the coefficient of readiness does not significantly affect the level of effectiveness of the system. The effectiveness of such a system basically depends on the number of failures during a given period of exploitation, that is, on the failure-freeness expressed by the mean time of operation until failure. Prerequisites for achieving the previously mentioned characteristics are a particularly high level of maintenance convenience and the preventive nature of the thermal power plant maintenance process, which ensures the complete restoration of the working capacity of components and systems. The selection and analysis of the effectiveness criteria of a specific subsystem or component or the thermal power plant system as a whole is usually done based on the impact of the failure of a given component on safety during operation, and the overall consequences of events caused by its unreliability (impact on personnel and the surrounding environment). In the statistical definition, the failure probability represents an empirical function of the failure distribution. Thus, the events, which consist of the occurrence or non-occurrence of cancellations, in working time t, are opposed. With logistic regression, predictions for the dependent variable based on the independent variables can be used. Thus, the percentage of variance in the dependent variable, which is defined by the independent variable, is determined. Logistic regression is used to predict the probability of an event by fitting the data to a logistic curve. Logistic regression is an analysis in which one dependent variable is dichotomous, i.e., binary that can have two outcomes, and takes the values 0 or 1, and has at least one or more independent variables. It predicts the probability of the event, and the data is adapted to the logistic curve that has the shape of the letter S. On the other hand, based on the small number of failure occurrences, as rare events that can occur during a certain interval of exploitation of the water supply system at the thermal power plant and lead to a loss of the system's working capacity as a whole, only an approximate estimate of reliability parameters is possible. Determining the legality of the distribution of the probability of failure-free working time in the considered case is practically very difficult, which means that the two-sided assessment of the required probability in engineering calculations and assessments becomes of great importance. The relationship between the value of the time of use or the duration of exploitation of the water supply system with the MTBFs, which are characteristic of modern water supply systems of TPP, indicates that the lower and upper limits of the rating of the required function of the probability of failure-free operation form a very narrow range of possible values.

2.3.1.2 Reliability models of some of the elements of complex technical systems

For the solution of tasks based on the assessment of reliability and prediction of the working capacity of objects, it is necessary to have a mathematical model, which is represented by an analytical expression. The basic way to obtain the model consists of the tests carried out, the calculation of statistical estimates, and their approximation of analytical functions. In the period of normal exploitation, the intensity of failure decreases and practically remains constant, whereas failures are due to wear and tear of a random nature and appear suddenly, first of all, due to non-compliance with the conditions of exploitation, random load deviations, unfavorable external factors, etc. Namely, this period was agreed on the basis of the time of exploitation of the facilities. The increase in the intensity of failure refers to the period of aging of the facilities and is caused by the increase in the number of failures due to wear, aging, and other causes associated with long-term exploitation. For the purposes of researching the reliability of complex systems (condensation thermal power plant pumping stations), mathematical models based on the analysis of diagnostic quantities in the assemblies of these systems are presented. In the creation of the model, several based methods were applied. The ROCC model (Receiver Operating Characteristic Curve Model) was used, where the sizes of errors on the bearings of machine systems were defined and the scale of the considered states of the diagnosed systems was determined. A binary logistic regression model was also used, which indicates a statistical method that allows for the prediction and assessment of the possibility of the occurrence of a probability when a certain defect occurs on a rolling bearing. On the basis of the research results obtained in laboratory and operational samples of pumps, fans, and compressors, with the application of appropriate mathematical algorithms, adequate mathematical models for complex mechanical systems were formed, which presented the behavior during exploitation (condition assessment) and maintenance from the aspect of the occurrence of malfunctions.

A system model based on the use of the ROCC curve (Receiver Operating Characteristic Curve Model). Classifier testing is based on determining the ROC (Receiver Operating Characteristic) curve. The ROC curve is a diagram that graphically shows the quality of the classifier depending on the decision parameter, Figure 2.3. Originally, ROC analysis was developed during World War II with the aim of detecting enemy aircraft using radar and was later applied in other areas, in psychology, medicine, biology, economics, shape recognition, etc. The area under the ROC curve (AUC) is used as a quality assessment of the two classifiers, and it is considered that the classifier is better if the AUC value is higher. For a certain set (False Positive rate – FPR, True Positive rate – TPR) of values, a classifier with a lower AUC value has better properties, e.g., in the picture, classifier A for FPR > 0.6 also gives a higher

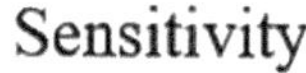

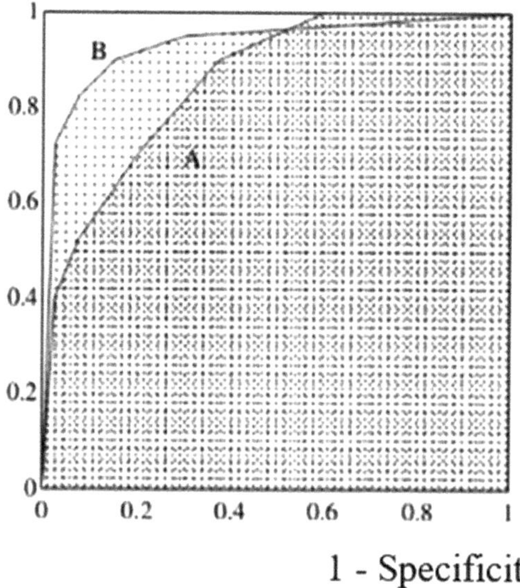

Figure 2.3 Example of ROC curve.

TPR value. Nevertheless, in practice, the AUC value gives a very good assessment of the quality of the classifier because the classifier with a larger AUC area contains the point that is closest to the coordinates (0,1). The ROC curve usually compares the results of all experiments and an additional best score scenario designed to validate and demonstrate the validity of the method for estimating the probability of adverse events. An example of using the ROCC model is most often used to identify errors in rolling and other types of bearings in complex machine systems. This method is unique in setting the tasks that the researcher exposed to the research should perform through conducting testing. In this way, researchers manifest their abilities, which are measurable through the use of certain techniques of determination (test) and measurement of the initial state and the situation created after the action of certain factors. In doing so, it is based on strictly established models of activity, which include certain checks (method of checking the ability of research, method of optimal operating modes during exploitation, method of optimal maintenance of the machine system, etc.). For example, by measuring vibrations and temperature on machine systems, it is possible to define the limits of permissible values for the field of application of bearings in exploitation (the limit that the bearings are in a pre-failure condition, which is still allowed, the tested bearings on the machine have no deformations, or are with small deformations, but in a condition that does not yet require their repair or replacement). Diagnostics can determine that the bearing is in a FAILURE state (unacceptable state) and that they have certain errors that cannot be repaired and represent scrap (exactly positive). After further testing, the bearings on the machine, which have no deformations, are in good condition (RUNNING condition), the so-called exactly negative fits). With further testing, the bearings on the machine have certain errors and these errors can be repaired, they are in an acceptable state (false negative). The test is, therefore, accurate to a certain extent (it correctly identifies damaged bearings and good bearings), and it is somewhat inaccurate (it wrongly classifies the good ones in the group of damaged ones, and vice versa). Sensitivity and specificity are measures related to their accuracy, precision, and predictive value.

ROCC curves represent a statistical technique that aims to determine the threshold value (limit value) of a certain test, in which there is the best ratio of sensitivity and specificity. The ROCC curve is a graphical representation of the ratio of sensitivity (*i*-axis) and "1-specificity" (*k*-axis). The final goal of a good test is to make its sensitivity as high as possible, and "1-specificity" as low as possible (the curve should be as close as possible to the upper left corner of the coordinate system). One of the goals of the analysis related to the ROCC curve is to examine whether it deviates statistically significantly from the reference line (that is, the one that passes through the middle of the graph). If the result is positive (it shows errors), we can say that a certain test has a statistically significant degree of diagnostic efficiency. More precisely, in the research it is important whether the area under the ROCC curve is statistically significantly different from the area under the reference line. The area under the ROCC curve is always a positive number, which, theoretically, ranges from 0 to 1, and can be understood as the sum of the test's discriminative accuracy.

A system model based on the use of logistic regression. Regression is an integral part of any data analysis that deals with looking for a relationship between dependent and independent variables. Its goal is to find a model that is best adapted to the data, that contains only those independent variables that have an impact on the outcome of the dependent variable and that well describes the relationship between the dependent and independent variables (Pešić 2016). Regression is used to indicate a statistical method that allows predicting and evaluating *one* variable based on the value of another variable or multiple variables. Logistic regression can be used to predict the dependent variable based on the independent variables, it determines the percentage of variance in the dependent variable that is explained by the independent variable. It is also used to rank the importance of independent variables, to evaluate the effects of interactions, as well as to determine the influence of the interval of controlling the independent variable. Logistic regression applies the maximum likelihood estimate after changing the dependent variable to a logistic variable (the natural logarithm of the likelihood of the independent variables appearing). In this way, logistic regression estimates the probability of the occurrence of a certain event. Logistic regression calculates changes in the logarithm of the probability of dependent variables, unlike regression (least value square), which analyzes changes in dependent variables. Using logistic regression requires serious and detailed knowledge of measurement scales, distributions, and interpretation of results.

Methods of ensuring the reliability of complex systems. To increase the reliability of complex technical systems in the conditions of exploitation, the activities can be divided into the following four groups: development of scientific methods of exploitation (include scientifically based methods carried out by technical maintenance, repairs and other activities to increase the reliability of complex technical systems in the process of their exploitation), collection of analyses and generalization of experience (determined by the type of product and the peculiarities of the exploitation of those products), the connection between design and production (the process of designing technical systems is reduced to the creation of complete technical documentation that defines their goal, gives the basic characteristics, composition and mutual connections between components and its functioning) and increasing the qualification of maintenance personnel (the performance of maintenance activities significantly depends on the availability, schedule and level of training of technical personnel for the maintenance of complex technical systems). A large part of the experience from the process of exploitation of complex technical systems is used to solve organizational and technical activities, as well as to create future designs with high reliability (methods of similarity or use of analog methods). Of great importance is the organization of gathering information about the dismissal. Possible sources of statistical data can be obtained from the results of various types of tests in exploitation, which are formed periodically in the form of reports on the technical correctness and reliability of the product. Studying the peculiarities of their behavior enables the possibility of using the accumulated data for the design of future products. In this way, the collection and generalization of data on product failures is one of the most important tasks, to which special attention should be paid.

2.3.2 Development of diagnostic methods for complex technical systems, their maintenance, and condition forecast

Checking the correctness, working capacity, and functionality of the technical system, along with locating the point of failure at the lowest hierarchical level, are the elements on the basis of which the assessment of the remaining life of use or the trend of occurrence of malfunctions is carried out. Significant economic effects and reduction of operating costs through timely detection of possible causes of failure of technical system components can be achieved through the application of technical diagnostic methods and tools. At the same time, diagnosis and definition of the cause can be realized during the exploitation of the system itself or within downtime and time for the overhaul of the plant and equipment, so the exploitation and overhaul of technical diagnostics are different, as integral elements of condition-based maintenance. A significant application of technical diagnostics is also in forecasting the short-term and long-term reliability of the technical system and its optimization, most often according to economic criteria (Milovanović 2011a). Technical diagnostics is a significant tool for increasing the reliability, economy, and safety of complex technical systems. The basic tasks of technical diagnostics of complex technical systems such as production systems, energy systems, are formulated as: forecasting and prevention of breakdowns, reduction of the number and duration of outages through forecasting, detection and monitoring of failure development, shortening the scope of planned and unplanned overhauls at the expense of improvement and application of technical diagnostic methods, prevention or elimination, in the process of exploitation of technical systems, of working conditions that are a generator of damage and failures, and computer-supported monitoring of working resources and the effectiveness of the production of mechanical technical systems. Measurement methods for determining the state of a technical system include a set of special procedures that define the relationships of some measured quantities. They can be absolute and relative, contact and non-contact, with or without material destruction, or differential and complex measurement methods. The absence of regulations on the permissible values of diagnostic criteria and the necessity for their use makes it difficult to formalize diagnostic procedures. Key factors, such as experience and professional knowledge of using diagnostic methods and equipment, significantly influence the comparability of results obtained by diagnostic methods on analog technical systems. The peculiarities of the operation of separate groups of complex systems in most cases do not allow the transfer of this experience to other objects. Under such conditions, it is especially important to form a single universal criterion assessment, and objective maintenance of the current state of the analyzed system.

2.3.3 Water supply systems at condensing thermal power plants

2.3.3.1 *Water balance of thermal energy plant*

The flow of cooling water for condensation of the produced steam is determined from the heat balance in the condenser. The produced steam of low humidity (%) usually enters the condenser, so the subcooling of the condensate in condensers of especially modern designs and with very low flow resistance on the steam side can be ignored. When drawing up the water balance of the power plant, it is necessary to consider the consumption of water for other purposes in the power plant. Table 2.6 gives approximate values of water consumption in a pure condensation power plant, where the flow of water for condensation is taken as 100%. In the case of a reverse system of hydraulic transport of slag and ash, water consumption can be reduced to 0.1–0.4% (depending on the consumption of coal, its ash content, the method of extracting ash from flue gases, and its internal and external transport). A similar distribution of water consumption given for the condensing thermal power plant is also valid for TPP, but only during the operation of their turbines in purely condensing mode of operation with nominal power. When operating in the heating mode (with removal of steam for heating, i.e., to the basic and peak network heater), the flow of steam into the condenser decreases, and correspondingly, the flow of cooling water into the condenser, while water consumption for other needs increases significantly (the share of water consumption for powering steam boilers is increasing, with the appearance of supplementary water consumption for powering the heating network) (Milovanović 2011a).

Table 2.6 Approximate values of water consumption in a pure condensation power plant

Indication of water consumption	*Water consumption, %*
Condensation of steam in the condenser at N_{nom}	100.0
Gas cooling of generators and large electric motors	2.5–4.0
Cooling of turbine oil and oil in turbocharging pumps	1.2–2.5
Cooling of bearings of auxiliary drive machines	0.7–1.0
Hydraulic transport of slag and ash	2.0–6.0
Additional water for the boilers of the condensing thermal power plant	0.04–0.1

When it comes to heating condensing turbines, full (nominal) electrical power is usually required when steam removals are turned off for external HCs (condensing mode of operation), so the flow of cooling water for their condensers defines their electrical power. In the case when condensing steam turbines are used to drive feed pumps (FP) or other auxiliary devices, it is necessary to add the cooling water flow for the auxiliary turbine condensers to the flow of cooling water for the main condenser of the steam turbine. The supply of cooling water is connected with the consumption of electricity, which is why we should strive to reduce it to a minimum (Milovanović 2003).

2.3.3.2 Water supply systems

The sources of water supply for the thermal power plant can be rivers, lakes, and seawater (in the case of its construction on the sea coast). The most commonly used water sources are rivers, with different cooling systems depending on the available average water flow and its change during the year (flow duration curve). Water supply systems can be performed as *flow (open)*, *reverse (closed)*, *mixed*, and *combined* (Milovanović 2011b). *The (open) cooling system*, in the case where there is a strong water source, the yield of which is significantly higher than the water consumption in the thermal power plant, provides the power plant with water directly from the river, lake, or sea (water intake), with water outflow downstream of the water intake, Figure 2.4.

Often, regardless of the fulfillment of the conditions regarding the abundance of the water flow of the river, the open system is not applied due to the lack of an adequate location of the power plant (possibility of flooding, extremely high banks, the existence of cultural and historical monuments, protection of plant and animal life, etc.). In the case where a lake is used as a source, it is necessary that it has sufficient size with flowing water, and

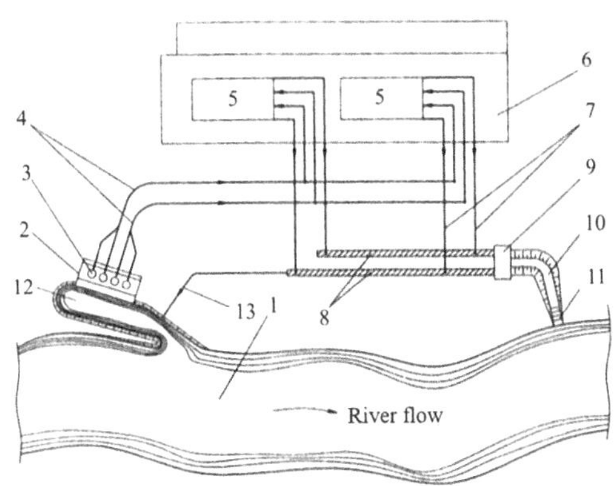

Legend:

1 - river,
2 – pumping station,
3 – circulating pumps,
4 – pressure pipelines,
5 – condensers,
6 – main power facility,
7 – drainage pipelines,
8 – cover drainage channels,
9 – a device for regulating the water level in covered drainage channels,
10 – open drainage channels,
11 - device for releasing water into the river,
12 - bay for taking water,
13 - pipeline for heating the water intake

Figure 2.4 Scheme of the flow (open) water supply system (Milovanović 2011b).

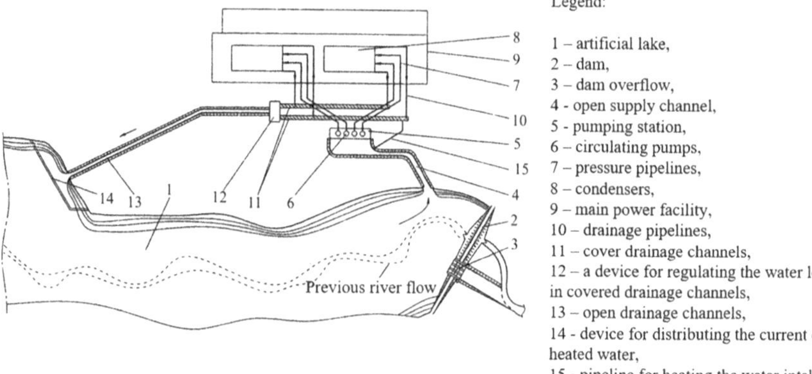

Figure 2.5 Scheme of a reverse (closed) water supply system with an accumulation lake (Milovanović 2011b).

when using seawater when the power plant is located on the sea coast, it is necessary to implement additional measures to protect equipment and pipelines from corrosion (salt water) (Milovanović et al 2017). This is especially important when choosing a capacitor, whose elements must be made of special alloys resistant to corrosion and with the application of special electrochemical protection. In the case when there is no river with sufficient flow near the planned micro-location of the power plant or it cannot be used, a *reverse (closed) water supply system* is applied. In such systems, the water is used more than once, and the absorbed heat is transferred by the water to special cooling devices (it circulates in a closed circuit). Water cooling in these cases is achieved in cooling lakes (Figure 2.5), in pools with water spray (Figure 2.6), or in cooling towers of different designs, Figure 2.7. Reverse systems require a smaller amount of water to cover the losses incurred in the process of its cooling in cooling devices. A *combined* water supply system is used when there is enough water in the river, but not for the application of a

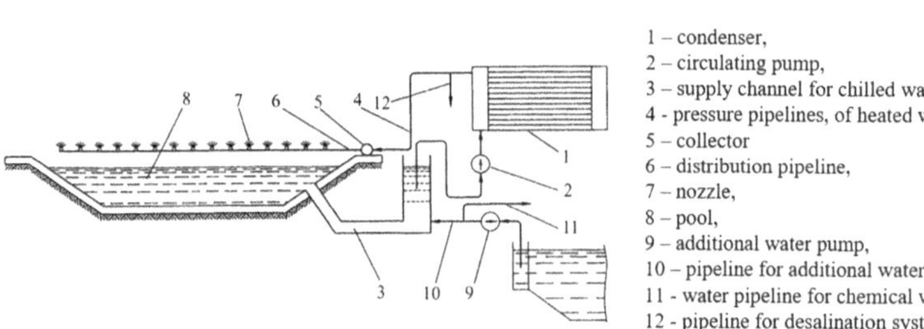

Figure 2.6 Scheme of a closed water supply system with sprinkler devices (Milovanović 2011b).

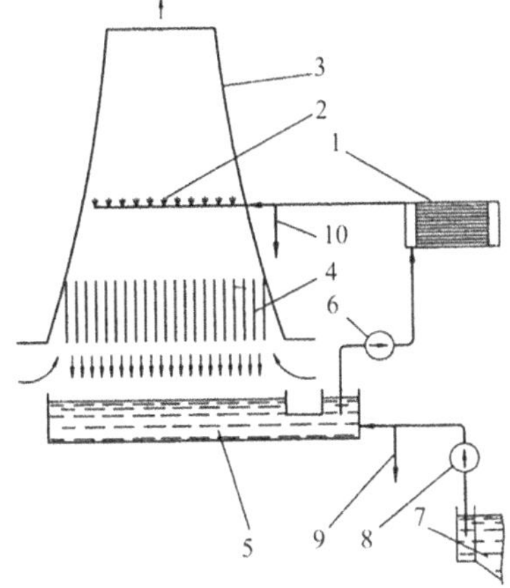

Legend:

1 – condenser,
2 – water distribution device with spray nozzles,
3 – cooling tower,
4 - sprinkler device,
5 – pool,
6 – cooling water circulation pump,
7 – accumulation lake,
8 - additional water pump,
9 - pipeline of additional water for the cycle,
10 - desalination pipeline

Figure 2.7 Scheme of a closed water supply system with a cooling tower.

flow water supply system, so a part of the heated water is mixed with colder river water. The combined water supply system is used in cases where only a part of the power plant's water needs can be provided with a flowing or mixed water supply system. Then it is necessary to combine a flow or mixed water supply system with a reverse system (cooling ponds, cooling towers, or pools with water spray). At the same time, it should be pointed out that spraying devices are very rarely used.

The initial temperature of the cooling water depends on the choice of the water supply system, and therefore the possible depth of the vacuum in the turbine condensers, as well as the cooling surface of the condenser (the price of the condenser). Related to this is the flow of cooling water through the condenser. Depending on the geographical position of the selected location of the power plant, that is, on the source of water supply, the initial temperature of the cooling water also changes for different water supply systems. For the conditions that were valid in the former Yugoslavia, the temperature of cooling water in flow systems was 8–14°C, while in reverse systems it was 12–16°C (reverse systems with cooling lakes) or 20–25°C (reverse systems with cooling towers). Approximate consumption of cooling water depending on the inlet temperature of cold water and pressure in the condenser is shown diagrammatically in Figure 2.8.

In the reverse water supply system with cooling towers, the pressure p_{con} also depends on the chosen cooling system, i.e., by the type of cooling tower (wet tower with natural draft, wet tower with forced or artificial draft, dry tower with natural draft and dry tower with artificial draft). It is necessary to design the water supply system considering the annual change in temperature and flow of the source, as well as possible daily and annual changes in the load of the power plant. For these reasons, it is necessary to carry out detailed climatological, topographical, hydrogeological, seismological, and other

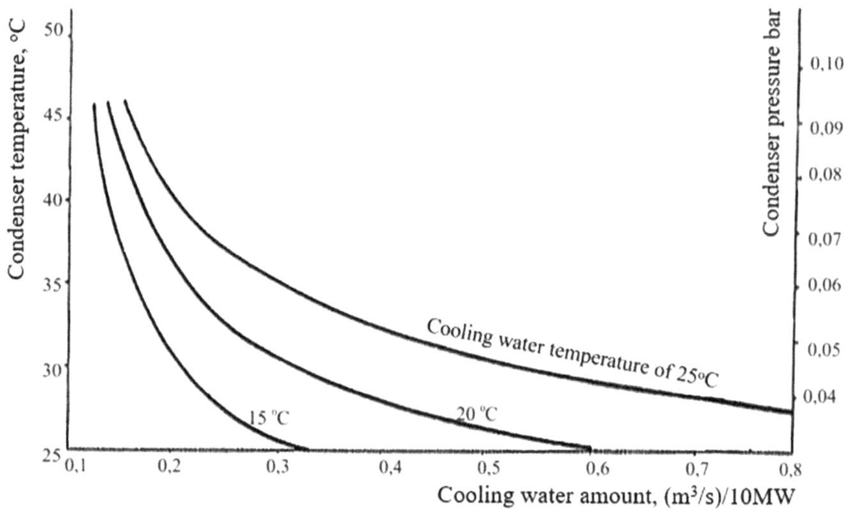

Figure 2.8 Required amount of cooling water $D_{cw} = f\left(t_{in}, p_{con}\right)$, depending on its temperature and required vacuum.

researches (quality of water, air, and soil in the observed macro location or micro location, etc.) before the design itself. The importance of achieving optimal pressure and temperature values can best be seen on the basis of Figure 2.8, which shows the dependence $p_{con} = f\left(t_{con}\right)$ for three different cooling systems (flow cooling and cooling on wet and dry cooling towers). The role of the cooling system in TPP is to transfer the unused heat in the steam cycle to the environment, preferably at the lowest possible temperature because the operating temperature of the cooling system is closely related to the steam condensation pressure (a lower condensation pressure means that the turbine can produce more useful work).

The diagram from Figure 2.9 shows the limiting temperature difference (underheating) of cooling water for three characteristic cooling systems, with marked intervals of limiting temperature differences and steam condensation lines, as a function of temperature and pressure with typical temperatures of the corresponding cooling system.

At the same time, the limiting temperature difference is a measure of the heat transfer efficiency in the condenser and expresses the dependence between the temperature of the steam and the temperature of the heated water at the outlet of the condenser, $\vartheta = t_{con} - t_{out}$. In flow-through systems, water is taken from a river or lake with temperatures that are usually lower than 25°C, then passing through a condenser it is heated to 8–15°C. With a limiting temperature difference of 2.5–5°C, the outlet condensation pressure is between 0.003 and 0.008 MPa. An artificial cooling lake is a closed system in which the cooling water is circulated and cooled. Since the cooling lakes are significantly smaller in capacity than flow systems, the transferred heat

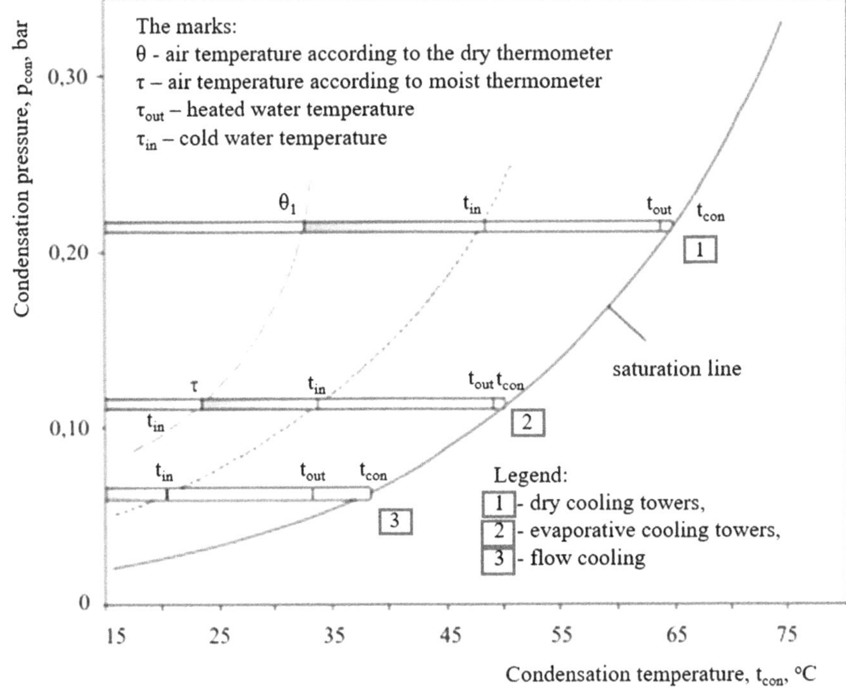

Figure 2.9 Display of dependencies $p_{con} = f\left(t_{con}\right)$ for three different cooling systems.

results in a significantly higher operating temperature, so the condensation pressure itself is somewhat higher (0.006–0.012 MPa). Evaporative cooling towers are also widely used in closed systems. In them, the largest amount of unused heat is transferred from the cooling water to the surrounding air by evaporation, so the limiting temperature in this cooling process is equal to the air temperature according to the wet thermometer.

The cooling limit, i.e., the difference between the temperature of the cooled water t_{in} and the air temperature according to the wet thermometer $\tau = t_{w.term}$, determines the extent to which the water cools in relation to the temperature (usual values of the cooling limit are 6–12°C). The maximum design values of the air temperature according to the wet thermometer for our meteorological conditions are 21–27°C and result in the highest condensation pressure of 0.007–0.015 MPa. Heat transfer in dry cooling towers is based solely on convection, so the air temperature according to the dry thermometer is the lowest temperature to which water can be cooled, that is, the condensation pressure is significantly higher and amounts to over 0.002–0.027 MPa. Therefore, when working with such high condensation pressures, the efficiency of the turbine is reduced, so when using dry cooling towers, this is one of the reasons for less economy in some areas, where the availability of cooling water is significantly lower.

2.3.3.3 Combined cooling system

The combined condenser cooling system is considered as a function of supplementing or increasing the cooling capacity of the basic cooling system (flow system or cooling lake). At the same time, in the function of auxiliary systems in combination with the basic cooling system, artificial lakes, pools with sprinklers, or cooling towers are used. The use of auxiliary cooling can also be a consequence of legal restrictions related to heat release or temperature rise in a lake or river, or as a result of implemented optimizations according to economic criteria (high temperature in the primary cooling system). When considering these systems, the following characteristic combinations are most often observed: combined system of flow and reverse cooling (Figures 2.10 and 2.11) and combined system for reverse cooling using dry and wet cooling towers (Figure 2.12).

Combined flow and reverse cooling system. In the case when additional cooling is necessary, a combined cooling system is applied or a special auxiliary system with reverse cooling is added, depending on the required investments, annual fuel costs, and the efficiency of the system itself. The improvement of the basic flow-through cooling system with the supplementary reverse cooling system with wet towers with artificial draft is shown in Figure 2.10. The increase in investment resulting from the investment in the supplementary system is compensated by increasing the degree of heat utilization, that is, by increasing the capacity of the cooling system and the power plant as a whole, especially in the case when the flow decreases significantly, and the temperature of the water in the river increases.

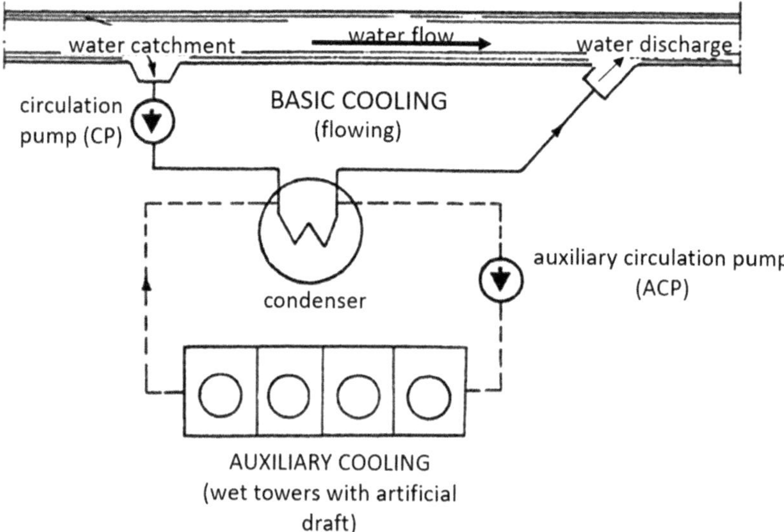

Figure 2.10 View of the combined system for auxiliary cooling: Flow (basic) and reverse (auxiliary) cooling.

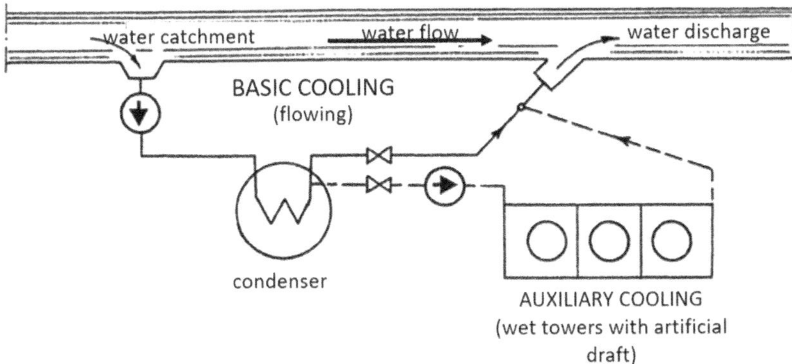

Figure 2.11 Combined cooling system with supplementary system to protect the river from excessive heating.

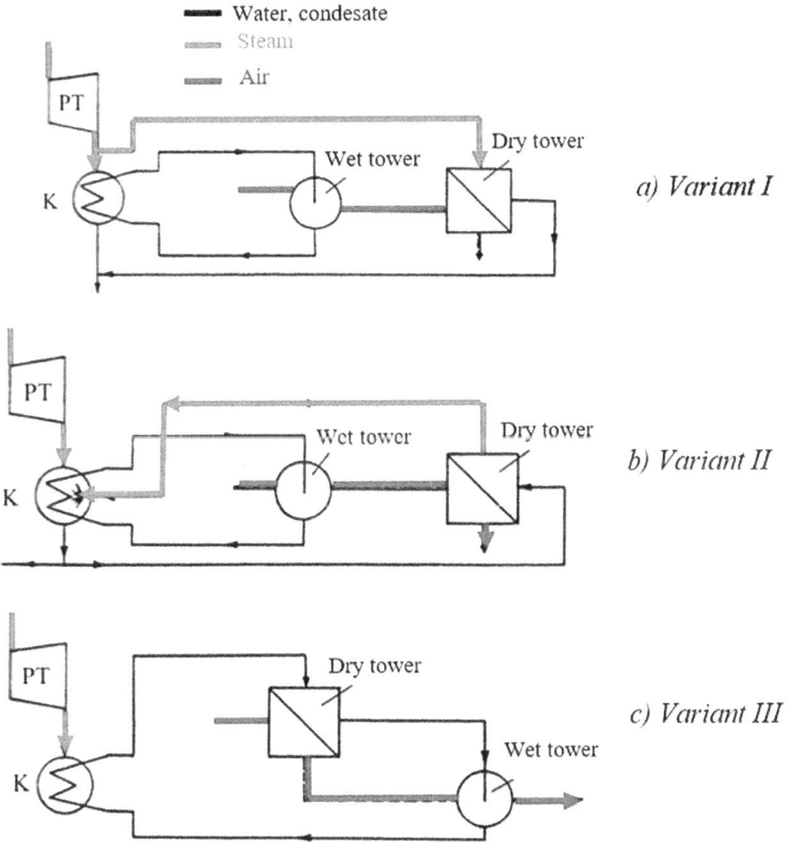

Figure 2.12 Combined system for reverse cooling (dry and wet cooling tower): a) Variant I, b) Variant II and c) Variant III.

High investments, required by the realization of a wet cooling tower with natural draft, represent a basic obstacle for their use in combined cooling systems for small- or medium-capacity power plants. This is the main reason why wet towers with forced draft (ventilator wet towers), splash basins, or smaller watercourses are most often used for additional cooling in a combined system. Additional cooling is also applied in cases of the need to protect the water flow from thermal overload, i.e., unauthorized heating. Additional cooling by sections of fan wet towers can be switched on before the condenser (thus achieving a better effect) or alternatively to cool the water before evaporating into the river, Figure 2.11. A combination of flow-through as a basic and reverse as an auxiliary or supplementary cooling system in the case where in the summer period, it cannot achieve flow cooling without limiting the power of the power plant, and it can be considered technically and economically justified in our conditions. Before any decision on the possibility of applying a reverse cooling system, it is necessary to consider the possibility of introducing combined cooling, considering specific location conditions.

Combined system for reverse cooling. In situations where there is not enough water to replenish losses in wet cooling towers, and it is necessary to improve the cooling effect of dry cooling towers, it is necessary to consider combined cooling in reverse systems. The combination of a dry and a wet cooling tower enables cooling at base load with a dry cooling tower, while a wet tower serves to cover the peak load. Figure 2.12 shows three switching schemes of such a combination. This combination will only be used if the cost of additional cooling water is high enough to justify the investment in this system. Therefore, for the choice of a combined system, the size of the "frozen" investments in the supplementary cooling system while it is not in operation and the costs of its operation while it is in operation are of primary influence, so for each specific case, complex technical and economic analyses of applicability and the fairness of such solutions. Each specific situation determines the most favorable water supply solution and power plant cooling system. It should always be borne in mind that the turbine, condenser, and cooling system are mutually dependent systems and determine the properties and price of the thermal power plant. For these reasons, the optimization of the choice of equipment and plant of the power plant is performed primarily for the turbine, condenser, and cooling system as a whole, which operates in the real energy and economic conditions of a higher hierarchical electrical energy system. The result of optimization of the integral system ("cold end") carried out in this way can provide a solution that will have high reliability in its operation and which will also be economical during the entire lifetime of the thermal power plant. When choosing a cooling system and optimizing the "cold end" of the power plant, it is necessary, along with other influencing factors, to consider specific investments in thermal power plant cooling systems. Thus, the optimization of the vacuum is a function of the duration of use, the cost of fuel during the entire life of the power plant, natural environmental factors, percentage of equipment

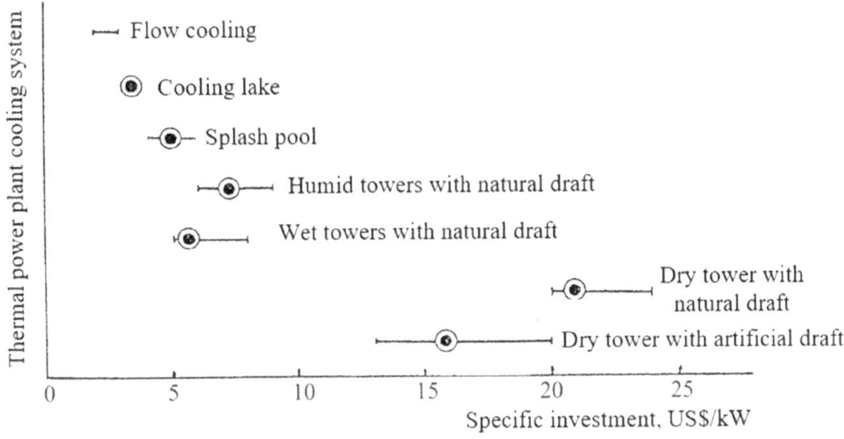

Figure 2.13 Approximate specific investments in cooling systems at thermal power plants.

depreciation (maintenance costs), steam parameters, turbine dimensions, and a number of other parameters.

Figure 2.13 shows the relative relationships and orientation-specific investments in different thermal power plant cooling systems.

2.4 RELIABILITY OF THE WATER SUPPLY SYSTEM AT CONDENSATION THERMAL POWER PLANTS

2.4.1 Terms and definitions

Reliability is the property of a system to work without failure under certain conditions (modes of operation, personnel, and environment) for a precisely defined period of time (Sandler 1983). Reliability theory is a science that deals with the study of the failure of technical systems and their constituent elements. In the case of technical systems where there are requirements for a high level of reliability and availability with continuous operation of the plant, a maintenance strategy based on reliability (Reliability – Centered Maintenance – RCM) has been developed. The basic prerequisites for providing this maintenance strategy are: determining the current state and level of reliability of the technical system, developing a methodology for determining the limits of desired reliability, analyzing current and desired reliability, predicting and analyzing the consequences of changes in reliability and defining reliability values in the future. Maintenance according to reliability uses a simple methodology that enables making correct and applicable decisions and activities that support the operation of the technical system (Milovanović and Branković 2021).

The reliability R(t) of the technical system, in addition to the readiness or availability of G(t) and the functional convenience of the FP technical system, represents the third characteristic that describes the behavior of the technical system in terms of effective commissioning. Reliability is the probability that the system will successfully perform the objective function in the designed duration and given environmental conditions after repeated use, i.e., the probability that the system will successfully perform the design task in the given time and given environmental conditions after one-time use. As one of the important components of the quality of technical systems in modern conditions, the reliability of technical systems plays a significant role. Starting from this fact, the production of the most reliable systems and their elements is sought today. In order to achieve this, it is necessary to approach reliability in the development phase, that is, in the phase of designing a new system, and reliability analysis in the laboratory and exploitation. It is necessary that the system being designed has certain predetermined characteristics and reliability indicators (Todorović and Zelenović 1994).

Defining the reliability parameters as one of the most important characteristics of the system requires a good knowledge of the system. Since reliability depends on stochastic processes, the selection of its indicators is based on the probability of occurrence of an event. The selection process itself is carried out according to the differential (first it is determined what properties the reliability indicators should have, and then concrete indicators are established according to certain norms) and integral (a list of all possible reliability indicators is established, without distinguishing them by properties) scheme. The optimal solution is obtained by analyzing and excluding indicators that, according to the selected criteria, are not subject to standardization. For each scheme, a number of alternative solution variants are formed for the evaluation of each variant and the selection of the best solution.

Failure or malfunction is the cessation of the element's ability to perform its function, whereby the failure of one element does not have to simultaneously represent the failure of the system as a whole (e.g., the element on which the failure occurred is of peripheral importance). On the contrary, if the element is of vital importance, then its failure causes the failure of the system or the reduction of its productivity or degree of useful action. The state in which the technical system is capable of performing its assigned function, without certain parameters of functioning not exceeding the limit values, represents the operational capability of the observed object. From the point of view of the theory of reliability and the theory of probability, the failure of an object is a random event, which consequently leads to a loss of working capacity. The time of correct operation until failure (the effective time of operation of the object until the occurrence of failure without considering possible interruptions in operation) includes the time period of correct operation of the object until the occurrence of failure. Sometimes this time coincides with the calendar time, in the case when the object is affected by a continuous process independent of interruption of operation (e.g., corrosion, action of external temperature factors, and hydraulic shock). The time of correct work is most often expressed in hours of work, but it can also be the same it can be expressed also in numbers of revolutions, oscillation circles, path length, etc. The time of correct operation can also be defined as the time of reaching the limit value of any functioning parameter (Adamovic 1998).

Technical systems as sets of elements and relations between them and their characteristics are connected to each other in a way suitable for performing useful work or performing a certain service activity. This means that for the functioning of the system, in addition to the quality of the elements as a whole, a certain connection between them is necessary. Each complex system combines a greater or lesser number of constituent elements, and its reliability can only be judged if the reliability of each detail is analyzed and analytically included. The final working connection and ability of the block as a whole depends on the work and behavior during the exploitation of each of the elements from the scheme. Starting from the applicable methods for these analyses and criteria for single failures, it is possible to determine the reliability and availability of complex technical systems by breaking them down into component elements, with the determination of appropriate parameters of reliability and availability using statistical analysis and by observing and defining the interconnections or the influence of individual elements on the system as a whole. Starting from the initial stage of development, design, and production of a certain type of thermal energy equipment in order to fulfill all requirements without restrictions, which arise from its purpose, the designer is faced with a multivariate choice, with the need for optimization according to certain already adopted algorithms. The goal is to create such a plant that has a satisfactory structure in terms of reliability indicators, with minimal maintenance costs during the expected working life. Defining the plant that best meets the set requirements related to reliability and the process of exploitation and maintenance itself must be the result of the implemented optimization process (most often based on the chosen criterion of the minimum investment for the desired effectiveness of the technical system) (Papić and Milovanović 2007).

Some of the auxiliary equipment in TPP related to the process of changing the state of the working fluid (water-steam) are pumps (condensate pumps, drain and drainage pumps, FP, oil pumps, cooling water pumps, pumps for technological or operational cooling, mains water pumps or heating water, additional water pumps, etc.). Of all the installed pumps, the most commonly used are FP for transporting feed water to the steam boiler (steam generator), condensate pumps that reverse condensate from the condenser to the system, oil pumps, and many others for auxiliary systems. Steam generator FP are used to pump water into the steam generator. Most often, they are high-pressure, centrifugal, or reciprocating and use a condensate reverse suction system, while boiler FP are centrifugal type, with several stages in horizontal execution. Dosing pump aggregates are used in the feed and additional water preparation system at TPP for dosing catalysts, inhibitors, or other media of acidic, basic, or neutral composition. Where it is necessary to provide a constant flow of droplet working media at high pressures, piston pumps are used. Multistage pumps are intended for the transport of clean and partly impure water, cold oils, liquid fuels, etc., while thermal oil pumps are used for the transport of thermal oil, which has the role of heat energy transfer. The construction of the pumps is determined by the operating conditions within the thermal scheme of the thermal power plant. When visually monitoring the operation of pumping plants, you should pay attention to whether the pump is working normally, i.e., within the limits of the prescribed characteristics of the pump and the motor load (checking the pump effort and motor current, bearing temperature and lubrication oil level, noise and possible vibrations during pump operation). In case of any deviation in the operation, additional diagnostic controls of the pumps are performed.

2.4.2 Thermal scheme and equipment structure of the thermal power system within the power system

The generated thermal power plant equipment within the power system is located at a lower hierarchical level, where, together with hydropower plants, nuclear power plants, and a smaller part in industrial power plants (IPP), it forms the basic structural forms of the electricity production system (production capacities within the electroenergy system). In principle, two types of basic technological schemes or technological structures are distinguished. Accordingly, power plants are classified into *block* and *non-block thermal power plants*. Modern TPP with intermediate superheating of steam are performed according to the block system: each turbine is supplied with steam from the boiler, which is only connected to it in a block connection – *monobloc*. In some cases, the turbine is connected with two boilers – *double-block*. With the block structure of the thermal power plant, there is no connection between the blocks with pipelines of fresh and intermediate superheated steam, as well as feed water. Those with such schemes are cheaper and easier to manage and automate. The power plant is assembled as a set of separate energy blocks, and they have common facilities and connections for water, fuel, etc. Condensing power plants with lower initial pressure (9 MPa and below) and TPP with pressures of 13 MPa and below, as well as other types without steam reheating (e.g., IPP), are usually not built in a block structure. Such non-block structures have common mains (collectors) of fresh steam and feed water for all boilers. Steam turbines receive steam from these mains in a centralized or sectional scheme. In principle, there are two groups of equipment that are technologically connected to each other by fundamentally different physical and chemical production processes: electrotechnical and thermomechanical. In the first group, mechanical work is transformed into electricity, and in the second group, the chemical energy of the fuel or the heat obtained during the splitting of atoms or the kinetic energy of falling water is transformed into mechanical work.

The largest and most basic variable cost within the complex system of a thermal power plant is the cost of fuel, so its economy has an exceptional importance on the total costs and the price of the product unit. Other more important costs relate to water preparation, spare parts and materials for plant maintenance (reliability level maintenance), salaries of employees, legal obligations to the state, etc. A special and characteristic cost of a thermal power plant is its own consumption of electricity, which is used in the power plant to drive auxiliary facilities (mills, fans, pumps, chemical water preparation plant [CWPP], etc.) and lighting. Setting a feasible task for determining reliability properties is not a straightforward process. It usually goes in two independent different directions. The basic postulate of the first direction is based on the fact that the component equipment and facilities of a complex energy system represent a uniquely determined whole, the reliability of which is evaluated based on the results of the fulfillment of assigned functions without the participation of physical and chemical processes. In that case, ratings or forecasts of reliability properties are given using a standard probabilistic array of indicators. Within the second group of tasks, the flow of changes in the primary form of energy is considered. Reliability is determined by deterministic indicators of changes in the properties of the materials from which the objects are built, the interaction of the working environment, as well as other factors that affect changes in primary energy. This approach in the general case does not exclude the properties of reliability indicators obtained on the basis of solving the first group of tasks. The assessment of the level of reliability, as well as the management based on reliability itself, leads to the analysis and synthesis of patterns of changes in the value of reliability indicators, which are characterized by reliability, suitability for overhaul, longevity (basic operational life), a possible revitalized period of exploitation (supplementary operational life). The reliability scheme of a complex thermal energy system in the general case is reduced to the connection of its most important elements: steam boiler, steam pipeline (steam pipeline), heat exchanger, steam turbine, electric generator, and FP.

In order to have a more detailed look at the working process of a thermal power plant, it is necessary to have an idea of the technological scheme and the composition of the equipment of such a power plant. We will observe such a scheme in the example of a condensing thermal power plant (KO-TPP) in which solid fuel (coal) is used, Figure 2.14.

The technological scheme shows two flows of materials that participate in the technological process: fuel (e.g., coal, natural gas, or fuel oil) – air – combustion products (flue gases, slag, ash) and feed water (turbine condensate, additional chemically prepared water, internal reverse condensate, external reverse condensate from HCs) – fresh steam – condensate. It can be said that there is also a third separate flow, the respective cooling water circulation circuit. As you can see, the first flow of matter is open, while the second is a closed circulation system. From the scheme given in Figure 2.14, you can see which facilities and objects are included in the technological composition of a thermal power plant (heating plant or power plant). It is also suitable from the aspect of understanding the choice of criteria and indicators for determining reliability, as well as giving forecast ratings and defining possible forecast errors.

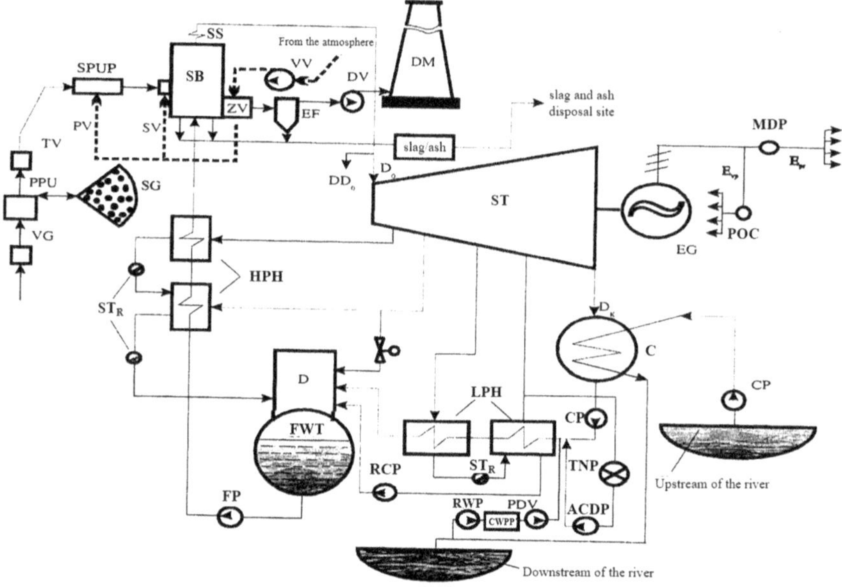

Figure 2.14 Principle technological scheme of a thermal power plant (Milovanović 2011b).

The technological scheme represents the sequence of flow (route) of fuel, water, steam, and electricity of the thermal power plant, which delivers electricity and thermal energy to external consumers. The first important technological complex of the thermal power plant is the tract that begins with the reception of fuel in the thermal power plant. Another important technological complex of the thermal power plant is the *water-steam tract*, which consists of the steam-water part of the steam boiler (SB), steam turbine (ST), condensation plant including the condenser (C) and condensate pump (CP), then the low-pressure heater (LPH), feed water tank with degasser (FWT + D), feed pump (FP) and high-pressure heater (HPH). This complex includes facilities for technical water supply and chemical water preparation, as well as all water and steam pipelines. The obtained steam in the steam boiler (SB) and the superheated steam in the steam superheater (SS) are fed through steam pipes to the steam turbine (ST), which drives the electric generator (G). The obtained electrical energy (Epr), through the block-transformer and the main distribution plant (MDP), is transferred to the power system via high-voltage transmission lines (110, 220, 380 kV, and more). Part of the electrical energy (Evg) produced on the generator is transformed from the generator voltage to 6 kV for self-consumption with high-voltage motors (6 kV). Low-voltage drives for self-consumption are supplied from the low-voltage plant, which consists of a 6/0.4 kV transformer, cells, and drains to substations in the thermal power plant. This part of the electrical plant and equipment represents the power plant's own consumption (POC). The produced steam in the turbine (Dk) is directed to the condenser (C), where it transfers its remaining heat to the cooling water (Qk). The resulting condensate is pumped by condensate pumps (CP) into the degasser (D) and feed water tank (FWT). In the degasser (D), dissolved gases in the water (O_2, CO_2) are separated, in order to protect the pipe system from corrosion. In addition to technical degassing, FWT is also performed chemical degassing, i.e., binding of oxygen with specially dosed chemicals.

The feed pump takes feed water from the FWT and feeds it to the boiler (steam generator) through feed pipes through FP. On the line for transporting condensate to FWTs and feed water to the boiler, they are heated by the steam taken from the turbine. LPHs and HPHs were installed for this purpose. Condensate, created in the heaters (LPH and HPH) by condensation of extracted steam, is reversed to the basic condensate line or directly to FWTs. For this purpose, steam traps (ST_R) are installed, which provide drainage of condensate without steam leakage. In addition to condensing boilers, auxiliary condensate drainage pumps (ACDP) are also used. In the condenser (C), steam is condensed using cooling water that circulates through the condenser from the cooling system. The cooling system is made of *flowing (open)* and *closed type*, depending on whether there is a large watercourse or lake near the power plant. In the flow type, water is taken from the river upstream of the thermal power plant, and is released into the river downstream after passing through the condenser (Figure 2.14). There is an ecological limitation here: the water in the river must not be heated by more than 2°C in order to protect the river fauna (it is desirable that this value be as small as possible, on the order of 0.5 or a maximum of 1°C). With a closed system, cooling towers are built in which the water is cooled by the temperature rise gradient in the condenser ($\Delta t = 8 - 10$°C). It is, in other words, a closed circulation circuit in which water is added as a compensation for evaporated losses in the cooling tower. For example, in the case of a 300 MW power unit, 36,000 m^3/h of water circulates through the condenser in the cooling system, the evaporation losses amount to about 500 m^3/h, and cooling by $\Delta t = 9$°C is achieved in the tower.

Theoretically speaking, the working process in the thermal power plant is closed, i.e., it is assumed that all the amount of steam entering the turbine from the boiler reverses to the boiler in the form of condensate. However, a certain loss of steam and condensate is unavoidable, which should be compensated with additional demineralized water. Additional water is added according to the CWPP, and the raw water for preparation is taken directly from the watercourse after physical purification or from the reverse circulation line using the raw water pump (RWP). The chemical preparation of water consists of a *decarbonization plant* (removal of carbonate hardness up to 3–4 °dH) and a *demineralization plant* (complete desalination in ion exchangers). The most common procedure (especially with a closed cooling system) is to decarbonize the entire raw water (partially soften it), so it is used as additional water for the cooling system, and the rest is used for demineralization. A part of the steam, which partially provides energy to the turbine, can be used for heating needs or as technological steam in another type of production. The steam is taken from the turbine and taken to the exchangers of the HC. Condensate is fully or partially reversed to the thermal power plant by means of reverse condensate pumps (RCP) and included in the condensate regeneration system for boiler feed water, Figure 2.14.

Condensate, feed and network pumps. These pumps are used to transport condensate from individual headers and condensers to the degasser (condensate pump), i.e., to deliver feed water to the steam boiler and ensure the set pressure of fresh steam at the outlet (feed pump), i.e., to supply mains water through the heating line from mains heaters to installed consumer heat exchangers (network pumps, line I) or in the case of using a peak water heating boiler at lower outside temperatures (network pumps, line II). The construction of condensation pumps is determined by the operating conditions within the thermal scheme of the plant. They serve to drain the condensate from the turbine condenser and deliver it through the LPH system (most often surface type) to the degasser (contact or mixing type exchanger).

2.4.3 An example of the analysis of the power supply system for TPP Gacko with an installed capacity of 300 MW

Within this work, more detail will be dealt only with the scheme of an energy block with a flow boiler and supercritical steam parameters in a block with a turbine power of 300 MW, type K-300-240 (LMZ), which are installed in TPP Gacko and TPP Ugljevik, which uses lignite, brown coal. A simplified thermal scheme for the release of such a block is given in Figure 2.15.

At KO-TPP and TPP-HP with subcritical parameters of fresh steam, FP with electric drive and hydraulic coupling are used, while at power units with supercritical steam parameters, high-speed FP with steam turbine drive (turbo FP) are used. The principle scheme of installing the feed pump is shown in Figure 2.16, while the representation of the typical design of the turbo feed pump for power units with supercritical steam parameters is given in Figure 2.17.

Mains pumps are most often centrifugal, horizontal type, one or two-stage, and with a rotation speed of 1500 and 3000 r/min. The feed plant intended for feeding the boiler with water at TPP Gacko consists of three booster pumps that supply water under pressure from the deaerator to the suction in the FP, ensuring safe cavitation-free operation of the feed pump. Three FP are installed on the block, two of which are electrically driven, and one is turbo-driven. Electric FP are made at half capacity and are used when starting the block. The electric pump is driven by an electric motor via a hydro coupling and an accelerating reducer. The turbo feed pump is used as a basic feed pump and provides boiler power in the load range of 35–100% of the nominal value. The drive of the main feed pump is a special turbine OP – 12 PM with a high number of revolutions, which is fed by steam from the third subtraction of the turbo generator. The feed water is circulated by means of one feed pump through a group of HPH – 6, 7, 8. The feed water is heated

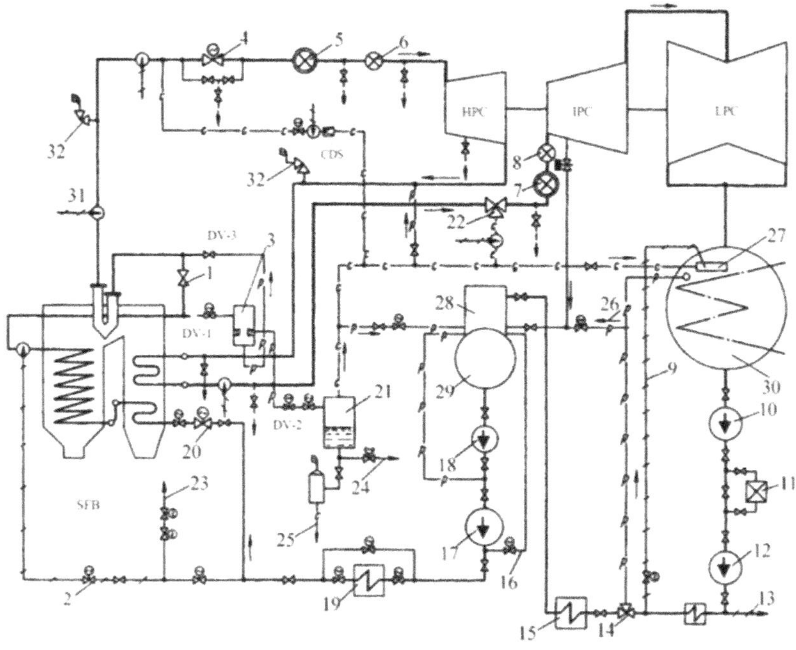

Legend:
1 - start shutter; 2 - control valve for injection of water in steam coolers; 3 - built-in start separator; 4 - main steam shutter (MSS); 5 - stop valve HPC; 6 - control valve HPC; 7, 8 - stop and control valves IPC or combined stop-regulating valve IPC; 9 - line of injection of condensate into the steam-receiving capacitor device; 10 - condensate pump I degree; 11 - ion exchangers for purification of turbine condensate; 12 - 2nd stage condensate pump; 13 - condensate drainage system BRRS; 14 - condensate recirculation valve; 15 - low pressure heating group (LPP); 16 - line of recirculation of supply water in water tank (WT); 17 - a supply pump; 18 - pre-pump or buster pump; 19 - high pressure heating group (HPP); 20 - control supply valve (supply head); 21 - Starting expander; 22 - a quick-acting vacuum valve for releasing steam from the inter overheating system to the condenser; 23 - return of excess water to the deaerator; 24 - discharge of condensate from the starting expander into the deaerator; 25 - discharge of condensate from the starting expander, alternatively: into the neutralizing basin of contaminated water, or to the cooling conduit or the reserve condensate tank; 26 - the supply of steam from a foreign source: a joint station for the distribution of low pressure steam, an auxiliary boiler room or a near block in operation; 27 - steam-receiving device in the condenser; 28 - deaerator; 29 - a water tank (WT); 30 - turbine condenser; 31 - injecting a refrigerator of fresh steam; 32 – manual control; DV-1 ÷ DV-3 - damping valves on the old appliance.

Figure 2.15 Simplified principled heat transfer scheme for the operation of a 300-MW monoblock with a supercritical once-through boiler and a one-bypass scheme (Milovanović et al. 2018).

in the HPH from 165 to 270°C with a flow rate of 950 t/h. The pipe system inside the HPH is under full pressure from the FP, and the steam side is under pressure from the turbine. It is HPH-6 (III subtraction), HPH-7 (11th subtraction), and HPH-8 (I subtraction). Structurally, the heater is made in the form of a vertical cylindrical welded casing with a height of 8860 mm and a pressed dancette. Inside there is a pipe system consisting of six spiral columns. The steam drainage scheme for heating HPH was carried out in cascade, from HPH-8 to HPH-7, and HPH-6. From HPH-6 steam comes to the deaeration

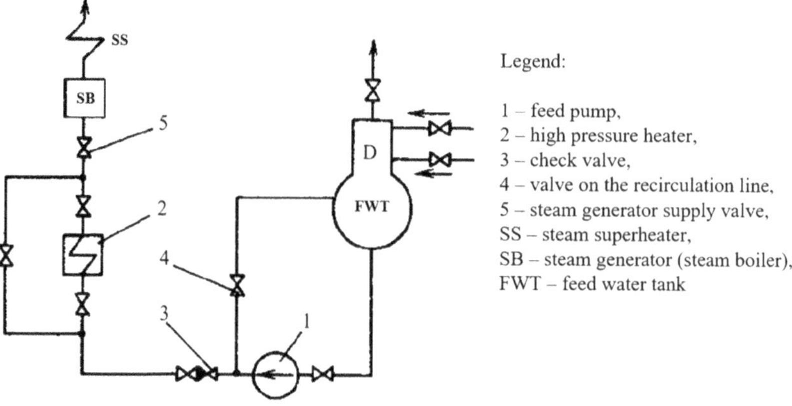

Figure 2.16 Principle scheme of installing a feed pump within the TPP.

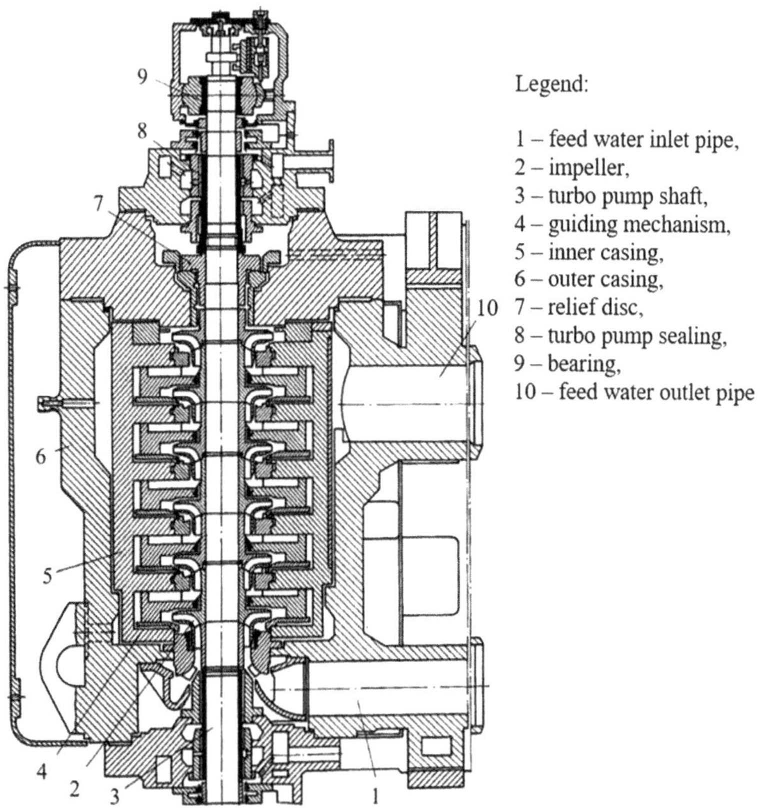

Figure 2.17 View of a turbocharger pump for power units with supercritical steam parameters.

tank. When reducing steam pressure in subtractions, the scheme provides that the drainage from HPH-6 comes to HPH-4 and from HPH-7 to the deaerator. Non-condensable gas exhausts were also carried out at HPH in cascade from HPH-8 to HPH-7 and HPH-6, and from HPH-6 to LPH-4. At the start of the block and when it reaches the nominal load for 48 hours, it is allowed to increase the specific electrical conductivity of the steam up to 50%, as well as the content of combined sodium and silicic acid in it, and in the feed water – specific electrical conductivity, general hardness, content of sodium compounds, silicic acid, iron, and copper.

In the first 24 hours, the content of iron compounds and silicic acid is allowed up to 50 μg/dm^3 for each of these compounds. At the start of the block after capital and intermediate overhaul, an increase of these values up to 50% is allowed for a duration of 96 hours. At the same time, in the first 24 hours, the content of iron compounds and silicic acid is allowed up to 100 μg/dm^3 for each of these compounds.

Two PE-600-300-1 Type electric FP with a capacity of 600 m^3/h, with a discharge pressure of 330 bar, are installed at TPP "Gacko", the electric motor AB-8000/6000 is a two-pole, asynchronous, short-circuited, direction of rotation in the direction clockwise if viewed from the drive end of the valve, Figure 2.18. The electric motor has sliding bearings.

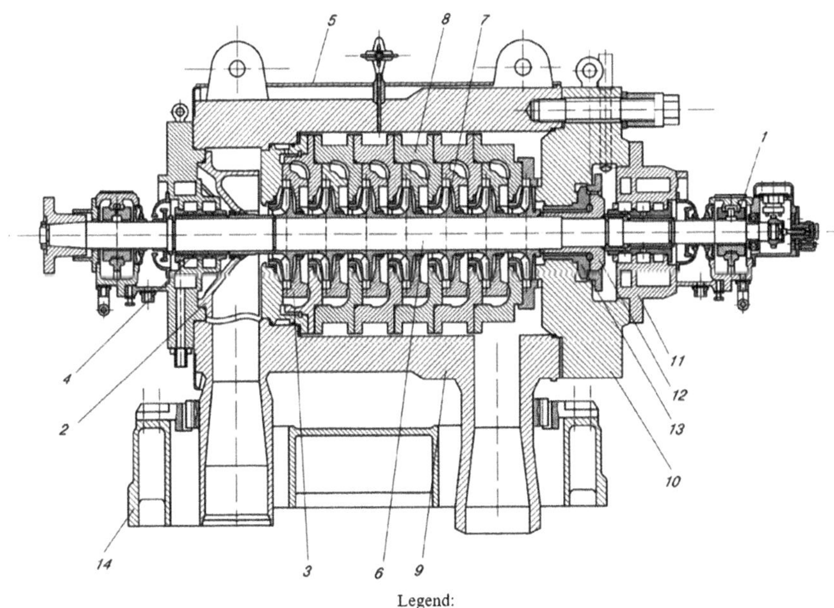

Legend:

1 – bearing, 2 – suction part, 3 – guiding mechanism, 4 – sealing shell, 5 – pump housing, 6 – rotor shaft, 7 – impeller, 8 –section, 9 – outer casing, 10 – thrust part, 11 – end seals, 12 – relief disc, 13 – relief disc shell, 14 – pump stand

Figure 2.18 Section of feed pump PE 600-300-2.

The electric feed pump (EFP) PE 600-300-2 is a centrifugal, horizontal, double-cased pump with inlet and discharge ports directed downwards and with seven stages. The pump rotor consists of a forged shaft with seven impellers mounted on it and a relief disk. The rotor rests on two slide bearings with forced lubrication from the general turbine lubrication system. The axial force on the pump rotor is completely balanced with the hydraulic force acting on the relief disk, due to the pressure difference on both sides of the disk. From the side of the working wheels, water is brought under pressure to the disc, which is taken after the last stage. 2/3 of the chamber behind the water disc is drained to the pump suction through an overflow pipe. The pressure difference on both sides of the disk is determined by the resistance to the movement of the liquid in the axial gap between the relief disk on the shaft and the heel cushion in the housing. The end seal is of the labyrinth type and is used to seal and cool the shaft by supplying cold condensate and draining water to the deaerator and condenser. To control the wear of the sides of the relief disk and hydropet cushion, and also to limit the movement of the rotor on the thrust side, a rotor buffer with an automatic and visual indicator of axial displacement is placed on the rear bearing. Behind the first stage of the feed pump, feed water is taken at a pressure of 80 bar and a quantity of 80 t/h to regulate the temperature of the intermediate superheating steam.

The hydraulic coupling type MG2L-650 is intended for changing the number of revolutions of the pump while the number of revolutions of the driving electric motor remains unchanged. The gearbox is intended for transmission of rev of the torque of the electric motor to the pump with an increase in the number of revolutions from 2970 to 6300 r/min. The gearbox is made as a single-stage with two gears. The toothing is hypoid, which ensures the balancing of axial forces. The gears of the gearbox are housed in a cast steel housing with a horizontal position. Special protrusions are made on the sides of the housing, where the radial bearing inserts are located. The outer walls of the housing and the cover of the gearbox are equipped with ribs that increase the hardness of the structure and also improve heat dissipation. The oil is supplied to the toothed gear via pipes attached to the gearbox housing from the inside. Oil is supplied to each half of the gearbox through special holes drilled on the sides of the gearbox housing. The high-speed gearbox gear has an internal opening through which a flexible torsion shaft is threaded. The torsion shaft is connected to the gear on one side by a toothed coupling, and on the other by a rigid coupling to the pump. The torsion shaft reduces vibration when de-centering the gearbox and pump.

The electric motor cooling system is combined: the rotor winding is cooled by water directly, the stator core together with the part of the windings that are placed in the grooves is cooled indirectly, and the front parts of the stator winding are cooled by air. Air circulation is carried out according to a closed contour, and circulation is provided by the electric motor's own fan. A water air cooler type BPT-108-100 is used for air cooling. The flow of cooling water with a temperature of 33°C through the cooler is 15 m^3/h. At the same time, the temperature of the cooled air must not exceed 45°C. The water system for cooling the rotor and stator of the electric motor of the EFP is autonomous. In the system, there is a tank with a volume of 2 m^3 from which, with the help of one of the pumps (the other is in reserve), water passes through the filter and heat exchanger and leads to the cooling of the electric motor. After passing through the cooling system of the electric motor, the water comes back into the tank. The tank is supplied with demi-water: Pump characteristics: TYPE – 4MS-10, flow rate – 60 m^3/h, pressure – 66 mVS. Characteristics of the electric motor are: TYPE – A-262-2, voltage – 380 V, number of revolutions – 2950 (r/min)

The flow of water at a temperature of 33°C is: (a) in the rotor system – 40 m^3/h, (b) in the stator system – 6.5 m^3.

2.4.3.1 EFP protection and blocking system

Blockages that condition the inclusion of EFP:

- at a pressure in the lubrication system 1.2 bar
- at oil pressure in front of the hydro-coupling regulating valve 0.7 bar
- when the water pressure at the intake is 16 bar
- when the flow of cooling water to the rotor is 25 t/hour
- when the flow of cooling water to the stator is 3 t/hour
- when the recirculation valve is open
- when supplying power to protection devices
- when the valve is open on sealing 655 more than 50%

Protections that shut down the pump:

- when reducing the pressure in the lubrication system to 0.5 bar with a delay of 3 seconds,
- at axial displacement (0.8 mm),
- when reducing the suction pressure to 14 with a time delay of 3 seconds,
- when the discharge pressure is reduced to 200 bar,
- when reducing the flow of cooling water to the rotor to 25 t/h with a delay of 3 minutes,
- when reducing the flow of cooling water to the stator to 3 t/h with a delay of 3 minutes.

When preparing the pump for start, it is necessary:

a. Check that all repair works have been completed.
b. Make sure of the correctness of the complete equipment, as well as fittings with drives, by careful inspection.
c. Check the correctness of control and measuring instruments and turn them on.
d. Check the sufficient flow of oil in each of the bearings using inspection glasses. Inspect the oil pipelines and make sure that there are no oil leaks through the seals.

e. Open the valves on the suction of the electric power supply pump in order to fill it with water and warm it up, as well as to release air from the pump housing through the vent valve.
f. Make sure that the EFP recirculation valve diagram is ready and that it is fully open.
g. Check the tightness of the fittings on the supply water withdrawal line behind the first stage.
h. Check the level in the 2 m^3 tank, turn on the pump for cooling the electric motor EFP, regulate the flow for cooling the electric motor, and check the AUR EFP.
i. In the test procedure, check EFP protections and interlocks, and try the EFP emergency button.
j. Open the valves at the inlet of the cooling water to the air cooler of the electric motor EFP.
k. Make sure that the feed pump is warmed up, that is, the temperature difference of the feed water at the EFP intake must not exceed 50°C.
l. Make sure that the oil temperature after the oil coolers is ≈ 40°C.
m. Close the regulating valve of the hydraulic coupling, having previously made sure of the correctness of the manual and remote drive.
n. Turn on all protections. Starting the EFP with the protections off is prohibited.
o. Prepare EFP in a working position.
p. It is mandatory that two booster pumps must be in operation before turning on the EFP.
q. Switch on the EFP and check the current load (electric motor and pump have opposite directions of rotation).
r. By slightly opening the hydraulic control valve from the control block or on the spot, establish a pressure of 320 bar.
s. Make sure on the spot and based on the display of instruments, that the EFP is working normally: that there are no large vibrations, that the temperature of the oil at the outlet from the hydro-coupling does not exceed 75°C, that the temperature difference of the feed water in the overflow pipe and at the intake does not exceed 30°C, and when the cooling water temperature reaches 35°C, open the circulation water supply to the heat exchanger.
t. Open the valve on the supply water withdrawal line after the first stage of the pump, for injection into the intermediate heating.
u. When the flow of feed water increases above 130 t/h, the recirculation valve closes due to the blocking effect, and its closure must be checked on-site.
v. During EFP exploitation, control the pressure drop on the EFP suction screen, which must not exceed 3 bars.
w. Continuously monitor EFP operation, aggregate vibrations, electric motor load, bearing lubrication system, absence of leaks, and sealing. When making any changes in the operating mode of the power supply device, carefully examine the readings of the control-measuring instruments and make sure that their readings are within the permitted limits.

x. Do not allow an increase in the EFP flow rate above 700 m^3/h.
y. Monitor the load of the electric motor EFP, the current must not exceed 900 A.
z. In the event of a leak in the cooling system of the rotor and stator, electric motor and water reaching the winding of the electric motor, immediately turn off the EFP and disconnect its electrical circuit; re-starting can only be done after removing the leakage, checking the insulation of the electric motor and a careful inspection.

Turbocharger pump (TNP) PN-1135-340-10. The pump type is PN-1135-340-10, capacity of 1135 m^3/h, discharge pressure of 340 bar, and suction pressure of 20 bar, Figure 2.19. The pump is driven by an OP-12 M turbine, power 12500 kW. Number of revolutions 6000 r/min. Feed pump PN-1135-340-10 is a centrifugal type with six pressure levels, it is structurally similar to an electronic feed pump. The pump is connected to the turbine by a gear coupling. The oil for the lubrication of the drive turbine pump bearings and the gear recess of the coupling is supplied from the oil system of the main turbine.

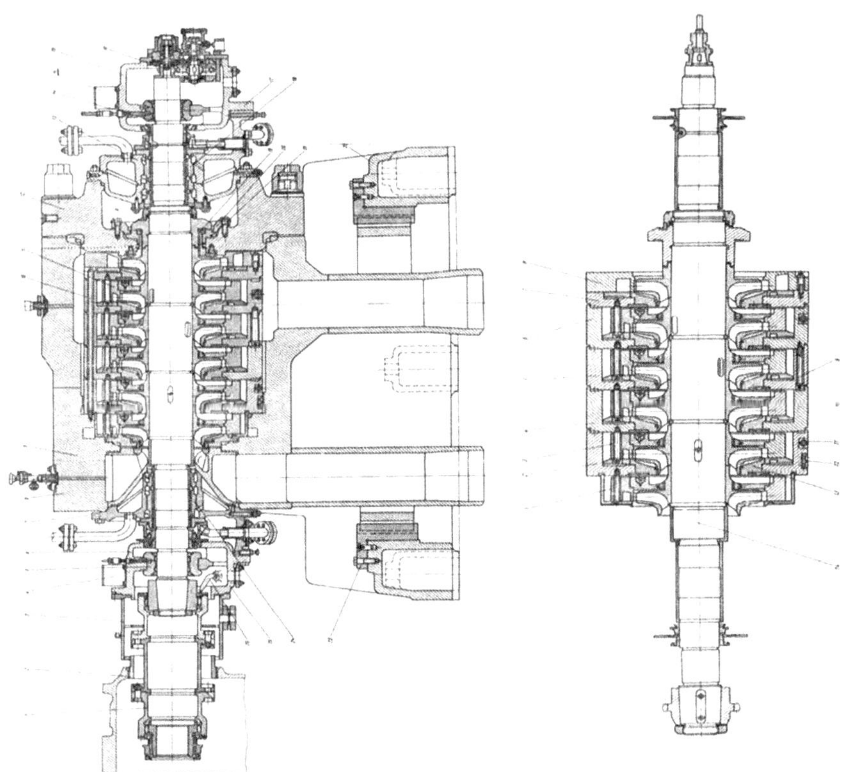

a) section TNP PN-1135-340-10 b) flow part TNP PN-1135-340-10

Figure 2.19 Turbocharger pump PN-1135-340, a) section and b) flow part.

2.4.3.2 *Operation of condensate pumps, drainage pumps LPH, ejector pumps, and booster pumps*

The following pumps are installed at TPP "Gacko":

1. Pumps of the condensate treatment plant – PCT (KP-I stage) which are intended for the supply of condensate from the condenser to the PCT. The pump is centrifugal, vertical, three-stage. The constructive design ensures relief of a significant part of the axial load that occurs during pump operation. The upper bearing is axial-radial, spherical, lubricated with turbine oil that fills the oil tub and is cooled by technical water. The lower bearing is made of special bronze and is lubricated with water. The pump is sealed with a seal, it has a water supply for cooling and sealing the pump shaft. The type of pump is KCB-500-85 where 500 – flow m^3/h, 85 – thrust in mVS, number of revolutions per minute 985, motor power 200 kW, voltage 380 V; direction of rotation – clockwise if we look at it from the electrical side of the engine.
2. Condensate pumps of the second stage (KP-II) KSV-500-150: flow – 500 m^3/h, pressure – 150 mVS; number of revolutions per minute – 1480; power – 272 kW; voltage – 6000 V; direction of rotation – clockwise as seen from the engine side. They are intended for supplying purified condensate from KCB to the deaerator, through low-pressure regeneration. The pump is vertical type, with inner and outer housing, centrifugal, four-stage. The constructive design of the pump ensures relief of a significant part of the axial load that occurs during pump operation. The upper bearing is axial, spherical, and lubricated with turbine oil, which fills the oil bath of the bearing. The bearing housing has water cooling. The lower bearing is radial, made of special bronze, and lubricated with water. The pump seal is a soft seal, between the layers of which are placed rings that form a chamber into which condensate is fed under a pressure of 3–5 bar to prevent air suction and to cool the seal. The pump housing has a vent in the upper part for air release.
3. Drainage pumps LPH KCB-200-220: flow – 200 m^3/h pressure – 220 mVS; in terms of technical performance, they are absolutely the same as the previous ones, except that they differ in the power of the electric motor. The purpose of LPH sump pumps is to, at Ne = 80 MW, drain the heating steam drainage from LPH – 2 into the basic condensate line behind LPH – 2. Ejector pumps are intended for the supply of circulation water from pressure circulation pipelines to the basic turbine ejectors, ejectors for PS-115, and filters F-250.
4. The ejector pump is centrifugal; single-stage; with double-sided suction. The pump type is – 32 D – 19; flow – 4700 m^3/h, pressure – 33 mVS; number of revolutions per minute – 730; electric motor power – 630 kW; voltage – 6000 V; direction of rotation – counter-clockwise if viewed from the side of the engine. The pump housing is made of cast iron. The bearings are spherical, and their lubrication is annular. The oil is poured

through the openings in the upper part of the bearing housing. There is a "snake" in the bearing housing for water cooling. The oil level is controlled by the level indicator. Drainage of dirty oil is done through the lower plug for taking samples. The sealing of the pump valve is sealed, the seal is a tallow braid. Between the rings of the cord, there is an annular chamber into which water is fed from the pump discharge channel for cooling and sealing the seal.

5. Buster pump PD 650 – 160 is intended for supplying feed water from the deaerator to the FP of the block. Flow – 650 m^3/h; thrust – 158 mVS; engine power – 500 kW; voltage – 6000 V; number of revolutions per minute – 2975. The pump is centrifugal, horizontal, spiral-type, single-stage with a double-sided suction impeller. The housing is cast from iron with a horizontal composition. The pump rotor consists of a stainless-steel shaft and impeller, the pump rotor rests on two sliding bearings. The axial load of the rotor is transferred to the axial-radial double-row ball bearing. Lubrication of the bearings is annular, with turbine oil. "Snakes" are installed in the bearing housing to cool the oil. The end seals are sealed.

During the exploitation of the mentioned mechanisms, special attention should be paid to the vibrations of the mechanism, their heating, condition and temperature of the bearings, oil level and their cleanliness, and pressure differences. The pump must be turned off immediately with the emergency button in the following cases: if the extension of the pump's operation endangers people's lives, if there is a fire or smoke from the electric motor or the pump bearing, and if there are strong vibrations or metal impacts that endanger the equipment. In all cases of emergency shutdown of the pump, the turbine operator is obliged to check that the backup pump is turned on by the AUR or turn it off manually. After turning off the pump, the assistant turbine engineer is obliged to make sure that the pump has not started to rotate in the opposite direction, inspect the pump, whether it is turned on by the AUR effect, and inform the block engineer about the condition of the pumps. The pump must be turned off with prior activation of the backup pump in the event of a sudden increase in temperature to 75°C, one of the bearings, if there is large heating of the seal, if water indirectly reaches the electric motor.

2.4.4 Preparations for the overhaul of condensate pumps (CP) and booster pumps (BP)

Preparations for the overhaul of condensate pumps (CP) and booster pumps (BP) include:

a. Preparation for overhaul of rotary mechanisms should be carried out in accordance with the conditions of the works. In doing so, the mechanism should be switched off, the voltage from the electric motor and the electric drive of the armature should be removed and the coupling should be separated. The locking armature of the mechanism (valves, shutters) must be in a position that ensures the safety of the work. Actuators for switching on the armature must be secured by means of chains or other devices or accessories. Warning signs prohibiting the supply of voltage must be placed on the disconnected drives and the starting device of the mechanism. CP and BP during preparation for overhaul must be safely turned off on the thrust, suction side, and drained, and for BP the pump cooling must be turned off, after which the valve on the condensate supply to seal the pump is closed. During the overhaul of the CP, after the pressure valve is closed, the suction valve is closed. Then the pressure in the pump should be controlled and if there is a significant leakage of the pressure valve after the suction valve is completely closed, the pump remains under full pressure, which leads to damage to the pump housing.

b. During the trial switching on the rotary mechanism, the button for emergency shutdown of the electric motor of the mechanism must be in operation. Next to the emergency button, there must be a designated person who will follow and who will turn off the mechanism according to the order of the works manager.
c. Before turning on the rotary mechanism, as well as before testing, the coupling must be assembled, all protective parts of the rotary parts must be installed, warning signs removed, tools and materials collected, and people removed from the work site.
d. After switching on the mechanism, monitor its operation, listen, and make sure on the basis of the instruments that the pump generates working pressure, that there are no objections to the operation of the bearings and the cleanliness of the oil. If there are no objections to the operation of the given mechanism, it should be left to work for 2 hours under constant supervision, after which it can be switched off in reserve or left in constant operation.

2.4.5 Objectives and specifics of the development of the reliability model for the water supply system at the condensing thermal power plant

The main goal of the work is to determine, through the developed reliability model, the basic factors that contribute to the poor performance of the observed technical water supply system at the condensing thermal power plant. Also, during the operation of the joint components of this system, there is relative movement and interaction of their parts, whose surfaces can be in direct or indirect contact. Also, the possibilities of failure of the bearings of these systems are high. Analyses show that deviations range up to 20 times compared to the theoretical life of the bearing (Miličić and Milovanović 2010). According to research given in Ašonja (2005), 34% of installed roller bearings fail due to material fatigue, 36% due to inadequate lubrication, and the other causes are fatigue limit, improper assembly, excessive vibrations, influence of electricity, etc.

Due to the fact that the rolling element and the ring are not separated by the necessary thickness of the protective lubricant layer, material fatigue starts very early (Miličić and Milovanović 2010). In addition to the thickness of the lubricant, its contamination should also be considered (Ašonja 2005). The basic processes that occur during the relative movement of two solid bodies are friction on contact surfaces (loss of mechanical energy during the operation of a mechanical system) and wear of contact surfaces (changes in the surfaces of machine parts during exploitation). The surfaces of machine elements that are in contact with each other during movement also change their properties over time (change in geometry, size, structure, and properties of surface layers). The size and character of the changes are a function of the load conditions, the amount of movement, the nature of the material of the elements in contact, the composition and properties of the environment (moisture, air, etc.) and the composition and properties of the lubricant, etc. Wear, as a progressive process of material loss of elements or parts of machines that are in direct contact and are in relative motion, results in the development of wear and tear of machine parts, a decrease in the efficiency of their work, an increase in energy consumption, etc. Bearing wear on technical systems includes abrasion, material fatigue, and corrosion. Abrasive wear is a form of wear, that is, the removal of material from the surfaces in contact as a result of the penetration of higher hardness surface irregularities or solid particles of various origins into a lower hardness surface. Particles of material can be free and act as a third body, but can also be pressed into one or both surfaces. They can arise either within the tribomechanical system itself as a result of the surface wear process or they can be introduced by means of lubrication (Ašonja 2012). In order to prevent the entry of solid particles, good sealing of the corresponding machine elements must be ensured, and filtering devices must be installed in the lubrication or hydraulic systems. Abrasive wear can in a positive sense have a wide application in metal processing by grinding, scraping, polishing, etc. Wear of surfaces in contact, which is conditioned by the appearance of abrasive wear, can have the following forms: cloudy and bright marks, scratches, notches, wavy rolling paths, etc. This type of wear is present in rolling bearings that are exposed to the constant effects of vibrations in a corrosive environment, where the total friction on rolling bearings amounts to 97–99%. The occurrence of material fatigue wear is the most frequent occurrence of damage to rolling bearings, to which all its elements are exposed. Surface fatigue of the material occurs when the variable load volumes and the number of cycles exceed those that the material can handle on its own. The occurrence of material fatigue is particularly pronounced in the group of higher kinematic pairs, where the points of contact of the elements in relative motion are very small, and in addition, they are always exposed to high specific pressures. During operation, the bearing elements are exposed to severe fatigue of the entire system due to frequent sudden changes in load. The places where the first cracks appear are defined by the position of the maximum tangential shear stress, which reaches its maximum value immediately below the loaded surface, and the changing normal tensile stresses cause their further expansion. As the cracks spread further, small flakes are separated, and the surfaces become rough and uneven, and their geometry is disturbed. Further use of such bearings leads to a decrease in the accuracy of rotation, the appearance of vibrations, and a violation of the ergonomic conditions for human work in the maintenance process. The remaining service life of rolling bearings, where material fatigue wear is registered, amounts to another 10% of the calculated life. All machines in the production plant that contain rolling bearings, primarily loading and unloading transport aggregates, are exposed to this type of wear. This type of wear is not characteristic of machine elements where constant or very little variable load is present. It often occurs with gears, all types of rolling and sliding bearings, camshafts, etc. This type of wear can also be initiated by stress concentration of foreign harder particles pressed into the bearing materials, long-term rest of the bearing under static load, improper assembly, improper maintenance, long-term use, passage of electric current, sparking, etc. (Ašonja 2012). Corrosive wear is the process of damaging surfaces that slide in a corrosive environment, where the action of the corrosive medium can be chemical or electrochemical. Depending on the intensity and degree of surface involvement, it can be continuous or partially interrupted. If the surfaces were at rest, corrosion products could create secondary protective layers on them. By sliding, however, these layers are torn and damaged, so corrosion can spread further. The problem of corrosion wear is not dominant only in internal systems, where various combustions and evaporations have a destructive effect on the entire assembly, but also in poorly protected external systems, where external atmospheric influences have a dominant influence on the appearance of corrosion wear. These phenomena are particularly pronounced in industrial machines, especially in thermal energy plants such as pumps and compressors (Ašonja 2012). If preventive activities for the protection of rolling bearings are not taken during exploitation, corrosion and considerable wear of the rolling elements usually occur. The size of this wear depends on the quality of the construction solution, the physical and chemical properties of the material from which the bearing is made, the conditions of the working environment, the quality of assembly, the quality of oil and grease for lubrication, the operating mode, etc. Necessary protection, which can greatly reduce the dominant influence of aggressive and other substances, and therefore corrosion, is the use of lubricants capable of neutralizing them (Ašonja 2012).

2.4.6 Optimizing the reliability of water supply systems for condensing thermal power plants

The optimization of diagnostic procedures (centrifugal pump vibration) is obtained by different methods that define the criteria according to which the optimal solution is determined, which provides the highest readiness and availability of the components of the technical system. The performance of the water supply system of condensing TPP, as an important aspect of their

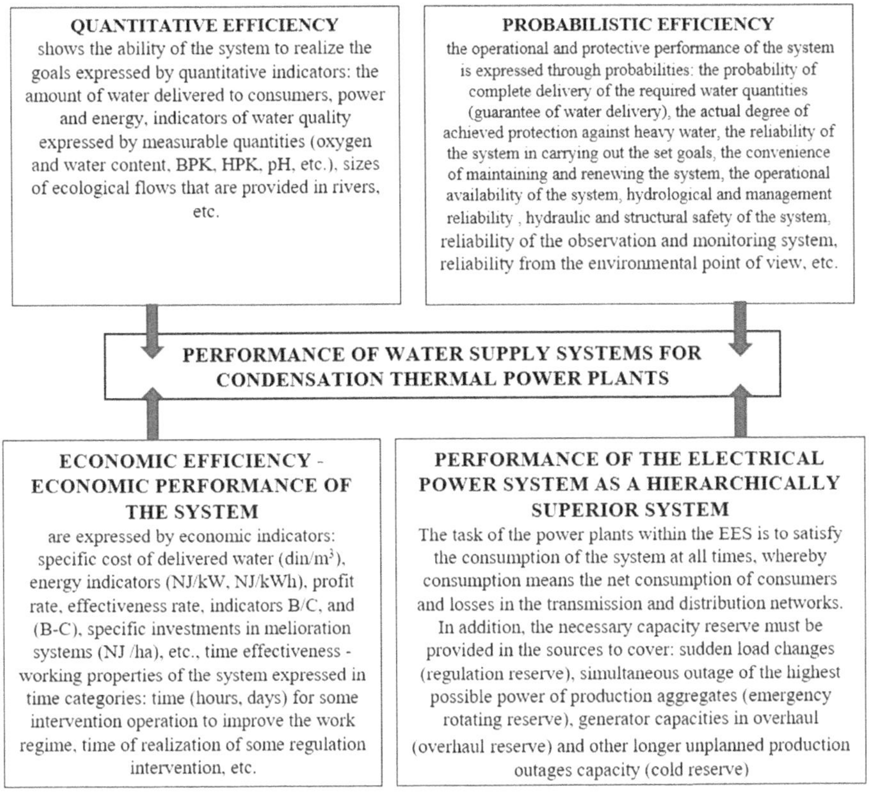

Figure 2.20 Systematization of the performance of water supply systems for condensing thermal power plants.

working and protective effectiveness, can be systematized into the categories given in Figure 2.20.

Analysis of the practice of planning and presenting the water supply system to the public indicates two possible omissions:

- most often, during the presentation of these systems to the public, they are quite wrongly named after only one of the primary users (e.g., a thermal power plant water supply system, a reversible hydropower plant water supply system, and an irrigation system), and it is not defined as an integral development project with the entire the complex target structure of the system, which should include all production and protection functions within the broader water management system of the given river basin, but also all other social, economic-developmental, urban, ecological, traffic and other goals that are achieved by building the planned system,

- when analyzing such a system, most often not all four categories of effectiveness of technical systems are evaluated, but only some of their working and protective performances, so the conclusion reached after such analyses cannot give all the answers about the functionality of the system as a whole (especially the category of probabilistic effectiveness is forgotten).

The lack of timely analysis of other important indicators of probabilistic effectiveness in the phase of designing and working out system dispositions often neglects the assessment of system reliability. In systems with water transport (pumping) over long distances with long pipelines and pumping stations with working and reserve pumping units, the following are of particular importance: operational readiness of the system, reliability of network systems with complex configurations, management reliability, etc. In order to ensure this, it is necessary to determine and analyze the functions of the intensity of failure (λ) of individual system elements and the system as a whole (Λ), analysis of the necessary intensity of repair (μ), etc., while choosing the configuration of the system and accompanying equipment. By looking at all dispositional aspects of the system, the necessary conditions are created so that the adopted disposition of the technical system enables unhindered preventive and corrective maintenance and restoration of the system. Otherwise, the economic performance of the system is not complete, damages due to outages from system functions, as well as costs related to ongoing and investment maintenance, have not been assessed. On the other hand, the time effectiveness is not nearly complete if the indicators of operational readiness of the system, outages, etc. are not established. As a result of these deficiencies in the phase of planning and selection of the disposition with the associated equipment and facilities, the appropriate measuring equipment required for quick detection and location of the fault is usually not planned in a timely manner in order to shorten the administrative time for its elimination, and in accordance with the size of the intensity of the repair (μ), acceptable from the point of view of the normal functioning of the system during exploitation.

The requirements for detailed analyses of all reliability categories gain even more importance when pumping stations are viewed as part of wider water management systems. The final choice of system configuration and facility parameters in the design phases (layout of water intakes, disposition of pumps and suction and pressure pipelines with accompanying equipment at pumping stations, along with solutions for machine buildings due to the convenience of maintenance, when solutions must also be found for appropriate measuring systems for "on line" monitoring the behavior of the system for adequate management and regulation during their use), determines the later use of the system, during which various categories of reliability (hydrological, hydraulic, facility reliability, operational readiness, etc.) must be periodically reviewed. From all of the above, it follows that the overall reliability of the system should be determined both during the design of new systems (sublimation of the aspect of reliability in the choice of system configuration) and in the case of systems in operation (analysis, verification, and optimization of system reliability).

2.5 RELIABILITY OF WATER SYSTEM OBJECTS ON FAILURES

The reliability of the system for continuously supplying the thermal power plant with feed water of a certain quality is defined by the probability that the system will not fail during exploitation. According to the cause of the failure of the system for the continuous supply of feed water to the thermal power plant, the following are distinguished: *primary failure of elements* (in normal operating conditions, they occur most often due to the aging of the system elements, and not due to some extraordinary events in the environment), *secondary failures* (caused by some influences from the outside or due to operation in overload conditions) and *failures due to incorrect management* (failure occurs due to a wrong command that put the device in some unplanned operating mode). Terminations caused by the error of the working staff or the influence of the environment are usually considered separately. Key reliability indicators are: reliability function $R(t) = P[T > t]$, failure density function $f(t) = dQ(t)/dt$, failure intensity function (failure hazard function) $\lambda(t) = f(t)/R(t)$, as well as the mean time of operation without failure T_0. In more complex systems composed of more plants and equipment, reliability depends on the way elements are connected into a functional unit (serial, parallel or network structure, combined connection). Although the serial connection is unfavorable from the point of view of reliability, it cannot be avoided in the case of pumping stations for supplying water to the thermal power plant. Parallel connection is significantly better from the point of view of reliability because when one element fails, the system continues to function, partially or completely, depending on the type of parallel connection by which the elements are connected. In complex systems, a number of elements are connected into a larger assembly (system) by various combinations of connections. It is necessary to analyze the reliability of such complex systems even when choosing the layout of the system. One of the possibilities is to use the *fault tree* graph method, or, bearing in mind the duality of the tree, the *success tree* (Đorđević and Milanović 1995).

2.5.1 Reliability of network systems under pressure

From a reliability point of view, pressurized water network systems are more complex than other technical systems. In order to create the conditions for their normal exploitation, two reliabilities must be satisfied simultaneously: the mechanical reliability of the pipeline and the hydraulic reliability of the system or the reliability of satisfying the hydraulic parameters (Đorđević and Milanović 1995). Mechanical reliability and reliability of hydraulic parameter satisfaction are combined into the total mechanical-hydraulic reliability of the system, which represents the probability that the required amount of water of the required pressure will be provided in a certain node of the distribution system, provided that there are sufficient amounts of water in the source nodes. For *safety-sensitive systems*, a requirement can be set a priori that the reliability should be higher than some required reliability R_z: $R \geq R_z$, whereby the *reliability allocation* problem is additionally analyzed. By using the failure tree method, the reliability of subsystems, subassemblies, assemblies, and elements is considered, and the weakest points of that complex system are strengthened with parallel connections, until the required reliability is achieved. This activity is realized in the design phase when the system layout is defined and basic and auxiliary equipment is selected. Therefore, it is necessary to additionally consider several variants that have been shortlisted for evaluation before the final selection (introduction of additional pump aggregates to ensure reliability, conversion of break chambers into reservoirs to enable disconnection of part of the pipeline for repairs without interrupting the supply of consumers, provision of suitable closures to facilitate repairs, "bypassing" some safety-delicate sections of supply systems in order to increase reliability, forecasting compensation basins in order to ensure guaranteed ecological flows and during repairs of parts of the system, etc.). Due to the requirements that may arise during the exploitation of the system to increase the reliability of the water supply system of the thermal power plant, they seek, even in the phase of designing and developing the system, the prediction and planning of the necessary disposal possibilities for this purpose (disposition of the machine building, pump or booster station that does not prevent later expansion, in order to it was possible to install an additional aggregate, which will increase the reliability of the system as a cold reserve of the system in a parallel connection, etc.). During the development of further project documentation (Conceptual design, Main design, Design for execution), more detailed reliability analyses are required: analysis of failure intensity (λ_i) of all vital elements (i) that can be separated as functional units at the thermal power plant (pumping stations, booster stations, pipelines) and failure intensity of the entire system (Λ), as well as consideration of other reliability indicators, e.g., determination of MTBF.

2.5.2 Convenience of maintaining and renewing the water supply system of thermal power plants

Thermal power plant water supply systems require adequate maintenance, reconstruction, revitalization, and restoration after the end of the basic working life period (25–30 years). For this reason, this form of probabilistic effectiveness is analyzed even in the phase of design, development, and selection of system equipment, with the aim of harmonizing the disposition and parameters of the system with those requirements. The disposition should enable the system to be maintained in operational condition through *preventive maintenance*, and after failure occurs, to reverse to working condition in the shortest possible time through *corrective maintenance.* It is of utmost importance that the thermal power plant water supply systems have provisions that allow the *repair intensity* (μ) and mean repair time MTTR to be in accordance with the requirements set in relation to reliability and operational readiness (Papić and Milovanović 2007). Key measures required for activities during system maintenance in the exploitation period include:

a. *preventive – ongoing maintenance*, enables the system to be maintained in operational condition by preventing failures (monitoring of work or monitoring, periodic controls and checks, planned servicing and maintenance, periodic diagnostics and testing);
b. *corrective maintenance*, reversing systems to working condition by repairing them after a failure (failure detection, cause diagnosis, repair or replacement, correctness check or trial operation);
c. *investment maintenance*, enables major overhaul operations on the system (reconstruction, revitalization, and modernization of the plant, verification of correctness or trial operation).

For water supply systems as a whole, as safety-sensitive systems, it is most often recommended to choose a maintenance strategy according to reliability or condition-based maintenance, for which a detailed study of possible failures and identification of the legality of their occurrence is necessary.

Convenience of maintenance, as the probability that will the system being maintained remain operational or will be reversed to it in a given time interval, requires the choice of a system disposition that allows preventive and corrective maintenance to be performed in accordance with the required mean repair time MTTR and repair intensity (μ). In addition to ensuring accessibility to the elements of the system that are being maintained, this includes the installation of *emergency closures* (in the event of adverse events in the system, parts of the system that may have more serious consequences are automatically protected), and the provision of *repair closures* (allowing the isolation of malfunctioning parts for repair), prediction of *reservoirs* along the pipeline route (enable maintenance of vital functions of the system during repair), selection of *measurement equipment* disposition and information system that enables failure detection in accordance with the required MTTR. In the case of more complex water management systems, the following are additionally considered: the choice of the disposition of *ring pipelines* with the supply of important consumers from several directions, the prediction of *compensation basins* for the provision of ecological flows and during the overhaul of facilities, the use of equipment of the same type within one system (more rational from the point of view of spare parts), the use of the same range and types of overhaul closures, in order to meet multiple hydro nodes with one set of closures, bridging some vital devices with a bypass pipeline to ensure some minimum system functions during maintenance and repairs, etc.

The revitalization, reconstruction, and modernization of the water supply system of the thermal power plant is carried out after the end of the basic working life, with the aim of further extending it in order to follow the extended working life of the thermal power plant for an additional 15–20 years. The great complexity and safety sensitivity of hydrotechnical systems require timely determination of the period of replacement and modernization of safety-vital parts of the thermal power plant water supply system (Milovanović and Miličić 2012). For these reasons, it is necessary to take care of the maintenance and renewal of the thermal power plant's water supply system, especially at the level of:

a. study considerations of the justification of the investment and the conceptual design: analysis of the system disposition from the point of view of maintenance convenience, analysis of the function of emergency and overhaul closures in all phases of system maintenance (especially important for systems with long pipelines, pumping and booster stations, tanks), evaluation of variant dispositions of the system from the point of view of convenience for maintenance,
b. main project: system maintenance project, with an analysis of the average and maximum planned times of corrective and preventive maintenance, information system project, and other systems to support system maintenance and renewal (expert system as an advisory system for maintenance needs, "on line" monitoring, etc.), analysis of the operability of system restoration and required contents (spare parts warehouses, access roads, etc.).

2.5.3 Availability of the thermal power plant water supply system

Availability (A) defines the preparedness (synonym: readiness) of the water supply system of the thermal power plant to be able to accept the execution of a continuous supply of feed water in the required quantity of appropriate quality when it is requested at some point in time t. It is determined by *the availability function A(t)*, which shows the probability that the system can accept the function of a continuous supply of water to the thermal power plant at the moment of time t. The size of the repair intensity (μ) significantly affects the readiness of the system, so it is considered in parallel with the analysis of maintenance convenience.

A system that is in two states is considered: "RUNNING State" with failure intensity λ, or "FAILURE State" with repair intensity μ. The availability function $A(t)$ for a certain repairable element of the system in the general case, by introducing the functions λ and μ is defined in the form:

$$A(t) = \frac{\mu}{\lambda + \mu} + \frac{\mu}{\lambda + \mu} \cdot e^{-(\lambda+\mu)\cdot t} \tag{2.1}$$

For the case when $t \rightarrow \infty$ the availability function, for one element of the system, tends to the size:

$$A = \frac{\mu}{\lambda + \mu} \tag{2.2}$$

The quantity A represents the essential performance of the system as a stationary coefficient of availability. For the case of the exponential distribution, $\mu = 1/\text{MTTR}$ is valid, where MTTR – Mean Time To Repair. Repair intensity μ affects the reliability and availability of complex systems in the following way:

a. in the case of serial connection of elements, the repair intensity μ has no effect on the reliability of the complex system, but significantly increases the availability of the system (very important for pipelines),
b. in the case of the system, the parallel connection of the elements (reserve pumps in the parallel connection of the thermal power plant pump station) significantly increases the possibility of repair and the availability and reliability of the system (one of the additional advantages of the system in parallel connection of the elements).

In the design phase at the conceptual design level, availability as an important aspect of reliability is included through the analysis of the availability of parts of the system and the system as a whole (for certain ranges of MTTR and repair intensity μ) and the analysis of the impact of system availability on the availability of the system and measuring equipment for monitoring the correctness of the system.

2.5.4 Hydrological reliability

Hydrological reliability represents the probability of calculated water to which some vital organs of hydrotechnical facilities are dimensioned, as well as water quality protection measures, which are determined in relation to calculated low water, often low monthly water with a probability of 95% (obligatory biological minimum flow). Analyses of hydrological reliability are usually done only during the design of facilities and systems, without subsequent verification, which is a big omission (insufficient biological minimum after taking water to feed the thermal power plant system). It is not a rare case that insufficiently long hydrological series of water flow data are used in the planning of budget water systems, which can lead to underestimated or overestimated budget waters. During the exploitation of the system for supplying the thermal power plant with feed water, hydrological information is additionally collected, so the available hydrological series are longer and more reliable. On the basis of supplemented hydrological series of data, it is necessary to periodically check the calculation of water on the basis of which the system was implemented. In the case of major deviations, corrections are necessary through the revitalization of the thermal power plant's water supply system. On the other hand, as a consequence of carrying out works in the basin (regulatory works with the exclusion of inundations in which there was the retention of water flow, the raising of embankments) but also due to the effects of climate change over time, there may be a significant change in hydrological regimes on larger rivers (e.g., an increase of large and decrease of small waters). Therefore, it is necessary to periodically check the hydrological reliability (through re-analysis of computational large waters, using new, extended hydrological series, etc.). Checking the budget waters, with which the systems are designed, is necessary for the safety of the system and the environment. A good example is the protective embankments along our largest rivers, especially in the areas of very sensitive and important coastal defenses (REK Kostolac, TPP Nikola Tesla A and B in Obrenovac, large coastal settlements). The embankments along the Danube, Sava, Tisa, and Velika Morava rivers were dimensioned based on hydrological series until around 1965 because they were built after the great floods of 1965 and during the construction of HPP Đerdap. In the meantime, numerous regulatory works were carried out on the upstream parts of those rivers, especially the Danube and the Tisa, which led to gradual changes in the high-water regime, which were manifested by the acceleration of the concentration and propagation of high-water waves and an increase in the Qmax value. Therefore, it is quite certain that in the current conditions, the embankments do not provide the required level of protection against high waters with a probability of 1%, with protective heights of about 1.5 m, as planned and realized. It is necessary to carry out checks on the hydrological reliability of those protective systems as soon as possible, along the entire length of the rivers (Đorđević and Dašić 2016). It is necessary to determine new calculation waters for all key water measuring stations on those rivers, determine the level lines for those waters, compare them with the embankment elevations, and based on that, establish reserve protective heights ("freeboard") in the current state and determine critical insufficiently protected sections, according to priorities. At the same time, priority is given to stocks that protect large cities and large technological systems (e.g., REK Kostolac, REIK Kolubara, TE Obrenovac). That is why, in the current conditions, a realistic assessment of the level of protection against high water has priority and should be connected with the obligation to create flood risk maps, in accordance with the relevant EU Directive (FRMD – *Flood Risk Management Directive 2007/60/EC*).

2.5.5 Structural reliability

Structural reliability is expressed by the probability (R_k) that the working load (s) of the structure of the object will be less than the critical load (r). That starting point can be seen in Figure 2.21, where the distributions of working (s) and critical load (r) are given, as well as the risk – the probability of failure of the structure, which is defined by the overlapping area of those two

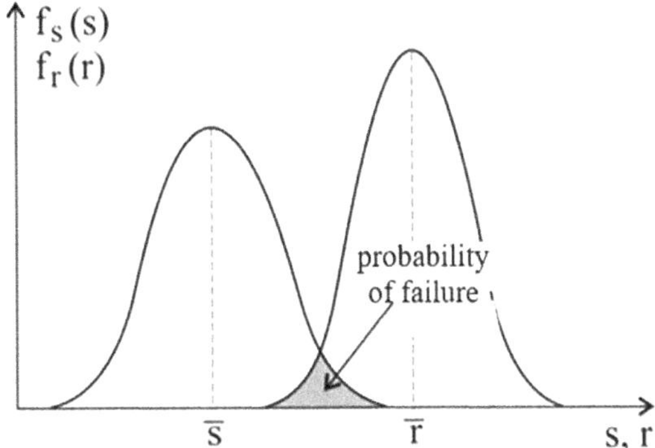

Figure 2.21 Distribution of working and critical load of the structure.

surfaces. The workload (s) is affected by a series of stochastic influences, so it is distributed by the distribution function $f_s(s)$. The limit-bearing capacity of the structure is also a stochastic category and is distributed according to the corresponding distribution $f_r(r)$.

Most often, in the analysis of the stability of the structure, the starting point is the assumed load scheme and the defined specific authoritative resistance of the structure. The relevant working load $s = S$, as well as the resistance value of the structure $r = R$, are extracted from the distributions $f_s(s)$ and $f_r(r)$, and the safety of the structure is defined through the safety coefficient. At the same time, the actual working load (s) can be higher than the selected value S ($s > S$), while the resistance (r) can be lower than the authoritatively selected value R ($r < R$), which is a consequence of the stochastic fact related to the deterministic starting point of uncertainty. The elimination of these uncertainties is realized through the adoption of a certain safety coefficient (k), greater than 1, most often $k > 1.25$. Since the random processes in which there are many mechanical, geophysical, and other influences as stochastic categories are superimposed within the random processes (s) and (r), according to the logic of stochastic processes (from a small size it can be even smaller, and from a large one even larger), it is clear that the structural reliability will certainly be less than 1 (regardless of the adopted safety coefficient). There remains the problem of not knowing the real reliability of the construction in that deterministic approach (disguised by an apparently high safety coefficient). Risk, as the probability of breaking the stability of the structure, given by the area marked with hatching in Figure 2.21, which shows the probability of failure in reality, i.e., the probability that the working load will still exceed the critical carrying capacity (resistance) of the structures at some point). Therefore, it is possible to obtain the reliability of the object exposed to the effect of the working load from the condition of existence of the probability of failure-free operation if the resistance of the element (r) is not exceeded by the working load (s), and the condition of failure-free operation is $r > s$. From this condition for the known distributions $f_s(s)$ and $f_r(r)$, a general equation for the structural reliability (R_k) of the elements or the object as a whole is derived:

$$R_k = \int dR = \int_{-\infty}^{+\infty} f(s) \left[\int_{s}^{\infty} f(r) dr \right] \cdot ds \qquad (2.3)$$

2.5.6 Hydraulic reliability

Hydraulic safety (R_h) represents one of the forms of structural safety of those objects whose safety is threatened in the case when the level or flow in some part of the system becomes greater than some critical value, so the object overflows. It is especially important for secondary water supply systems, such as embankment dams and embankments. During the design, this type of safety was implicitly included. By designing for some authoritative water (level and/or flow), as well as adopting an additional protective height of the structure (Δh), this form of structural safety was implicitly included at the stage of design and development of project documentation. By using hydrological models, the function of the frequency of extreme flows $f(Q)$ is first defined, and then the project high water Q_{pv} is determined, according to the previously adopted probability of exceeding the annual maxima. The design flow is transformed via the flow curve $H = H(Q)$ into the corresponding authoritative design level $H_{rv} = H(Q_{rv})$. Additional protective height ("freeboard") Δh increases the safety of the facility due to hydrological-hydraulic uncertainties. The project elevation of the embankment crown $Z_{kn} = H_{rv} + \Delta h$ was obtained. System failure occurs when $H(Q) > Z_{kn}$, and P_f – the probability of system failure is equal to: $P_f = P[H(Q) \succ Z_{kn}] \prec P_E$. It is smaller than P_E due to the effect of the protective height Δh, which can be seen in Figure 2.22.

Designing based on the probability of exceeding is only acceptable at the lowest level of development of project documentation (level of general and/or

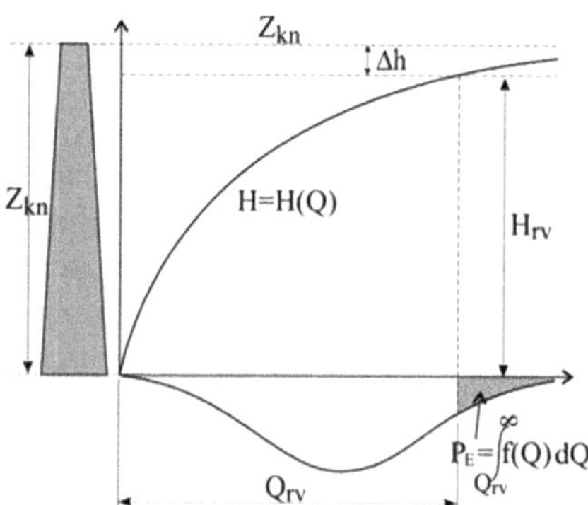

Figure 2.22 Design concept based on probability of exceedance.

preliminary designs). Determining the protective height Δh to the height of the embankment crown is particularly sensitive. Most often, in the case of water management systems, it is taken in the range of 1–2 m, depending on the additional consideration of uncertainties that he failed to consider in the initial phase of the documentation development (reliability of determining the calculated high water for which the embankment is dimensioned (e.g., $Q_{1\%}$), the height of the waves from the wind that can be expected on that section, etc.). At higher levels of design, the protective – safety performance of the system will have to be expressed by reliability values R.

Then the hydraulic reliability can be calculated in principle in the same way as in the case of structural safety, only the nature of the working load and critical bearing capacity (resistance) changes. In the case of hydraulic safety, the working load (s) is the high-water levels, together with the supplementary influence (wave heights from the wind), which are stochastic categories. The resistance of the system (r) represents the effective height of the objects (crown of embankments and dams, etc.), together with the additional overhang Δh, which is also distributed according to some distribution (reduction to the case shown in Figure 2.21, only for other physical quantities (s) and (r)).

2.5.7 Management reliability of the system

In addition to primary (occurring under normal operating conditions, without environmental influences) and secondary (occurring due to some adverse external influences or due to work in overload conditions) failures of the water supply system, there are also *failures due to incorrect management* (the result of operational errors personnel, to whom the plant was put into some irregular mode of operation). Conducted analyses of human errors in energy facilities showed that man is the weakest link in the management chain (Đorđević 1996), and some indicators of unexpectedly high unreliability of people were also pointed out in Đorđević and Milanović (1996), Dašić and Đorđević (2003) and Đorđević and Dašić (2015). The bottom line is that the human ability to analyze and use information in normal working conditions is in the range of about 5–10 bit/s, in short intervals of strenuous work it can increase to about 15 bit/s. However, under conditions of stress, due to the "panic control block", this ability to perceive and use information is reduced to only about 0.5 bit/s. Research has established that the average frequency of human error is 0.2–0.3 if a procedure should be performed in a precisely defined order, or, with a probability of 0.5, the operator (if not warned by persistent sound signals) will not notice the light warning on the control panel that some vitally important switches are in an illegal position. The probability of successful execution of set management tasks during a certain management time represents management reliability. It implies the probability that a person as an element of the system will not make management mistakes in the time interval t. Research has shown that man as a management organ behaves, from the point of view of reliability, completely analogous to all other systems, so the man's management reliability is R_c:

$$R_c(t) = \exp\left[-\int_0^t \lambda_c(t) \cdot dt\right] \tag{2.4}$$

where λ_c – the intensity of human errors in the management process (analogous quantity with the intensity of system failure λ), which in the case of human errors is best approximated with the Weibull distribution. The results of the conducted analyses show that the magnitudes λ_c intensity of human errors during management are surprisingly high. When designing control devices and defining procedures for performing safety-sensitive operations, it is necessary to include this segment. In order to eliminate the unreliability of man in the process of managing the water supply system, it is necessary to additionally use expert systems, as a management support system and to the greatest extent possible, the exclusion of man from the process of direct independent operational management in all crisis situations (flood protection, emergency situations). Man, as an irreplaceable management organ, with provided management support in the form of software, much more easily and studiously analyzes possible unfavorable scenarios that may happen, and for which he finds the best management responses for acting in crisis situations. Aspects of management reliability are solved during the development of the main project when the management system – its hardware and software parts – is designed. In project tasks, the management reliability of control devices should be analyzed and planned (according to position, grouping, start-up procedures, etc.) so that the required management reliability is achieved, and then management support (control software, expert systems, etc.) should be defined to increase reliability at to system management (Đorđević and Dašić 2016).

2.5.8 Reliability of the monitoring system

With the increase in the complexity of hydrotechnical systems for water supply and the increase in the number and safety sensitivity of their interactions with the environment, systems for observing (monitoring) objects are becoming more and more important. In some water supply systems, which have longer pipelines, the realization of the repair process depends on the speed of detection and localization of the failure. This aspect of reliability should be considered in the phase of creating the main project when designing the monitoring system and choosing the equipment for it. The selection of monitoring system equipment should be made according to reliability criteria. In parallel with this, it is necessary to perform: failure tree analysis, optimization of monitoring system reliability allocation, analysis of functions of the availability of its parts and the system as a whole, for certain ranges of MTTR and repair intensity μ (for repairable devices and parts of the system during exploitation), as well as analysis of monitoring of the system from the point of view of convenience of maintenance and the influence of the availability of the system on the disposition of its individual parts.

2.5.9 Reliability from an environmental point of view

As water supply systems have two-way interactions with the environment, environmental influences on the systems are independent of random geophysical phenomena (earthquakes, extremely high water, etc.). Also, they are appropriately included in structural and hydraulic reliability analyses. As some influences on the environment come from the direction of the system (the occurrence of so-called induced seismicity in the zone of large accumulations, as well as checking the stability of geotechnical formations in the zone of stagnation, etc.), it is necessary to additionally determine these forms of reliability. For the purpose of their determination, adequate methods are used.

2.6 DISCUSSION OF RESEARCH RESULTS

In TPP, water is used for cooling (through recirculation and flow) and for the transfer of heat energy, and surface water is the main source of technological water for TPP. The water used in the thermodynamic cycle at TPP must be completely clean. The methods of preparing and purifying water that are used today are very diverse. It is necessary to keep in mind that a very large amount of water is needed for the normal operation of the thermal power plant. About 93% of the total water is needed for cooling the steam in the condenser, and about 7% for other needs. The complete need for fresh water considering the three main consumers in the thermal power plant can be divided into: cooling water (used for cooling the condenser and cooling auxiliary equipment at the thermal power plant), service water (used as sanitary water, flushing water, water for manipulation with ash and slag, added water for the desulphurization plant and for sealing pumps) and water of high degree of purity (required for the water-steam system – feed water of the circular process, as a means of adding chemicals, for the regeneration of ion exchangers and as laboratory water). Based on the data from exploitation, the total durations of out-of-operation states of the units are calculated, and at the level of the entire observed time period, the following basic data: number of outages and downtimes, average duration of downtimes, causes of outages and downtimes, unavailability of component units and the thermal power plant as a whole, total undelivered electricity energy. Statistical processing of the recorded data calculates the reliability indicators of the thermal power plant system, as well as its individual units. We call the set of data created by recording driving events and the results of their statistical processing "statistics of driving events". The statistics of operating events are also used for comparison with other analog systems and evaluation of the business performance of the company that manages the thermal power plant, as well as for planning studies and probabilistic simulations of the operation of the system as a whole. Stoppages of units and components of the thermal power plant system can be considered random events to which a certain probability is attached. Before any forecast of the failure intensity of the entire water supply system at the thermal power plant, it is necessary to know or be able to calculate the failure intensity for each of its individual components, as well as the way of their mutual connection in the disposition scheme (serial, parallel or combined connection). In the case of the impossibility of predicting the severity of failure of individual components in the water supply system of the thermal power plant, it is necessary to try to use structural and parametric analysis to determine the components, that is, the forms of failure that most significantly affect the failure of the entire assembly. The results of the monitoring of the water supply system of TPP are also used to define the state of the level in the watercourses from the point of view of regulating the watercourses and protection against the harmful effects of water, including forecasts for the implementation of flood protection.

2.7 FINAL CONSIDERATIONS

The safety of the functioning of the water supply system of the thermal power plant and its accompanying energy equipment is determined by a large number of different factors. The construction quality of the materials used, manufacturing technology, assembly quality, operating modes and conditions of exploitation, maintenance strategies, along with the quality of the feed water during system operation, are the most common causes of complete or partial loss of functional properties, i.e., system failure. Ensuring the continuous supply of water for the thermal power plant, reliability in the operation of the water supply system for the thermal power plant, along with the convenience of maintenance and renewal, and operational readiness, represents important segments of the fulfillment of the primary function of supplying water to thermal power facilities. At the same time, they must meet additional reliability requirements from the point of view of failure (breakdowns), as well as the convenience of maintenance and renewal, on which the availability of the auxiliary water supply system of the thermal power plant depends on the wider hierarchical power system. The analysis of the provision of certain aspects of reliability, starting from the stage of elaboration and design and production of technical documentation, then through the creation of individual segments, their assembly, and trial operation within the thermal power plant, significantly affects the choice of achieving the required operational reliability and availability of the thermal power plant system as a whole. On the other hand, hydrological and hydraulic reliability as dynamic categories that, in addition to the analysis in the design phase, must be additionally checked in the exploitation phase of the object. Experience gained on the hydraulic behavior of thermal power plant water supply facilities needs to be supplemented with elements of management reliability with the aim of possible human error in operational management. Additional monitoring during use, as well as selection of the content of the monitoring system and verification of its reliability, create the necessary conditions for the full operation and functionality of the water system for supplying the power plant with water (Milovanović 2000).

REFERENCES

Adamovic, Ž. 1998. "*Technical diagnostics.*" Institute for Textbooks and Teaching Aids. Belgrade.

Adamović, Ž., Ašonja, A. 2014. *Methodology of Scientific and Research Work: Science – Methodology – Technology*. Serbian Academic Center, Novi Sad.

Asadzadehd, S.M., Salehib, N., Firoozic, M. 2015. "Condition-based maintenance effectiveness for series-parallel power generation system – A combined Markovian simulation model." *Reliability Engineering & System Safety*, 142, 357–368.

Ašonja, A. 2005. "Damage to rolling bearings on agricultural machines." *JUMTO – Journal of Scientific Society of Power Machines, Tractors and Power Machines*, 10(4), 120–125.

Ašonja, A. 2012. "Diagnostics of the state of rolling bearings and its influence on the reliability of cardan shafts on agricultural machines." Ph.D. Thesis, University of Novi Sad, Technical Faculty Mihajlo Pupin Zrenjanin, Serbia.

Ašonja, A., Adamović, Ž., Gligorić, R., Pastuhov, A.G., Mikić, D., Desnica, E. 2012. *Technical Solution: Laboratory-Experimental Table for Testing the Reliability of Agricultural Cardan Shafts, Model: "ANA", Type: 23-26-26-04*. Serbian Academic Center, Novi Sad.

Belitser, E., Serra, P., Zanten, H. 2015. "Rate-optimal Bayesian intensity smoothing for inhomogeneous Poisson processes." *Journal of Statistical Planning and Inference*, 166, 24–35.

Branković, D. 2018. "Reliability optimization of the production system for hygiene paper manufacturing by using the concept of condition based maintenance." Ph.D. Thesis, University of Banja Luka, Faculty of Mechanical Engineering, Banja Luka.

Branković, D., Milovanović, Z., Janičić-Milovanović, V. 2023. "Chapter 11: Maintenance and safety of industrial systems: Developed model for assessing the criticality of elements of technical systems." *Advances in Reliability Science Reliability Modeling in Industry 4.0*, edited by: Mangey Ram, Liudong Xing, 327–380. Elsevier, Amsterdam, Netherlands.

Branković, D., Milovanović, Z., Papić, L. 2021. "Chapter 10 – Analysis of the technical system reliability assessment with the application of technical diagnostics." *The Handbook of Reliability, Maintenance, and System Safety through Mathematical Modeling*, edited by: Amit Kumar, Mangey Ram, 373–417. Academic Press, UK.

Concawe. 2010. *Performance of European Cross-Country Oil Pipelines*. Report No. 4/10. Brussels, Belgium. www.concawe.be.

Dašić, T., Đorđević, B. 2003. "Method for determining the reliability of complex water management systems (NETREL)." *Vodoprivreda*, 35, 203–204.
Đorđević, B. 1996. "Human reliability as part of a complex management system." *Vodoprivreda*, 28(3–4), 181–190.
Đorđević, B., Dašić, T. 2015. "Expert systems for planning and operational implementation of flood defense." *Vodoprivreda*, 47, 276–278.
Đorđević, B., Dašić, T. 2016. "Reliability categories that must be checked during the planning and use of water management systems." *Vodoprivreda*, 48, 279–281.
Đorđević, B., Milanović, T. 1995. "Security of complex water management systems and the possibility of its allocation in the planning phase." *Vodoprivreda*, 27, 153–155.
Đorđević, B., Milanović, T. 1996. "Failure trees as an efficient method for reliability analysis of complex systems." *Vodoprivreda*, 28, 159–160.
He, Z., Gong, W., Xie, W., Zhang, J., Zhang, G., Hong, Z. 2016. "NVH and reliability analyses of the engine with different interaction models between the crankshaft and bearing." *Applied Acoustics*, 101(1), 185–200.
Henneaux, P. 2015. "Probability of failure of overloaded lines in cascading failures." *International Journal of Electrical Power & Energy Systems*, 73, 141–148.
Lee, M., Sohn, K. 2015. "Inferring the route-use patterns of metro passengers based only on travel-time data within a Bayesian framework using a reversible-jump Markov chain Monte Carlo (MCMC) simulation." *Transportation Research Part B: Methodological*, 81(1), 1–17.
Liu, Y., Li, C. 2016. "Complex-valued Bayesian parameter estimation via Markov chain Monte Carlo." *Information Sciences*, 326(1), 334–349.
Liu, Z., Liu, Y., Cai, B., Zhang, D., Zheng, C. 2015. "Dynamic Bayesian network modeling of reliability of subsea blowout preventer stack in presence of common cause failures." *Journal of Loss Prevention in the Process Industries*, 38, 58–66.
Marsland, S. 2014. *Machine Learning: An Algorithmic Perspective*, second edition. *CRC Machine Learning & Pattern Recognition Series*. Chapman & Hall, USA, p. 452.
Mikić, D. 2016. "Modeling of mechanical technical systems by using matrix of transformation", Ph.D. Thesis, University of Novi Sad, Technical Faculty Mihajlo Pupin Zrenjanin, Serbia.
Miličić, D., Milovanović, Z. 2010. *Monograph of Energy Machines – Steam Turbines*. University of Banja Luka, Faculty of Mechanical Engineering Banja Luka, Banja Luka.
Milošević, D. 2015. "Models for ensuring the reliability of complex plants in thermal power plants." Ph.D. Thesis, University of Novi Sad, Technical Faculty Mihajlo Pupin Zrenjanin, Serbia.
Milovanović, Z. 2000. "Modified method for reliability evaluation of condensation thermal electric power plant." Ph.D. Thesis, University of Banja Luka, Faculty of Mechanical Engineering Banja Luka, Banja Luka.
Milovanović, Z. 2003. *Optimization of Power Plant Reliability*. University of Banja Luka, Faculty of Mechanical Engineering Banja Luka, Banja Luka.
Milovanović, Z. 2011a. *Monographs: Energy and Process Plants, Volume 1: Thermal Power Plants – Theoretical Foundations*. University of Banja Luka, Faculty of Mechanical Engineering Banja Luka, Banja Luka.
Milovanović, Z. 2011b. *Monographs: Energy and Process Plants, Volume 2: Thermal Power Plants - Technological Systems, Design and Construction, Exploitation and Maintenance*." University of Banja Luka, Faculty of Mechanical Engineering Banja Luka, Banja Luka.

Milovanović, Z., Branković, D. 2021. "*Maintainability of industrial systems.*" DQM Monograph Library Quality and Reliability in Practice, The Research Center of Dependability and Quality Management DQM, Book 11, Prijevor, Serbia.

Milovanović, Z., Branković, D., Janičić-Milovanović, V. 2023. "Chapter 10: Efficiency of condensing thermal power plant as a complex system – An algorithm for assessing and improving energy efficiency and reliability during operation and maintenance." *Advances in Reliability Science Reliability Modeling in Industry 4.0*, edited by: Mangey Ram, Liudong Xing, 233–326. Elsevier.

Milovanović, Z., Dumonjić-Milovanović, S., Milašinović, A., Knežević, D. 2018. "Efficiency of operation of 300 MW condensing thermal power blocks with steam parameters in sliding pressure mode." *Thermal Science*, 22(Suppl. 5), S1371–S1382.

Milovanović, Z., Papić, L., Dumonjić-Milovanović, S., Milašinović, A., Knežević, D. 2017. "Sustainable energy planning: Technologies and energy efficiency." *DQM Monograph Library Quality and Reliability in Practice*, Book 9, University of Banja Luka, Faculty of Mechanical Engineering Banja Luka, Prijevor.

Milovanović, Z., Miličić, D. 2012. *Steam Turbines for Cogeneration Energy Production, Library of Monographs, Energy Machines 3*. University of Banja Luka, Faculty of Mechanical Engineering Banja Luka, Banja Luka.

Zdravko Milovanović, Ljubiša Papić, Valentina Janičić-Milovanović, Snježana Milovanović, Svetlana Dumonjić-Milovanović, Dejan Branković. 2020. "Methods for prognosis and optimization of energy plants efficiency in starting step of life cycle." *Advances in Reliability Analysis and Its Applications. Springer Series in Reliability Engineering*, edited by: M. Ram, H. Pham, 95–148. Springer, Switzerland.

Milovanović, Z., Papić, L., Milovanović, S., Janičić Milovanović, V., Dumonjić-Milovanović, S., Branković, D. 2021. "Chapter 8 – Qualitative analysis in the reliability assessment of the steam turbine plant." *The Handbook of Reliability, Maintenance, and System Safety through Mathematical Modeling*, edited by: Amit Kumar, Mangey Ram, 179–313. Academic Press, UK.

Očević, H. 2015. "Model procjene vjerojatnosti neželjenih događaja u informacijskom sustavu upotrebom Bayesovog teorema." Ph.D. Thesis, Josip Juraj Strossmayer University in Osijek, Faculty of Electrical Engineering, Osijek, p. 125.

Papić, L., Milovanović, Z. 2007. "Maintenance and reliability of technical systems." *DQM Monograph Library Quality and Reliability in Practice*, The Research Center of Dependability and Quality Management DQM Book 3, Prijevor, Serbia.

Paunjorić, P. 2016. "Maintenance methods and their influence on the reliability of complex machines in surface mines." Ph.D. Thesis, University of Novi Sad, Technical Faculty Mihajlo Pupin Zrenjanin, Serbia.

Pešić, M. 2016. "Upotreba logističke regresije u modeliranju verovatnoće bankrota preduzeća." Master Thesis, University of Novi Sad, Faculty of Science, Department of Mathematics and Informatics, Novi Sad, Serbia, p. 79.

Petrović, Z., Radičević, B., Vukičević, M., Bjelić, M. 2009. "Classification of activities based on RCM on the example of a pumping plant." *IMK-14 Research and Development*, 15(1–2), 185–201.

Radionov, A., Evdokimov, A., Petukhova, O., Shokhina, G., Yabbarova, L. 2016. "Vibrodiagnostic surveying of industrial electrical equipment." *IEEE NW Russia Young Researchers in Electrical and Electronic Engineering Conference*.

Sandler, G.H. 1983. *System Reliability Engineering*. PrenticeHall, Englewood Cliffs, NJ.

Shina, J., Junb, H. 2015. "On condition based maintenance policy." *Journal of Computational Design and Engineering*, 2(2), 119–127.

Šimunović, V. 2012. "Mikrobiološki poticana korozija zavarenih spojeva nehrđajućih čelika u vodi." Ph.D. Thesis, Faculty of Mechanical Engineering and Shipbuilding, University of Zagreb.
Todorović, D., Zelenović, D. 1994. *Efektivnost sistema u mašinstvu*. Science Book, Belgrade.
Ugechi, C., Ogbonnaya, E., Lilly, M., Ogaji, S., Probert, S. 2009. "Condition-based diagnostic approach for predicting the maintenance requirements of machinery." *Engineering 2009*, 1, 177–187.
Wu, X., Wu, X. 2015. "Extended object-oriented Petri net model for mission reliability simulation of repairable PMS with common cause failures." *Reliability Engineering & System Safety*, 136, 109–119.
Yi, C., Bao, Y., Jiang, Y., Xue, Y. 2015. "Modeling cascading failures with the crisis of trust in social networks." *Physica A: Statistical Mechanics and Its Applications*, 436(15), 256–271.

Chapter 3

Semi-Markov analysis of systems with Weibull interface

Veeachamy Mariappan, Ajit Srividya, and Milind Sakhardande

3.1 INTRODUCTION

Many real-life phenomena are time oriented and governed by random mechanisms. The entire queuing theory is the best example to support this fact. Such a process, known as a stochastic process, is a sequential arrangement of random variables $\{X_t\}$, where $t \in T$ is a time or sequence index. The range space for X_t is also known as state space, which may be either discrete or continuous. The applications in reliability engineering demand discrete state space. The stochastic process is in exactly one of n + 1 mutually exclusive and exhaustive states, at a given time 't', wherein the states are normally labeled 0, 1, 2, …, n. The random variables X_1, X_2, … might represent the number of customers awaiting service in a queue at times 1 minute, 2 minutes, and so on after the booth opens. It may be borne in mind that at time t, X_t is a random variable. As we advance in time, it will represent another random variable X at the advanced time. The process is called discrete stochastic or continuous stochastic, based on whether the sequence index or time is discrete or continuous.

A stochastic process is said to have Markovian property when and only when

$$P\{X_{t+1} = j \mid X_t = i\} = P\{X_{t+1} = j \mid X_t = i, X_{t-1} = i_1, X_{t-2} = i_2, \ldots, X_0 = i_t\} \quad (3.1)$$

For t = 0, 1, 2, …, and every sequence j, i, i_1, …, i_t.

Eq. (3.1) can be written as P_{ij} and extended to continuous Markov chain (process), with minor change in the terminology as t is on a continuous scale as $\{X(t+s) = j \mid X(s) = i\} = P_{ij}(t)$. In reliability engineering the continuous time Markov chain is more applicable. We denote the entire scenario as $P_n(t)$ which means the probability that the system being in state 'n' at t.

3.2 FAILURE INTERACTION

In reliability engineering, for system reliability estimation, the reliability of the components should always be considered for analysis. However, in real life, the failure of one component will affect the failures of other components

DOI: 10.1201/9781003546214-3

depending on the system configuration. The Markov process concept helps in quantifying the failure interactions.

3.2.1 Reliability analysis using Markov model

To quantify the issue of failure interactions, the system configuration should be addressed in terms of states that are defined as a particular combination of operating and failed components as demonstrated in Table 3.1.

3.2.1.1 *Markov formulation*

The probability of an item being in state 'i' at time 't' is denoted as $P_i(t)$, using which the system reliability can be obtained as shown in Eq. (3.2) with boundary conditions as stated in Eq. (3.3).

$$R(t) = \sum_{i \in o} P_i(t) \tag{3.2}$$

$$P_1(0) = 1;\ P_i(0) = 0, \forall\ i \neq 1;\ \sum_i P_i(t) = 1 \tag{3.3}$$

3.2.1.2 *Procedure*

For easy calculations, the following steps are proposed:

1. Construct a state table enumerating all states that would be 2^N. Then scan and retain only the feasible states as per the description of the system.
2. Convert the state table into state transition diagram (STD).
3. Formulate the Laplacian state transition matrix (LSTM) equation from STD using the designated rules.
4. Solve the matrix equation so obtained for $\overline{P}_i(s)$.
5. Take Laplace inverse and get $P_i(t)$.
6. Find appropriate quantity by properly grouping/summing up $P_i(t)$.

Table 3.1 State table for three units

	State No.							
Component	*1*	*2*	*3*	*4*	*5*	*6*	*7*	*8*
A	O	X	O	O	X	O	X	X
B	O	O	X	O	X	X	O	X
C	O	O	O	X	O	X	X	X

O – an operational component (operable state) and X – a failed component (failed state).

Rules for LSTM:

Normally, formulating state transition differential equations (DEs) for a state is a cumbersome process. Further transforming those DEs into algebraic equations is time consuming and needs application of Laplace transforms (LTs). By following the proposed rules, it is easier to form Laplacian algebraic equations directly.

1. A row represents the state under consideration, whereas the different column cells pertaining to the row represent the transition from respective states to the state under investigation.
2. The pivotal element of the matrix is the location of state under study.
3. Plug in Laplace operator, s to all the principal diagonal elements only.
4. Weights of arrow leaving a state are considered to be positive and those entering to the state under study as negative.
5. Assign these signed weights appropriately to the cells of the matrix.
6. Thus, the matrix formed is the coefficient matrix in parlance of Linear algebra and is denoted by A.
7. Let $\overline{P}_i(s)$ as column vector form variable matrix and is denoted by X.
8. $[1\ 0\ 0 \ldots 0]^T$ forms the constant matrix (row vector) and is denoted by C.

3.2.1.3 *Case I: Two-unit system*

Let us take a case of two units for illustration and realize the advantages of the LSTM rule. The total no. of states is $2^2 = 4$, and since nothing is mentioned about the system whether series or parallel, all the four states enumerated are feasible states. Therefore, the state table and STD are shown in Table 3.2 and Figure 3.1. The calculations reveal the advantages of LSTM.

Formulation of DE for State 1:

At 't + Δt' probability of being in State 1 = (at t at State 1 and remain in State 1 only during Δt), being no other possibility, which gives

$$P_1(t+\Delta t) = P_1(t)\left[1-(\lambda_a+\lambda_b)\Delta t\right]$$

$$\lim_{\Delta t \to 0} \frac{P_1(t+\Delta t) - P_1(t)}{\Delta t} = -(\lambda_a+\lambda_b)P_1(t)$$

Table 3.2 State table for two-unit system

	State No.			
Component	*1*	2	3	4
A	O	X	O	X
B	O	O	X	X

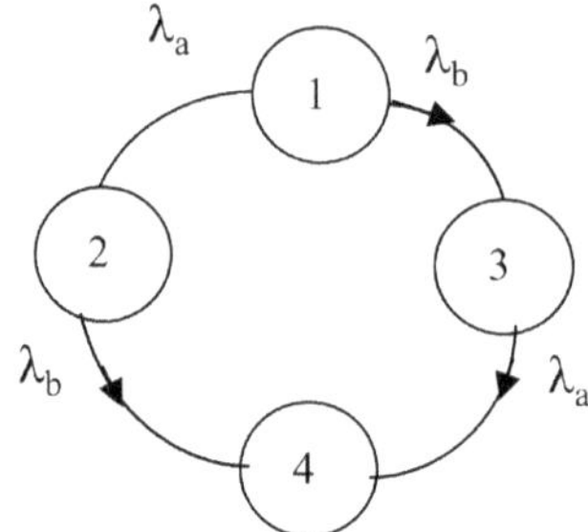

Figure 3.1 STD for two-unit system.

Using the first principles of calculus

$$\frac{d}{dt}P_1(t) = -(\lambda_a + \lambda_b)P_1(t) \tag{3.4}$$

Solving

$$dP_1(t) = -(\lambda_a + \lambda_b)P_1(t)dt$$
$$\frac{dP_1(t)}{P_1(t)} = -(\lambda_a + \lambda_b)dt$$

Indefinitely integrating both sides will give

$$\int \frac{dP_1(t)}{P_1(t)} = \int -(\lambda_a + \lambda_b)dt$$

$$lnP_1(t) + C_1 = -(\lambda_a + \lambda_b)t + C_2$$

$$P_1(t) = e^{-(\lambda_a+\lambda_b)t+C_3}$$

$$P_1(t) = Ce^{-(\lambda_a+\lambda_b)t} \tag{3.5}$$

Applying the boundary conditions, $P_1(0) = 1$
$1 = C.1 \Rightarrow C = 1$, thus Eq. (3.5) becomes

$$P_1(t) = e^{-(\lambda_a+\lambda_b)t} \tag{3.6}$$

Similarly for State 2:

At ‘t + Δt’ probability of being in State 2 = (at t at State 2 and remain in State 2 only during Δt) or (at t at State 1 and migrate to State 2 during Δt), which gives

$$P_2(t + \Delta t) = P_2(t)[1 - \lambda_b \Delta t] + P_1(t)\lambda_a \Delta t$$

$$\lim_{\Delta t \to 0} \frac{P_2(t+\Delta t)-P_2(t)}{\Delta t} = -\lambda_b P_2(t)+\lambda_a P_1(t)$$

Using the first principles of calculus

$$\frac{d}{dt}P_2(t) = -\lambda_b P_2(t)+\lambda_a P_1(t) \tag{3.7}$$

Instead of investing type of ordinary DE (ODE) to solve, we can apply LT to transform DE to algebraic equation and more importantly it has boundary conditions to apply which makes utility of LT more appropriate. The required fundamentals to deal with Laplace's transformation are given in the following section.

Taking LT on first order differential gives with the boundary conditions $P_i(0) = 0$ for $i \neq 1$, we get

$$s\bar{P}_2(s) - \cancel{P_2(0)} = -\lambda_b \bar{P}_2(s) + \frac{\lambda_a}{s+\lambda_a+\lambda_b}$$

$$(s+\lambda_b)\bar{P}_2(s) = \frac{\lambda_a}{s+\lambda_a+\lambda_b}$$

$$\bar{P}_2(s) = \frac{\lambda_a}{(s+\lambda_b)(s+\lambda_a+\lambda_b)} \tag{3.8}$$

Using partial fraction, we can split the above fraction after which simple Laplace inverse can be taken.

$$\frac{\lambda_a}{(s+\lambda_b)(s+\lambda_a+\lambda_b)} = \frac{A}{(s+\lambda_b)} + \frac{B}{(s+\lambda_a+\lambda_b)}$$

Taking LCM

$$\frac{\lambda_a}{(s+\lambda_b)(s+\lambda_a+\lambda_b)} = \frac{A(s+\lambda_a+\lambda_b)+B(s+\lambda_b)}{(s+\lambda_b)(s+\lambda_a+\lambda_b)}$$

From the above equation

$$A(s+\lambda_a+\lambda_b)+B(s+\lambda_b) = \lambda_a$$

Substituting $s = -\lambda_b \Rightarrow A = 1$
Now substitute $s = -(\lambda_a + \lambda_b) \Rightarrow B = -1$

$$\therefore \frac{\lambda_a}{(s+\lambda_b)(s+\lambda_a+\lambda_b)} = \frac{1}{(s+\lambda_b)} - \frac{1}{(s+\lambda_a+\lambda_b)}$$

By taking Laplace Inverse, we get

$$L^{-1}\bar{P}_2(s) = L^{-1}\frac{\lambda_a}{(s+\lambda_b)(s+\lambda_a+\lambda_b)} = L^{-1}\frac{1}{(s+\lambda_b)} - L^{-1}\frac{1}{(s+\lambda_a+\lambda_b)}$$

$$P_2(t) = e^{-\lambda_b t} - e^{-(\lambda_a+\lambda_b)t} \tag{3.9}$$

Thus, it is seen evidently that formulation of DE, thereby solving either by ODE or Laplace is cumbersome and time consuming. These difficulties are solved by LSTM. Let us apply and realize the advantage.

LSTM equation:

$$\begin{bmatrix} s+\lambda a+\lambda b & 0 & 0 & 0 \\ -\lambda a & s+\lambda b & 0 & 0 \\ -\lambda b & 0 & s+\lambda a & 0 \\ 0 & -\lambda b & -\lambda a & s \end{bmatrix} \begin{bmatrix} \bar{P}_1(s) \\ \bar{P}_2(s) \\ \bar{P}_3(s) \\ \bar{P}_4(s) \end{bmatrix} = \begin{bmatrix} 1 \\ 0 \\ 0 \\ 0 \end{bmatrix} \tag{3.10}$$

As we have got triangular matrix, by mere matrix multiplication will render the solution.

$$(s+\lambda a+\lambda b)\bar{P}_1(s) = 1$$

$$\bar{P}_1(s) = \frac{1}{(s+\lambda a+\lambda b)} \Rightarrow P_1(t) = e^{-(\lambda a+\lambda b)t}$$

This is the same as Eq. (3.6) presented in a more simplified form. The formulation and solution of ODE is taken care of by the rules running at the back end. Second-row matrix multiplication will provide $P_2(t)$.

$$-\lambda a\bar{P}_1(s) + (s+\lambda b)\bar{P}_2(s) = 0$$

Substituting

$$\bar{P}_2(s) = \frac{-\lambda a}{(s+\lambda a+\lambda b)(s+\lambda b)}$$

Splitting into partial fractions as done Eqs. (3.8)–(3.9), we get the same answer as

$$P_2(t) = e^{-\lambda_b t} - e^{-(\lambda_a+\lambda_b)t}$$

Similarly, by interchanging λ_a and λ_b, we can get

$$P_3(t) = e^{-\lambda_a t} - e^{-(\lambda_a+\lambda_b)t} \tag{3.11}$$

$$P_4(t) = 1 - \left[\cancel{e^{-(\lambda_a+\lambda_b)t}} + e^{-\lambda_b t} - \cancel{e^{-(\lambda_a+\lambda_b)t}} + e^{-\lambda_a t} - e^{-(\lambda_a+\lambda_b)t} \right]$$

$$P_4(t) = 1 - \left[e^{-\lambda_b t} + e^{-\lambda_a t} - e^{-(\lambda_a+\lambda_b)t} \right] \tag{3.12}$$

If the two units are connected in series, then $P_1(t)$ will yield R(t). It may be borne in mind that for series system, in the feasible state space, State 4 will be eliminated from the total state space. Thus, the entire argument holds good for any type of system description. For instance, in case of active parallel system, R(t) is sum of $P_1(t)$, $P_2(t)$, $P_3(t)$, or alternatively, $1–P_4(t)$.

3.2.2 Availability analysis using Markov model

The method can be extended on repairable systems to carry out the availability analysis. The availability analysis will involve the random variables of failure time and repair time.

3.2.2.1 *Transient state availability of single unit system*

The state table of a single unit system is as shown below:

	State	
	1	2
A	0	X

Let,

λ = Failure rate, i.e., hazard rate of exponential
μ = Repair rate exponential

LSTM:

$$\begin{bmatrix} s+\lambda & -\mu \\ -\lambda & s+\mu \end{bmatrix} \begin{bmatrix} \bar{P}_1(s) \\ \bar{P}_2(s) \end{bmatrix} = \begin{bmatrix} 1 \\ 0 \end{bmatrix}$$

Row transformation $[R_1/(s + \lambda)]$ on row 1,

$$\begin{bmatrix} 1 & -\dfrac{\mu}{s+\lambda} \\ -\lambda & s+\mu \end{bmatrix} \begin{bmatrix} \bar{P}_1(s) \\ \bar{P}_2(s) \end{bmatrix} = \begin{bmatrix} \dfrac{1}{s+\lambda} \\ 0 \end{bmatrix}$$

Row transformation $R_2 + \lambda.R_1$ on row 2,

$$\begin{bmatrix} 1 & -\dfrac{\mu}{s+\lambda} \\ 0 & (s+\mu)-\dfrac{\lambda\mu}{s+\lambda} \end{bmatrix} \begin{bmatrix} \bar{P}_1(s) \\ \bar{P}_2(s) \end{bmatrix} = \begin{bmatrix} \dfrac{1}{s+\lambda} \\ \dfrac{\lambda}{s+\lambda} \end{bmatrix}$$

Considering the second row,

$$\left[(s+\mu)-\frac{\lambda\mu}{s+\lambda}\right]\bar{P}_2(s)=\frac{\lambda}{s+\lambda}$$

$$\bar{P}_2(s)=\frac{\lambda}{s(s+\lambda+\mu)}$$

$$P_1(t)=L^{-1}\left[\frac{\lambda}{s(s+\lambda+\mu)}\right]$$

Solving by partial differentiation,

$$P_2(t)=\bar{A}(t)=\frac{\lambda}{\lambda+\mu}-\frac{\lambda}{\lambda+\mu}e^{-(\lambda+\mu)t}$$

$$\text{Availability} = A(t) = 1-\bar{A}(t)$$

$$A(t)=1-\frac{\lambda}{\lambda+\mu}-\frac{\lambda}{\lambda+\mu}e^{-(\lambda+\mu)t}=\frac{\mu}{\lambda+\mu}-\frac{\lambda}{\lambda+\mu}e^{-(\lambda+\mu)t} \tag{3.13}$$

The first term of Eq. (3.13) represents transient state availability which is function of time, while the second term represents steady state availability which is time independent.

3.2.2.2 *Steady state availability analysis*

In reality, steady state availability alone is more meaningful than combining it with transient availability. The above procedure is modified to simplify the analysis.

The modified procedure is as follows:

The state transition matrix (STM) is worked out as follows:

1. Remove the Laplace operator 's' from the principal diagonal of the coefficient matrix.
2. Conversely, the weights of incoming arrows to the row under study will be positive and outgoing arrows will be negative.
3. Variable matrix is state probability column vector.

4. Constant matrix is a null vector.
5. However, the various columns of a given row represent the transition from respective state to the state under consideration and the principal diagonal, which is the state under consideration, will remain the same. Also all incoming arrows to the state under study will be summed up and will occupy the principal diagonal element.

STM:

$$\begin{bmatrix} -\lambda \mu \\ \lambda - \mu \end{bmatrix} \begin{bmatrix} P_1 \\ P_2 \end{bmatrix} = \begin{bmatrix} 0 \\ 0 \end{bmatrix}$$

As above equations are dependent, we have to replace one of the equations by $P_1 + P_2 = 1$.

Revised STM will be

$$\begin{bmatrix} 11 \\ \lambda - \mu \end{bmatrix} \begin{bmatrix} P_1 \\ P_2 \end{bmatrix} = \begin{bmatrix} 1 \\ 0 \end{bmatrix}$$

Row transformation $R_2 - \lambda . R_1$ on row 2,

$$\begin{bmatrix} 11 \\ \lambda - (\mu + \lambda) \end{bmatrix} \begin{bmatrix} P_1 \\ P_2 \end{bmatrix} = \begin{bmatrix} 1 \\ -\lambda \end{bmatrix}$$

Solving,

$$P_1 + P_2 = 1$$

$$-(\mu + \lambda) P_2 = -\lambda$$

$$P_2 = \frac{\lambda}{\lambda + \mu}$$

and

$$P_1 = \frac{\mu}{\lambda + \mu}$$

Steady state availability function, $A(\infty)$, is being in the state of 'O', i.e., P_1.

$A = \frac{\mu}{\lambda + \mu}$ which is the same steady state availability component represented by Eq. (3.13) using LSTM approach.

Dividing by $\lambda\mu$,

$$A(\infty) = A = \frac{1/\lambda}{\frac{1}{\mu} + \frac{1}{\lambda}} \tag{3.14}$$

The above can be described in text as

$$A = \frac{mean\ UP\ time}{Mean\ UP\ time + Mean\ DOWN\ time} \tag{3.15}$$

The steady state availability of the system is the ratio of up time to total time.

3.2.3 Limitations of Markov process

The limitations of Markov process are as follows:

1. The set of DEs evolved through modeling are difficult to solve.
2. The Markov model can only be applied when the state transitions follow exponential distribution.

Assurance sciences demand stochastic modeling to resolve real-life problems in the field of engineering. The Markov process as a simplified solution methodology finds its place predominantly in the field of Reliability and maintenance engineering and management. Due to its simplicity versatility of its application is curbed. At times it becomes difficult to obtain the solution when violating the assumption underlined by the Markov process. This is particularly when a failure process characterizes non-exponentiality, which is quite rare in mechanical systems involving moving parts. To make it furthermore severe downtime analysis involves repair time distribution which normally characterizes lognormality deviating significantly from exponential. So, the non-exponentiality of failure and repair phenomenon needed to be addressed in the Markov process framework.

3.3 SEMI-MARKOV PROCESS

The process becomes non-Markovian if the underlying distribution is non-exponential. The methods suggested in the earlier sections are no longer appropriate, and a new approach must be used to solve the issue. It is worthwhile looking into the assumption of exponential distributions a little bit more carefully before talking about these extra techniques.

The majority of reliability models presumptively assume that component uptime and downtime have an exponential distribution. The Markov model with constant interstate transition rates results from this assumption. In these situations, the analysis is comparably easier, and solutions are accessible more readily.

When the state transitions do not follow exponential distribution, the stochastic process becomes a semi-Markov process.

The next section deals with method of stages that addresses the semi-Markov process.

3.3.1 Method of stages

The non-exponential state transitions can be mapped to exponential form by introducing stages that behave exponentially as an equivalent Markovian model without losing the connectivity from initial to final states. This process is called 'Method of stages'. The prime objective of method of stages is to find the number of intermediate states, whether connected in series or parallel so that the sub-transitions follow the Markovian process, keeping the main transition non-Markovian.

3.3.1.1 *Erlang distribution*

Consider the pdf of a random variable, X

$$f(x) = \frac{a^b}{\Gamma b} x^{b-1} e^{-ax}, x \geq 0, a \ \& \ b > 0 \tag{3.16}$$

Moment generating function (MGF), $M_X(t)$ of the random variable is

$$\begin{aligned} E(e^{tx}) &= \int_0^\infty \frac{a^b}{\Gamma b} x^{b-1} e^{-ax} e^{tx} dx \\ &= \int_0^\infty \frac{a^b}{\Gamma b} x^{b-1} \mathrm{e}^{-x(a-t)} dx \end{aligned} \tag{3.17}$$

Put x(a−t) = y and dx = dy/(a−t) limits of t become 0 and ∞:

$$\begin{aligned} E(e^{tx}) &= \frac{a^b}{\Gamma b} \int_0^\infty \left(\frac{y}{a-t}\right)^{b-1} e^{-y} \frac{dy}{(a-t)} \\ &= \frac{a^b}{(a-t)^b \Gamma b} \int_0^\infty y^{b-1} e^{-y} dy \end{aligned} \tag{3.18}$$

As $\Gamma n = \int_0^{\infty} x^{n-1} e^{-x} dx$ Eq. (3.18) becomes

$$
\begin{aligned}
E(e^{tx}) &= \frac{a^b}{(a-t)^b \Gamma b} \Gamma b \\
&= \frac{1}{\left(1-\frac{t}{a}\right)^b} = \left(1-\frac{t}{a}\right)^{-b}
\end{aligned} \tag{3.19}
$$

Important property of MGF:

X is linear combination of 'n' independent random variables:

$$X = a_o + a_1 X_1 + \ldots + a_n X_n$$

$$E(e^{tx}) = \int_{-\infty}^{\infty} e^{tx} f(x) dx$$

$$E(e^{tx}) = \int_{-\infty}^{\infty} e^{t(a_o + a_1 x_1 + \ldots + a_n x_n)} f(x_1, \ldots, x_n) dx$$

From the property of marginal distribution,

$$E(e^{tx}) = \int_{-\infty}^{\infty}\int_{-\infty}^{\infty} \ldots \int_{-\infty}^{\infty} e^{a_o t}\, e^{a_1 x_1 t} \ldots e^{a_n x_n t} f(x_1) f(x_2) \ldots f(x_n) dx_1 dx_2 \ldots dx_n$$

$$E(e^{tx}) = e^{a_o t} \int_{-\infty}^{\infty} e^{(a_1 t) x_1} f(x_1) dx_1 \ldots \int_{-\infty}^{\infty} e^{(a_n t) x_n} f(x_n) dx_n$$

$$M_X(t) = e^{a_0(t)} \prod_{i=1}^{n} M_{X_i}(t) \tag{3.20}$$

Applying this property, T can be inferred that random variable in Eq. (3.19) represents linear combination of 'b' independently and exponentially distributed random variables each with parameter a. i.e., $X = \Sigma X_i$, where X_is follow exponential with parameter λ. Keeping 'a' as λ and b as r Eq. (3.16) can be rewritten as

$$
f(x) \begin{cases} = \dfrac{\lambda^r}{\Gamma r} x^{r-1} e^{-\lambda x}, x \geq 0, \lambda \ and \ \ r > 0 \\ = 0, \text{ otherwise} \end{cases} \tag{3.21}
$$

and is what is known as Gamma distribution as it basically involves Gamma function. It has two parameters λ and r. This represents a random variable which represents 'r' repeated rare event random occurrences.

Repeated occurrences of rare events are common in engineering life. To be specific, linear sum of many random variables each has exponential density of same parameter. For example, a system has time to failure of exponential fails once and after having failed, repaired, and put back to working condition. This means that the system is restored to as good as original new state, i.e., with the same parameter λ. Then it fails time and again and the procedure is repeated. If one is interested in the total life of the system which is linear sum of life given every time, then the variable follows gamma distribution. If 'r' is an integer, Γr can be expressed as (r-1)! and the random variable is known to follow the Erlang distribution. The whole concept in encapsulation is put in the form of convolution theorem, which states as follows:

Convolution theorem:

If a random variable, X, is the sum of independent, exponentially distributed random variables each with parameter λ, then X has a Gamma density with parameters λ and r. When r is integer, it is popularly known as Erlang distribution having extensive application in 'queueing theory' and method of stages in the semi-Markov process. For this distribution the salient measures, mean, and variance are obtained by simply applying E and V operators on $X = \Sigma X_i$.

$$E(X) = \frac{r}{\lambda}; V(X) = \frac{r}{\lambda^2}$$

The first and second moments of X, m_1 and m_2, become

$$E(X) = \frac{r}{\lambda} = m_1 \tag{3.22}$$

$$V(X) = \frac{r}{\lambda^2} = E(X^2) - [E(X)]^2$$

$$\rightarrow m_2 = \frac{r}{\lambda^2} + \frac{1}{\lambda^2} = \frac{r+1}{\lambda^2} \tag{3.23}$$

The application of this technique involves the following steps.

3.3.1.2 Stage combination identification

This section describes a number of simple stage combinations and their properties. The features of the given distribution or data should be compared to those of the stage combinations before selecting an appropriate combination. Because the Weibull family is responsible for all useful distributions in reliability and maintenance engineering, dealing with the Weibull interface with failure and repairs is far more meaningful.

3.3.1.3 Parameter determination

After deciding on a stage combination, the following step is to derive its parameters from those of the distribution. A moment matching technique can be used to accomplish this.

After deciding on a model to approximate a distribution, the following step is to determine the model parameters that will fit the distribution. There are no explicit formulae for directly calculating the approximate stage model parameters from the distribution parameters. The parameters that will best characterize an empirical distribution are not always known. However, the moments may always be calculated using either exact or approximate methods for any distribution.

The next sections offer a method for estimating the parameters for approximation stage models by matching the first r moments of the model with the distribution. This method is widely used in the parameter estimation of inferential statistics.

The stage model's parameters are non-linear and implicit functions of its moments. The first r moments for the stage combinations, on the other hand, can be simply determined from the parameters.

- Evaluation of the moments
- Evaluation of the parameters using the moments

The method of stages refers to the process of splitting a system state into sub-states, each of which is characterized as a stage. Weibull distributions with $\beta > 1$ can be represented as a set of stages connected in series with constant transition rates using the method of stages, as mentioned in the Erlang distribution.

3.3.1.4 Stages in series

The Erlang distribution discussed in Section 3.3.1.1 can be expressed in linear sum of identical exponential distribution, which gives way to express the Weibull interface into identical exponential. This will help us in converting semi-Markov into Markov and hence we can apply the methodology to carry out reliability and availability analysis of a system not having Markovian property. This is depicted in Figure 3.2. The Erlang distribution discussed has parameters λ and r where λ is the no. of stages to break the given distribution into 'r' no. of identical exponential distributions. In the analysis of availability as failure rate λ and repair rate μ are involved as per failure and repair distributions, it is better to use different notations for these parameters, which are α in place of r and ρ in place of λ. Hence, the Erlang distribution can be expressed as

$$f(x) = \frac{\rho(\rho x)^{r-1}}{(\alpha-1)!} e^{-\rho x},\ x > 0;\ \alpha \ and \ \rho > 0$$

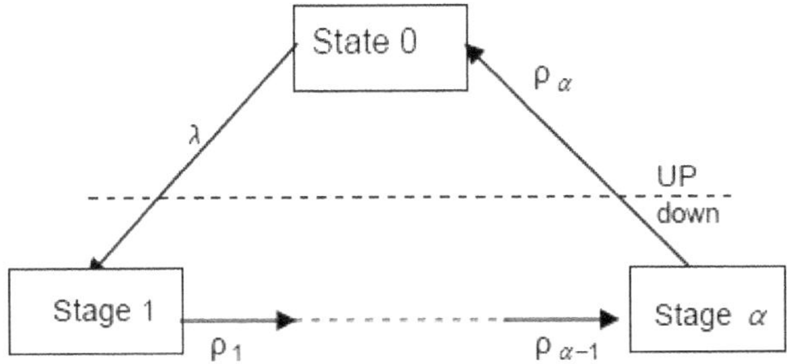

Figure 3.2 Stages in series.

If all the stages are identical with parameter ρ, the corresponding probability density function is the special Erlang distribution having

$$f(x) = \frac{\rho(\rho x)^{\alpha-1} e^{-\rho x}}{(\alpha-1)!} \tag{3.24}$$

$$m_1 = \alpha/\rho \tag{3.25}$$

and

$$m_2 = \frac{\alpha(\alpha+1)}{\rho^2} \tag{3.26}$$

Let M_1 and M_2 be the first two moments of the Weibull distribution being modeled. Using the method of moments to estimate the parameters, then

$$m_1 - M_1 = 0 \text{ and } m_2 - M_2 = 0 \tag{3.27}$$

Substituting Eqs. (3.25) and (3.26) into (3.27) gives

$$\frac{m_1(\alpha+1)}{\rho} = m_2$$

$$m_1^2 + \frac{m_1}{\rho} = m_2$$

$$\frac{m_1}{\rho} = m_2 - m_1^2$$

$$\rho = \frac{m_1}{m_2 - m_1^2} = \frac{M_1}{M_2 - M_1^2} \tag{3.28}$$

$$\alpha = m_1\rho = \frac{m_1^2}{m_2 - m_1^2} = \frac{M_1^2}{M_2 - M_1^2} \tag{3.29}$$

In the evaluation of Eq. (3.29), the value of α must be an integer and the calculated value of α should be rounded up or down as appropriate.

3.4 METHOD OF STAGES IN RELIABILITY ANALYSIS

Several efforts have been made to extend the Markovian approach to the study of reparable systems, during preceding years. The Markovian approach of modeling assumes distributions of exponential failure and repair, with constant rates of transition between the system states that do not depend on the in-transience time nor on the way in which it is arrived at. The assumption of exponential failure is used to model diverse situations of engineering problems. Particularly, those systems in which the components have been set under a previous process of running and are working in the stage of utility life, far from the zone of being wearing down.

3.5 FAILURE INTERACTION WITH SEMI-MARKOV MODEL

System reliability can be evaluated using Markov Models and evaluation techniques but with the following limitations:

1. In continuous processes, the models are administered by sets of DEs that can lead to difficulty in complex system applications.
2. The Markov modeling techniques are basically associated with constant hazard rates and hence can be used only for exponential distributions.

The first limitation can be overcome by the application of STM for solution development. The semi-Markov process can overcome the second limitation.

In the following sections, cases with the semi-Markov process have been analyzed, and the semi-Markov models are solved using method of stages.

3.5.1 Generic algorithm

1. Given data includes the parameters of Weibull distribution.
2. Draw STD.
3. Calculate the first and second moments.
4. Using the preceding equations, compute the number of stages (α) and the parameter of the exponential distribution (λ).
5. Create a new STD.
6. Adapt appropriate STM.
7. Solve the equations.

3.5.2 Reliability analysis for a single component system

Failure of the single component follows Weibull distribution with shape parameter, $\beta = 1.5$, with a Mean Life = 346. The STD is given in Figure 3.3.

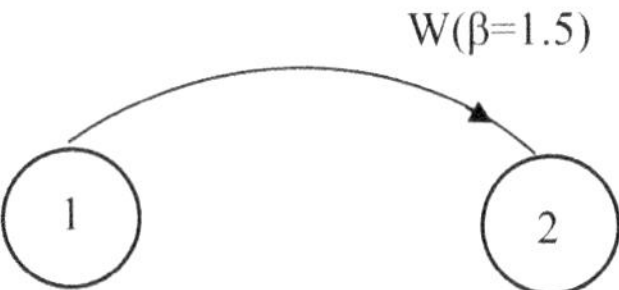

Figure 3.3 Initial STD for single component.

Here

$$E(T) = M_1 = \theta\Gamma\left(1+\frac{1}{\beta}\right) = 346$$

$$346 = \theta\Gamma\left(1+\frac{1}{1.5}\right) \rightarrow \theta = \frac{346}{0.9033} = 383.04$$

$$M_2 = \theta^2\Gamma\left(1+\frac{2}{\beta}\right) = 383.04^2 \times 1.188 = 174,302.934$$

$$\lambda = \frac{M_1}{M_2 - M_1^2} = \frac{346}{174302.934 - 346^2} = 0.006339$$

$$\alpha = M_1\lambda = 0.006339 \times 346 = 2.19 \sim 2$$

The revised STD is shown in Figure 3.4. In the revised STD, failure follows exponential distribution with parameter λ.

LSTM:

$$\begin{bmatrix} S+\lambda & 0 & 0 \\ -\lambda & S+\lambda & 0 \\ 0 & -\lambda & S \end{bmatrix}\begin{bmatrix} P_1(S) \\ P_2(S) \\ P_3(S) \end{bmatrix} = \begin{bmatrix} 1 \\ 0 \\ 0 \end{bmatrix}$$

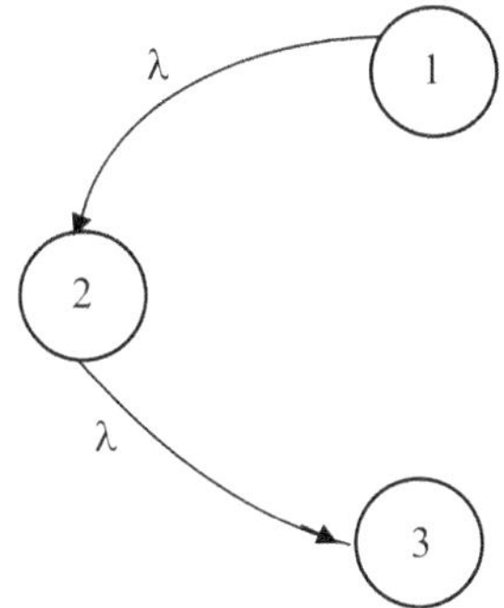

Figure 3.4 Revised STD.

Solving

$$(s+\lambda).\ \overline{P_1}(S)=1$$

$$\overline{P_1}(S)=\frac{1}{S+\lambda}$$

$$\therefore P_1(t)=e^{-\lambda t}$$

$$\frac{-\lambda}{(S+\lambda)}+(S+\lambda)\ \overline{P_2}(S)=0$$

$$\overline{P_2}(S)=\frac{\lambda}{(S+\lambda)\ (S+\lambda)}$$

$$P_2(t)=\lambda t\ e^{-\lambda t}$$

$$P_3(t)=1-e^{-\lambda t}-\lambda t.\ e^{-\lambda t}$$

$$\therefore R(t)=e^{-\lambda t}+\lambda t\ e^{-\lambda t} \tag{3.30a}$$

Assume t = 100, then

$$\begin{aligned} R(t) &= e^{-0.006339\times 100}\left[(100\times 0.006339)+1\right] \\ &= 0.5305\times 1.6339 \\ &= 0.8668 \rightarrow \textit{By Method of Stages} \end{aligned}$$

By conventional method

$$\begin{aligned} R(t) &= e^{\left(t/\theta\right)^{\beta}} = e^{\left(\frac{100}{383.04}\right)^{1.5}} = e^{-0.133393} \\ &= 0.8751 \rightarrow \textit{By Conventional Method} \end{aligned} \tag{3.30b}$$

The values attained by conventional reliability calculation and method of stages are comparable. Hence the method of stages appears to be good solution for semi-Markov models. By varying the parameters β and θ, the model was solved, for different values of T.

3.5.3 Availability analysis for a single component system

Given

Avg. no. of repairs = 10 = M_1

$$\beta=2$$

$$\lambda=0.003$$

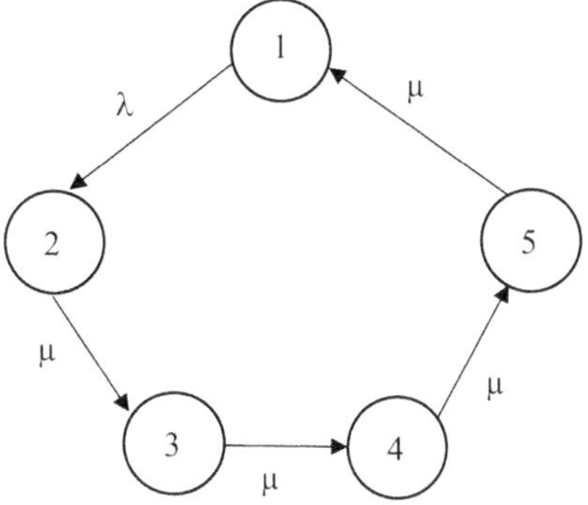

Figure 3.5 Revised STD for single unit availability analysis.

$$E(T) = M_1 = \theta\Gamma\left(1+\frac{1}{\beta}\right) = 346$$

$$10 = \theta\Gamma\left(1+\frac{1}{2}\right) \rightarrow \theta = 11.28$$

$$M_2 = \theta^2\Gamma\left(1+\frac{2}{\beta}\right) = 11.28^2\Gamma\left(1+\frac{2}{2}\right) = 127.24$$

$$\rho = \frac{M_1}{M_2 - M_1^2} = \mu = \frac{10}{127.24 - 10^2} = 0.367$$

$$\alpha = \frac{M_1^2}{M_2 - M_1^2} = \frac{10^2}{127.24 - 10^2} = 0.367 \cong 4$$

The revised STD is given in Figure 3.5.

STM:

$$\begin{bmatrix} -\lambda & 0 & 0 & 0 & \mu \\ \lambda & -\mu & 0 & 0 & 0 \\ 0 & \mu & -\mu & 0 & 0 \\ 0 & 0 & \mu & -\mu & 0 \\ 0 & 0 & 0 & \mu & -\mu \end{bmatrix} \begin{bmatrix} P_1 \\ P_2 \\ P_3 \\ P_4 \\ P_5 \end{bmatrix} = \begin{bmatrix} 0 \\ 0 \\ 0 \\ 0 \\ 0 \end{bmatrix}$$

Transforming the second row,

$$\begin{bmatrix} -\lambda & 0 & 0 & 0 & \mu \\ 1 & 1 & 1 & 1 & 1 \\ 0 & \mu & -\mu & 0 & 0 \\ 0 & 0 & \mu & -\mu & 0 \\ 0 & 0 & 0 & \mu & -\mu \end{bmatrix} \begin{bmatrix} P_1 \\ P_2 \\ P_3 \\ P_4 \\ P_5 \end{bmatrix} = \begin{bmatrix} 0 \\ 1 \\ 0 \\ 0 \\ 0 \end{bmatrix}$$

$R_1 / -\lambda$ *on Row* R_1

$$\begin{bmatrix} 1 & 0 & 0 & 0 & -\mu/\lambda \\ 1 & 1 & 1 & 1 & 1 \\ 0 & \mu & -\mu & 0 & 0 \\ 0 & 0 & \mu & -\mu & 0 \\ 0 & 0 & 0 & \mu & -\mu \end{bmatrix} \begin{bmatrix} P_1 \\ P_2 \\ P_3 \\ P_4 \\ P_5 \end{bmatrix} = \begin{bmatrix} 0 \\ 1 \\ 0 \\ 0 \\ 0 \end{bmatrix}$$

$R_2 - R_1$ *on Row* R_2

$$\begin{bmatrix} 1 & 0 & 0 & 0 & -\mu/\lambda \\ 0 & 1 & 1 & 1 & 1+\mu/\lambda \\ 0 & \mu & -\mu & 0 & 0 \\ 0 & 0 & \mu & -\mu & 0 \\ 0 & 0 & 0 & \mu & -\mu \end{bmatrix} \begin{bmatrix} P_1 \\ P_2 \\ P_3 \\ P_4 \\ P_5 \end{bmatrix} = \begin{bmatrix} 0 \\ 1 \\ 0 \\ 0 \\ 0 \end{bmatrix}$$

$R_3 - \mu R_2$ *on Row* R_3

$$\begin{bmatrix} 1 & 0 & 0 & 0 & -\mu/\lambda \\ 0 & 1 & 1 & 1 & 1+\mu/\lambda \\ 0 & 0 & -2\mu & -\mu & -\mu(1+\mu/\lambda) \\ 0 & 0 & \mu & -\mu & 0 \\ 0 & 0 & 0 & \mu & -\mu \end{bmatrix} \begin{bmatrix} P_1 \\ P_2 \\ P_3 \\ P_4 \\ P_5 \end{bmatrix} = \begin{bmatrix} 0 \\ 1 \\ -\mu \\ 0 \\ 0 \end{bmatrix}$$

$R_3 / -2\mu$ *on Row* R_3

$$\begin{bmatrix} 1 & 0 & 0 & 0 & -\mu/\lambda \\ 0 & 1 & 1 & 1 & 1+\mu/\lambda \\ 0 & 0 & 1 & 1/2 & (1+\mu/\lambda)/2 \\ 0 & 0 & \mu & -\mu & 0 \\ 0 & 0 & 0 & \mu & -\mu \end{bmatrix} \begin{bmatrix} P_1 \\ P_2 \\ P_3 \\ P_4 \\ P_5 \end{bmatrix} = \begin{bmatrix} 0 \\ 1 \\ -\frac{1}{2} \\ 0 \\ 0 \end{bmatrix}$$

$R_4 - \mu R_3$ *on Row* R_4

$$\begin{bmatrix} 1 & 0 & 0 & 0 & -\mu/\lambda \\ 0 & 1 & 1 & 1 & 1+\mu/\lambda \\ 0 & 0 & 1 & 1/2 & (1+\mu/\lambda)/2 \\ 0 & 0 & 0 & -3\mu/2 & -\mu(1+\mu/\lambda)/2 \\ 0 & 0 & 0 & \mu & -\mu \end{bmatrix} \begin{bmatrix} P_1 \\ P_2 \\ P_3 \\ P_4 \\ P_5 \end{bmatrix} = \begin{bmatrix} 0 \\ 1 \\ -\frac{1}{2} \\ -\frac{\mu}{2} \\ 0 \end{bmatrix}$$

$R_4 / \frac{-3\mu}{2}$ *on Row* R_4

$$\begin{bmatrix} 1 & 0 & 0 & 0 & -\mu/\lambda \\ 0 & 1 & 1 & 1 & 1+\mu/\lambda \\ 0 & 0 & 1 & 1/2 & (1+\mu/\lambda)/2 \\ 0 & 0 & 0 & 1 & (1+\mu/\lambda)/3 \\ 0 & 0 & 0 & \mu & -\mu \end{bmatrix} \begin{bmatrix} P_1 \\ P_2 \\ P_3 \\ P_4 \\ P_5 \end{bmatrix} = \begin{bmatrix} 0 \\ 1 \\ \frac{1}{2} \\ \frac{1}{3} \\ 0 \end{bmatrix}$$

$R_5 - \mu R_4$ *on Row* R_5

$$\begin{bmatrix} 1 & 0 & 0 & 0 & -\mu/\lambda \\ 0 & 1 & 1 & 1 & 1+\mu/\lambda \\ 0 & 0 & 1 & 1/2 & (1+\mu/\lambda)/2 \\ 0 & 0 & 0 & 1 & (1+\mu/\lambda)/3 \\ 0 & 0 & 0 & 0 & -\mu - [\mu(1+\mu/\lambda)/3] \end{bmatrix} \begin{bmatrix} P_1 \\ P_2 \\ P_3 \\ P_4 \\ P_5 \end{bmatrix} = \begin{bmatrix} 0 \\ 1 \\ \frac{1}{2} \\ \frac{1}{3} \\ -\frac{\mu}{3} \end{bmatrix}$$

Solving

$$P_5[-\mu - \{\mu(1+\mu/\lambda)/3\}] = -\frac{\mu}{3}$$

$$\therefore P_5 = \frac{\lambda}{4\lambda+\mu} \tag{3.31}$$

And

$$P_4 = P_3 = P_2 = \frac{\lambda}{4\lambda+\mu}$$

$$P_1 = 1 - (P_1 + P_2 + P_3 + P_4) = 1 - 4\frac{\lambda}{4\lambda + \mu} = \frac{\mu}{4\lambda + \mu}$$

$$\text{Availability of system} = A = \frac{\mu}{4\lambda + \mu} \tag{3.32}$$

$$= \frac{0.367}{0.367 + 0.003 \times 4} = 0.968$$

3.5.4 Availability analysis for two-unit redundancy with Weibull interfaced failures and repairs

The appropriate STD is shown in Figure 3.6.

Failure for Unit-1:

$$\mathrm{W}(\beta = 1.5,\ \theta = 10)$$

$$E(T) = M_1 = \theta\Gamma\left(1 + \frac{1}{\beta}\right) = 10\theta\Gamma\left(1 + \frac{1}{2}\right) = 9.03$$

$$M_2 = \theta^2\Gamma\left(1 + \frac{2}{\beta}\right) = 10^2\Gamma\left(1 + \frac{2}{1.5}\right) = 118.8$$

$$\lambda = \frac{M_1}{M_2 - M_1^{\;2}} = \frac{9.03}{118.8 - 9.03^2} = 0.2424$$

$$\alpha = M_1\lambda = 0.2424(10) = 2.24 \cong 2$$

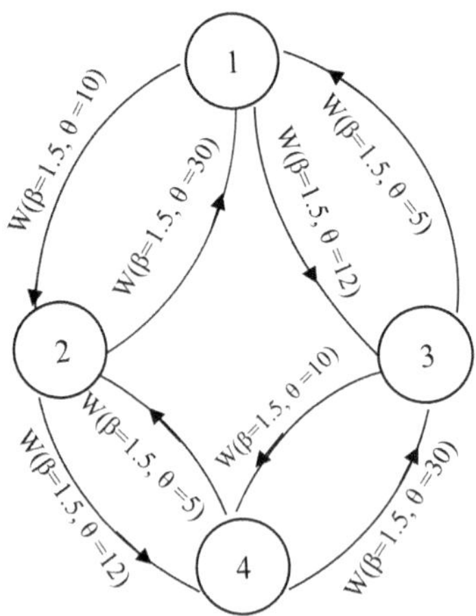

Figure 3.6 STD for two-unit redundancy with Weibull failures and repairs.

Failure for Unit-2:

$$W(\beta = 1.5,\ \theta = 12)$$

$$E(T) = M_1 = \theta\Gamma\left(1+\frac{1}{\beta}\right) = 12\theta\Gamma\left(1+\frac{1}{1.5}\right) = 10.8396$$

$$M_2 = \theta^2\Gamma\left(1+\frac{2}{\beta}\right) = 12^2\Gamma\left(1+\frac{2}{1.5}\right) = 171.072$$

$$\lambda = \frac{M_1}{M_2 - M_1^2} = \frac{12}{171.072 - 12^2} = 0.2023$$

$$\alpha = M_1\lambda = 0.2023(10) = 2.023 \cong 2$$

Repair for Unit-1:

$$W(\beta = 1.5,\ \theta = 30)$$

$$E(T) = M_1 = \theta\Gamma\left(1+\frac{1}{\beta}\right) = 30\Gamma\left(1+\frac{1}{1.5}\right) = 27.099$$

$$M_2 = \theta^2\Gamma\left(1+\frac{2}{\beta}\right) = 30^2\Gamma\left(1+\frac{2}{1.5}\right) = 1069.2$$

$$\mu = \frac{M_1}{M_2 - M_1^2} = \frac{27.099}{1069.2 - 27.099^2} = 0.0809$$

$$\alpha = M_1\mu = 0.0809(27.099) = 2.42 \cong 2$$

Repair for Unit-2:

$$W(\beta = 1.5, \theta = 5)$$

$$E(T) = M_1 = \theta\Gamma\left(1+\frac{1}{\beta}\right) = 5\Gamma\left(1+\frac{1}{1.5}\right) = 4.5165$$

$$M_2 = \theta^2\Gamma\left(1+\frac{2}{\beta}\right) = 5^2\Gamma\left(1+\frac{2}{1.5}\right) = 29.7$$

$$\mu = \frac{M_1}{M_2 - M_1^2} = \frac{4.5165}{29.7 - 4.5165^2} = 0.4856$$

$$\alpha = M_1\mu = 0.0809(30) = 2.42 \cong 2$$

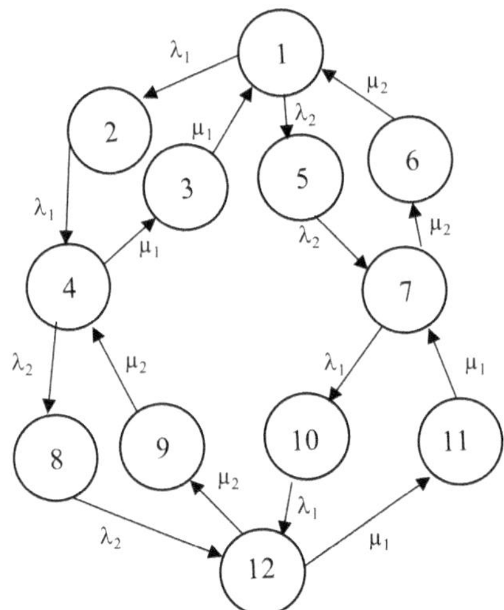

Figure 3.7 Revised STD for two-unit redundancy with Weibull interface.

The revised STD is shown in Figure 3.7.
Revised STM will be

$-\lambda_1-\lambda_2$	0	μ_1	0	0	μ_2	0	0	0	0	0	0
λ_1	$-\lambda_1$	0	0	0	0	0	0	0	0	0	0
0	0	$-\mu_1$	μ_1	0	0	0	0	0	0	0	0
0	λ_1	0	$-\mu_1-\lambda_2$	0	0	0	0	μ_2	0	0	0
λ_1	0	0	0	$-\lambda_2$	0	0	0	0	0	0	0
0	0	0	0	0	$-\mu_2$	μ_2	0	0	0	0	0
0	0	0	0	λ_2	0	$-\mu_2-\lambda_1$	0	0	μ_1	0	0
0	0	0	λ_2	0	0	0	$-\lambda_2$	0	0	0	0
0	0	0	0	0	0	0	0	$-\mu_2$	0	0	μ_2
0	0	0	0	0	0	0	0	0	$-\mu_1$	0	μ_1
0	0	0	0	0	0	λ_1	0	0	0	$-\lambda_1$	0
0	0	0	0	0	0	0	λ_2	0	0	λ_1	$-\mu_1-\mu_2$

Substituting the values of λs and μs in the matrix, and solving we get, –

$P_1 = 0.0394$; $P_2 = 0.0394$; $P_3 = 0.1959$; $P_4 = 0.1959$; $P_5 = 0.0394$; $P_6 = 0.0197$; $P_7 = 0.0197$; $P_8 = 0.1959$; $P_9 = 0.0784$; $P_{10} = 0.0784$; $P_{11} = 0.0197$; $P_{12} = 0.0784$;

$$A(t) = 1 - P_{12} = 1 - 0.0784 = 0.9816$$

The solution complexity increases with increase in a number of stages. A MATLAB code is used for the solution.

3.6 CASE STUDY

A case of an underground copper mine workplace with modern amenities for copper mineral handling is investigated in this section. The problem comprised the availability/reliability investigation of the critical components in humid milling process. The aforementioned process is an element of the concentrator and is made up of the pumps that carry pulp from the ball mills to the hydraulicons, the mill, the battery of hydraulicons, and the feeding transporters. These pumps are constructed of special steel and require strict maintenance procedures since they must withstand the highly abrasive action of the pulp of the copper ore. The equipment is the most significant part of the humid milling process, and there is a lot of drive for better system reliability and lower operating costs.

In the first step, the analysis was focused on the study of the availability/reliability of the centrifugal pump as a global unit, with the objective to dimension the problem situation, and then, to continue, studying the other components of the system. The time between failures data obtained from the machine history and downtime data put under inferential statistics gave the following information.

The average time between failures was 346 hours. Accordingly, the density function of the failure distribution for the centrifugal pump is

$$F(t) = \frac{1}{346} e^{-\frac{t}{346}} \tag{3.33}$$

where t, is the operation time expressed in hours. As the first approximation, the repair time was assumed to be following the Erlang distribution, with a mean of 24 hours and a standard deviation of 12 hours. In this case the stages generated through the method are exponential. If all the stages are identically distributed with the parameter ρ, the corresponding density function will be represented with the following Erlang distribution.

For the failure time, T, which follows exponential distribution:

$$\begin{aligned} E(T) &= 346\,\text{hrs} \\ &= \frac{1}{\lambda} \rightarrow \lambda = 0.003 \text{ failures/hr or } 26.28 \text{ failures/year} \end{aligned}$$

For repair time,

$$M_1 = 24\,\text{hrs.}$$

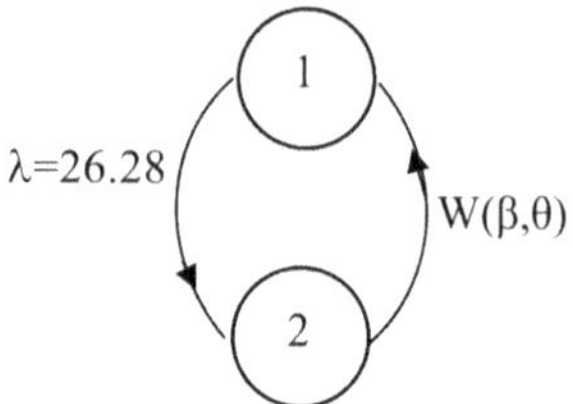

Figure 3.8 Initial STD for the centrifugal pump.

Therefore,

$$\alpha = \frac{M_1^2}{M_2 - M_1^2} = \frac{24^2}{720 - 24^2} = 4.0$$

$$\rho = \mu = \frac{M_1}{M_2 - M_1^2} = \frac{24}{720 - 24^2} = 0.167 \ repairs/hr = 1462.92 \ repairs/yr$$

The diagram of the state space would be represented in the following way. The initial STD is given in Figure 3.8, while the revised STD is given in Figure 3.9.

STM:

$$\begin{bmatrix} -\lambda & 0 & 0 & 0 & \mu \\ \lambda & -\mu & 0 & 0 & 0 \\ 0 & \mu & -\mu & 0 & 0 \\ 0 & 0 & \mu & -\mu & 0 \\ 0 & 0 & 0 & \mu & -\mu \end{bmatrix} \begin{bmatrix} P_1 \\ P_2 \\ P_3 \\ P_4 \\ P_5 \end{bmatrix} = \begin{bmatrix} 0 \\ 0 \\ 0 \\ 0 \\ 0 \end{bmatrix}$$

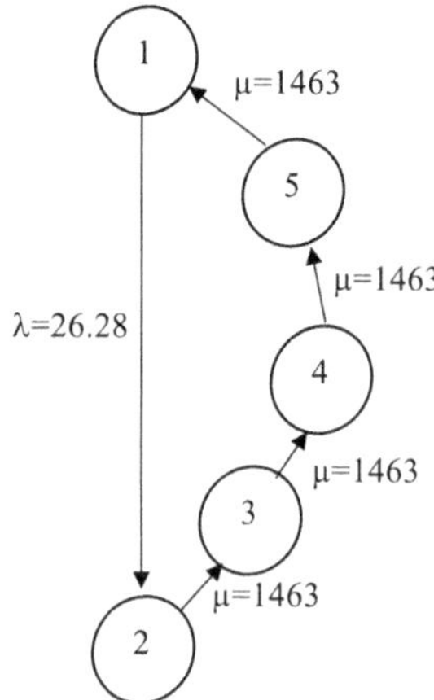

Figure 3.9 Revised STD for the centrifugal pump.

Solving as delineated in Section 3.5.3, which we get

$$P_5\left[-\mu-\{\mu(1+\mu/\lambda)/3\}\right]=-\mu/3$$

$$\therefore P_5=\frac{\lambda}{4\lambda+\mu}$$

And

$$P_4=P_3=P_2=\frac{\lambda}{4\lambda+\mu}$$

$$P_1=1-(P_1+P_2+P_3+P_4)=1-4\frac{\lambda}{4\lambda+\mu}=\frac{\mu}{4\lambda+\mu}$$

$$\text{Availability of system}=A=\frac{\mu}{4\lambda+\mu} \tag{3.34}$$

$$=\frac{1463}{1463+26.28\text{x }4}=0.93$$

NOMENCLATURE

α, ρ	parameters of Erlang distribution
$\bar{A}$	steady state unavailability
A	steady state availability
A(t)	availability function
β, 0	Parameters of Weibull distribution
DE	differential equation
E(T)	expected value or mean of random variable 'T'
f(x)	probability density function (pdf) of random variable 'T'
λ	failure rate
LSTM	Laplacian state transition matrix
MGF, $M_X(t)$	moment generating function
M_i	i^{th} moment
μ	repair rate
ODE	ordinary differential equation
$P_n(t)$	probability of being in state 'n' at time 't'
R(t)	reliability function
STD	state transition diagram
STM	state transition matrix
W	Weibull distribution

BIBLIOGRAPHY

Dhawalikar MN, Mariappan V, Srividhya PK, Kurtikar V (2018) Multi-state failure phenomenon and analysis using semi-Markov model. Int J Qual Reliab Manage 35:2080–2091. https://doi.org/10.1108/IJQRM-01-2016-0001

Karr AF (1990) Markov Processes. Elsevier Science Publishers B. V.

Lewis EE (1994) Introduction to Reliability Engineering, 2nd edn. John Wiley & Sons, Inc.

Mariappan D, Dhawalikar MD, Srividhya PK, Kurtikar V (2017) Semi Markov model for optimizing availability of testable systems. AcadpublEu 114:83–93.

Michael S, Mariappan V, Amonkar UJ, Telang AD (2009) Availability analysis of transmission system using Markov model. Int J Indian Cult Bus Manage 2:551. https://doi.org/10.1504/ijicbm.2009.025280

Smith JUM, Billington R, Allan RN (1994) Reliability Evaluation of Engineering Systems. Springer Science.

Yu S-Z (2016) Hidden Semi-Markov Models Theory, Algorithms and Applications. Elsevier Inc.

Chapter 4

Reliability assessment of a multi-state degraded system

Rajesh S. Prabhu Gaonkar, Vishwanath L. Shet Verenkar, and Sunay P. Pai

4.1 INTRODUCTION

Multi-state reliability handles and analyses the situation wherein the system and its components can have varied performance levels from perfect functioning to complete failure. Reliability assessment is to assess the probability that a system will perform a required function without failure under stated conditions for a stated period. System performance may degrade owing to the degradation of its components over time. Partial failures lead to a decrease in the performance of components and of the system as well. Once partially failed, components operate at a reduced performance rate and are eventually unable to perform their function on complete failure. The binary system is the simplest case of a multi-state system having only two extremely distinctive states.

Before the complete failure, most components undergo a degradation process. All components/systems undergo an ageing process that depends not only on time but also on the state of the systems. Components undergo intermediate states between perfect functioning and complete failure. This is attributed to fatigue, burn-in, vibration, efficiency, or failure of nonessential components. Some of the degradation models used in reliability engineering include statistical distributions, stochastic processes, semi-Markovian models, and probabilistic superposition models [1]. Components in a system may not only deteriorate due to a simple failure process but it can be because of multiple degradation processes and also correlative degradation effect of other associated components. With the increase in number of components, the complexity of the system increases. It is very crucial to identify the weak links in a system. The performance of such links affects the reliability of the system.

Yan-Fu Li et al. [2] used the universal generating function technique to improve the descriptive power and solution efficiency of analytical probabilistic models of multi-state stochastic systems.

Multi-state reliability models precisely represent complex and realistic difficult engineering systems. The classical reliability theory considers the component or system to be in one of the two extreme states. But, with today's complex engineering systems, the requirement of accurate reliability and optimal design necessitates considering multi-state systems [3–6]. The universal

DOI: 10.1201/9781003546214-4

generating function was used for the reliability analysis of the power system [7], and for evaluating the performance distribution of complex series-parallel multi-state systems [8]. Monte Carlo simulation is also used by researchers on several occasions [9–12]. Sinan Calik [13] studied the multi-state component subjected to two kinds of stresses. Chi Zhang et al. [14] used Monte Carlo simulation and proposed the method to assess the reliability of components having continuous distribution. With time, the system starts deteriorating gradually to a lower performance level or fails suddenly. A. Salmasnia et al. [15] proposed an approach to optimise maintenance/repair costs and system availability for a multi-objective multi-state degraded system. Li Yong et al. [16] put to light a basic approach to solving problems in reliability modelling and assessment using uncertainty reasoning. There is a clear definition of the need to model the multi-state reliability for understanding the degradation of the system. The chapter explains the usefulness of using the Bayesian network (BN) for multi-state modelling due to its Boolean approach to defining states of working, failure, and degradation. Concluding the subject the chapter has included examples that have expressed the importance of posterior probabilities in the use of BN.

4.2 RELIABILITY SIMULATION USING FAULT TREE ANALYSIS (FTA)

Reliability block diagram (RBD) and fault tree analysis (FTA) have been extensively used for calculating system reliabilities. A fault tree gives a diagrammatic representation of a system and insights into the ways of system failure in a specific mode. Through a Boolean logic system, FTA identifies and combines the probabilities of the events impacting the system. It provides a list of all the causes of a system failure. The events identified in the fault tree may not be exhaustive but represents the most obvious events identified by analysts for failures. Hence, FT does not model the system for all possible failures but considers only those events identified by the analysts. It is important to identify the basic events that contribute the maximum to the system failure. It is however very difficult to gauge the failure rates accurately of individual events and it becomes necessary to work with rough estimates of probabilities. Moreover, it is important to identify the critical events and also events with maximum uncertainty.

A fault tree if properly validated by a knowledgeable person for its completeness and accuracy before its evaluation can act as a valuable design tool, can anticipate probable accidents, and can eliminate the required design changes. It can also serve as a diagnostic tool for predicting system failure. Regardless of the complexity of the system, FTA can be used efficiently for the reliability assessment of the system.

System reliability using Functional Block Diagrams (FBD) and FTA can be calculated under two scenarios. Firstly, wherein the probability of failure of various events is static i.e. it remains constant with time. Such calculations are performed only in the analytical mode. Under the second scenario, failure probability is considered to be time-dependent. Simulation techniques can be used in such cases for calculations wherein random failure times for each component are generated. However, in traditional methods, we have limitations on assuming failure rates as constant (λ). On analysing data from real life, we find that they exhibit different failure patterns and fit over in various distributions. There is, therefore, a strong need to draw our conclusion on the reliability of the system based on observation of real-life data instead of simply assuming the failure to be constant. Data collected from the manufacturers' specifications is more reliable and becomes the preferred choice. In case of the unavailability of such data, modelling, and simulation techniques are used to derive the event probabilities. Qualitative data from subject experts are also used. Real-life data helps to identify the weaknesses in the system, know the reasons behind the failure of the system, redesign the parts to suit the system, and improve the reliability of the system. We know that the electronic components are best modelled by using the Exponential distribution as such if we try to model a mechanical failure it will give a result to an approximation of the obtained solution. There are various situations where the modelling of a system may have a few components following Normal distribution, a few following Lognormal, Weibull, etc.

We have therefore taken up the study of simulation to build such models having mixed distribution to model and analyse such cases of interest. A simulation technique is explored for constructing a fault tree using quantitative data. The task involved is building the model that includes fitting of the collected data.

4.3 CASE STUDY

M/s ABC is a supplier of Diesel Generator (DG) of 350 KW, 450 KW, and 500 KW to the Indian Navy and the Indian Coast Guard, and many other private parties. They have recently developed a Programmable Logic Control (PLC)-based DG control system (refer to Figure 4.1 and Table 4.1) to monitor and govern variation of load for the remote monitoring and control of the DG system.

These PLC-based systems are in use since 2011 and are functionally well-appreciated. They have an excellent interface with the remote control system (RCS) and have proved their worth in automatic power management systems (APMS). The company however has seen a large failure in its components and is keen to analyse the case. There are carefully recorded failure times to build up a case study to estimate the reliability of the system. The primary function of the DG control system is to control the speed with the changing load on the system. The monitoring system is a secondary function. The firm desires a complete analysis of the failure of the main system and the subsystems to get a basis to improve the system using better components to ensure a better brand reputation in the future.

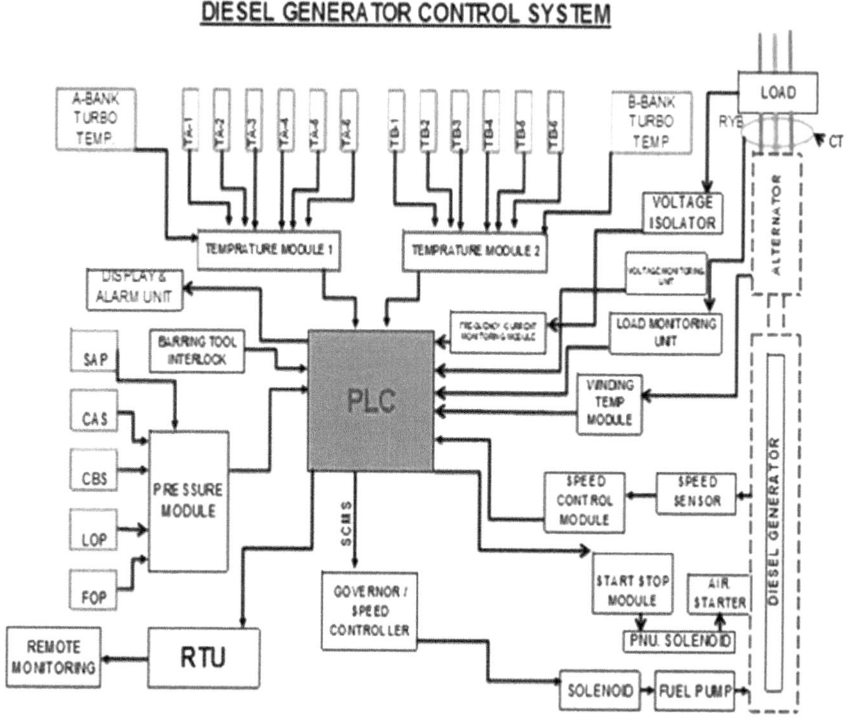

Figure 4.1 PLC-based DG control system.

Table 4.1 Components of PLC-based DG control system

Sr. no	*Component*	*Code*	*Quantity*
1	Thermo-couple temperature sensor for engine cylinders of the A-bank	TA-1–TA-6	06
2	Thermo-couple temperature sensor for engine cylinders of the B-bank	TA-1–TA-6	06
3	A-bank turbo temperature sensor	TAS	01
4	B-bank turbo temperature sensor	TBS	01
5	Temperature module 1 of A-bank	TM-1	01
6	Temperature module 2 of B-bank	TM-2	01
7	Starting Air pressure sensor	SAP	01
8	Crankcase pressure sensor A-bank	CAS	01
9	Crankcase pressure sensor B-bank	CBS	01
10	Lubricating oil pressure sensor	LOP	01
11	Fuel oil pressure sensor	FOP	01
12	Pressure module	PM	01
13	Display and alarm unit	–	01
14	Barring tool interlock	–	01

(Continued)

Table 4.1 (Continued)

Sr. no	*Component*	*Code*	*Quantity*
15	Programmable logic control		01
16	Remote transfer unit	RTU	01
17	Governor speed controller	–	01
18	Solenoid (electric)	–	01
19	Fuel pump	–	01
20	Air starter	–	01
21	Start stop module	–	01
22	Speed control module	–	01
23	Winding temp module	–	01
24	Load monitoring unit	–	01
25	Frequency/current monitoring unit	–	01
26	Voltage monitoring unit	–	01
27	Voltage isolator	–	01
28	Current transformer	CT	01
29	Pneumatic solenoid	–	01
30	Speed sensor	–	01

The Function of the DG control system is listed below.

1. Speed control system
2. Speed monitoring system
3. DG starting system
4. Temperature monitoring system
5. Pressure monitoring system
6. Load monitoring system

4.4 MODELLING AND SIMULATION OF THE PLC-BASED DG CONTROL SYSTEM

Step 1: The failure data of the various components of the DG recorded by the firm is used to fit in various distributions and cross-verified the same to obtain the failure distribution parameters. Refer to Table 4.2.

The data has been ranked and suitably verified using the method of maximum likelihood.

Step 2: A model of the PLC-based DG control system and its subsystems were built to study the reliability of the system.

Step 3: The parameters of various failure distributions estimated in Step 1 are assigned to the model at each component level.

Step 4: RBDs and FTAs are constructed to validate the model.

Step 5: The system is simulated for a time at 10 hrs, 100 hrs, 1,000 hrs, 10,000 hrs, 100,000 hrs, and 1,000,000 hrs of operation to draw and study the reliability curve.

Table 4.2 Estimation of parameters

Component	*Estimates from easy fit*	*Distribution*	*Estimates from (maximum likelihood)*	*Distribution*
Air starter	$\sigma = 0.97778$, $\mu = 8.1135$	Lognormal	$\sigma = 0.997141$, $\mu = 8.11346$	Lognormal
Pneumatic solenoid	$\sigma = 1{,}844.6$, $\mu = 4{,}784$	Normal	$\sigma = 1{,}844.555$, $\mu = 4{,}784$	Normal
PLC	NA (sample size > 5)		Mean time to failure = 6432	Exponential
Thermo-couple	$\sigma = 4{,}317.2$, $\mu = 5{,}910.9$	Normal	$\sigma = 4{,}317.214$, $\mu = 5{,}910.857$	Normal
Temperature module	$\sigma = 2{,}187.4$, $\mu = 6{,}115$	Normal	$\sigma = 2{,}187.44$, $\mu = 6{,}115$	Normal
Barring interlock	$\alpha = 1.7813$, $\beta = 6{,}307.4$	Weibull	$\alpha = 2.052$ $\beta = 6{,}349.171$	Weibull
CT	No failure reported Reliability = 1			
Fuel pump	NA (sample size > 5)		Mean time to failure = 6,888	Exponential
Winding temp. module	$\sigma = 3{,}219.9$, $\mu = 6{,}100.9$	Normal	$\sigma = 3{,}219.854$, $\mu = 6{,}100.888$	Normal
Voltage isolator	NA (sample size > 5)		Mean time to failure = 8208	Exponential
Governor speed controller	$\sigma = 0.54163$, $\mu = 8.1457$	Lognormal	$\sigma = 0.559391$, $\mu = 8.145731$	Lognormal
Speed sensor original	$\sigma = 2{,}890.0$, $\mu = 8{,}164.8$	Normal	$\sigma = 2{,}889.95$, $\mu = 8{,}164.8$	Normal
Speed sensor new	$\alpha = 2.7964$, $\beta = 2{,}648.5$	Weibull	$\alpha = 2.7964$, $\beta = 2648.5$	Weibull
Speed control module	$\sigma = 3{,}795.9$, $\mu = 6491.6$	Normal	$\sigma = 3{,}795.852$, $\mu = 6{,}491.612$	Normal
Display and alarm	$\sigma = 3{,}247.6$, $\mu = 7488$	Normal	$\sigma = 3{,}247.635$, $\mu = 7488$	Normal
RTU	NA (sample size > 5)		Mean time to failure = 8,208	Exponential
Freq. current monitoring module	$\sigma = 2{,}143.9$, $\mu = 7{,}569.6$	Normal	$\sigma = 2{,}143.913$, $\mu = 7{,}569.6$	Normal
Load monitoring unit	NA (sample size > 5)		Mean time to failure = 7,506	Exponential
voltage monitoring module	$\sigma = 68{,}1.28$, $\mu = 11{,}434.0$	Normal	$\sigma = 681.278$, $\mu = 11{,}433.6$	Normal
Pressure module	$\sigma = 1{,}163.0$, $\mu = 6{,}756$	Normal	$\sigma = 1{,}163.02$, $\mu = 6{,}756$	Normal

(Continued)

Table 4.2 (Continued)

Component	*Estimates from easy fit*	*Distribution*	*Estimates from (maximum likelihood)*	*Distribution*
Pressure sensor type 1	$\sigma = 0.16111$, $\mu = 8.8054$	Lognormal	$\sigma = 0.841591$, $\mu = 8.188469$	Lognormal
Pressure sensor type 2	$\sigma = 4{,}283.4$, $\mu = 7{,}210.3$	Normal	$\sigma = 4{,}283.374$, $\mu = 7{,}210.285$	Normal
Start stop module	$\sigma = 3{,}421.6$, $\mu = 5{,}292$	Normal	$\sigma = 3{,}421.553$, $\mu = 5{,}292$	Normal
Electric solenoid	$\sigma = 4{,}246.7$, $\mu = 5{,}512.8$	Normal	$\sigma = 4{,}246.712$, $\mu = 5{,}512.8$	Normal

4.4.1 Simulation model of the PLC-based DG control system

The modelling of the PLC-based system is done as shown in Figure 4.2. The same has been verified using FTA and RBD analysis. Refer to Figures 4.3 and 4.4.

Table 4.3 indicates the reliabilities at various times. A graph of reliability vs time shows the system nature of the DG control system (refer to Figure 4.4) The results are fairly poor, and there is a strong need for improvement. We carried

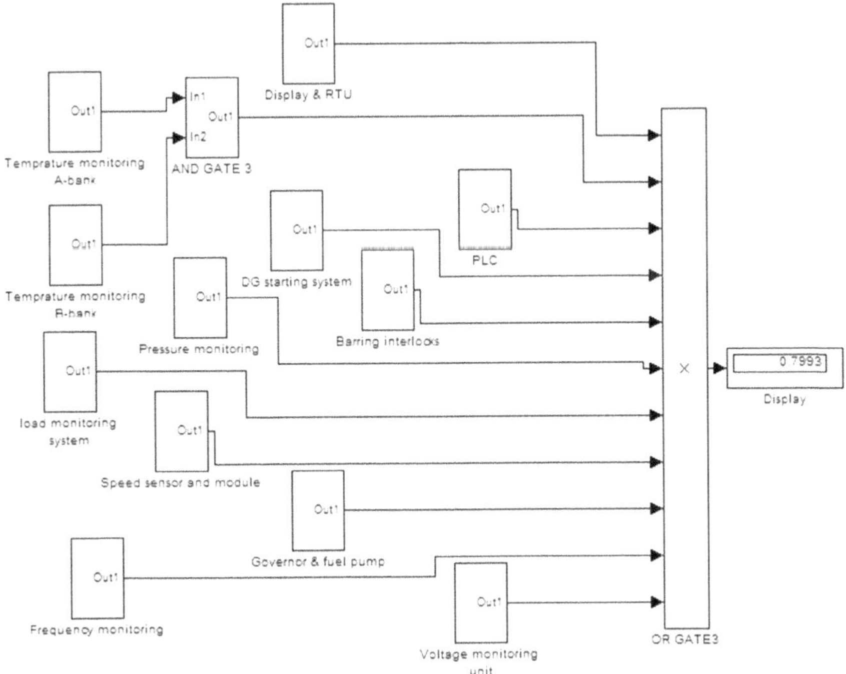

Figure 4.2 Model of PLC-based DG control system.

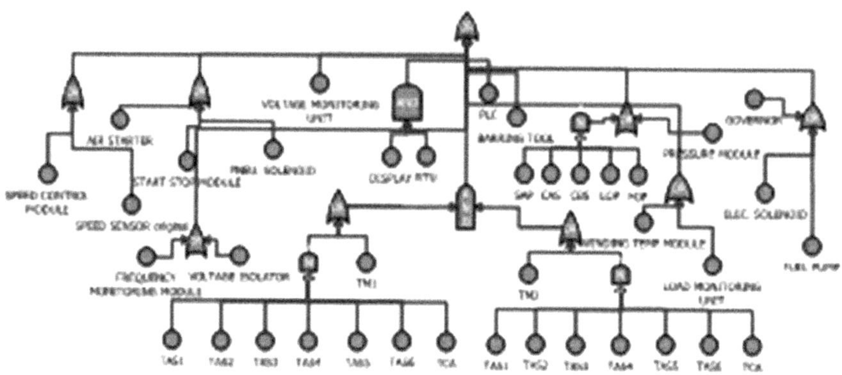

Figure 4.3 TA analysis of PLC-based DG control system.

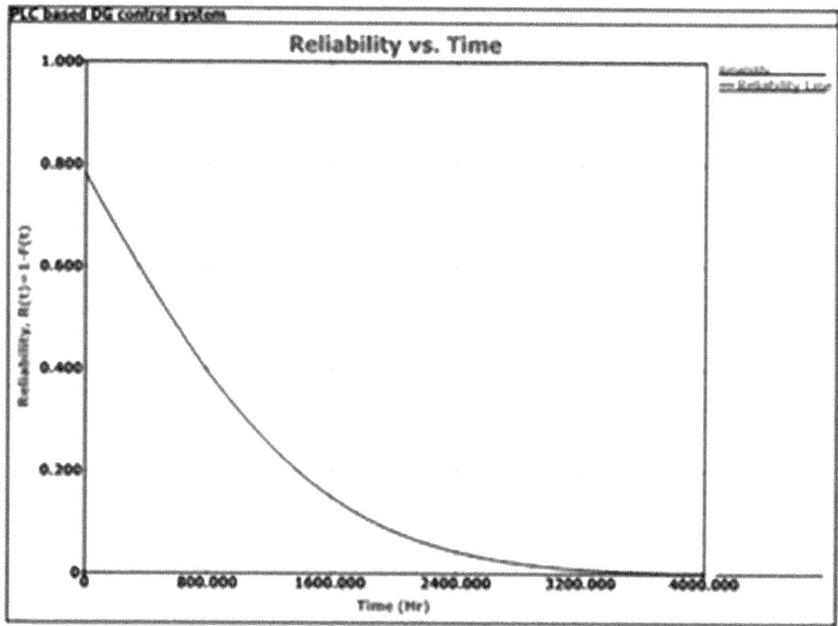

Figure 4.4 Graph of reliability vs time for DG control system.

Table 4.3 Simulation results of system reliability of PLC-based DG control system

		Time (hrs)					
		R(10)	*R(100)*	*R(1,000)*	*R(10,000)*	*R(100,000)*	*R(1,000,000)*
PLC-based DG control system	FTA (BlockSim®)	0.79928	0.75555	0.37752	8.67E-18	0	0
	FTA (Simulink®)	0.7993	0.7555	0.3775	8.67E-18	0	0

out a detailed system analysis of the pressure monitoring system, one of the subsystems of the DG control system, to find weak links in the system.

4.4.2 Simulation model of pressure monitoring system

The modelling of the pressure monitoring system is done and the same has been verified using FTA and RBD analysis refer to Figures 4.5–4.7.

A graph of reliability vs time shows the system nature of the pressure monitoring system (refer to Figure 4.8).

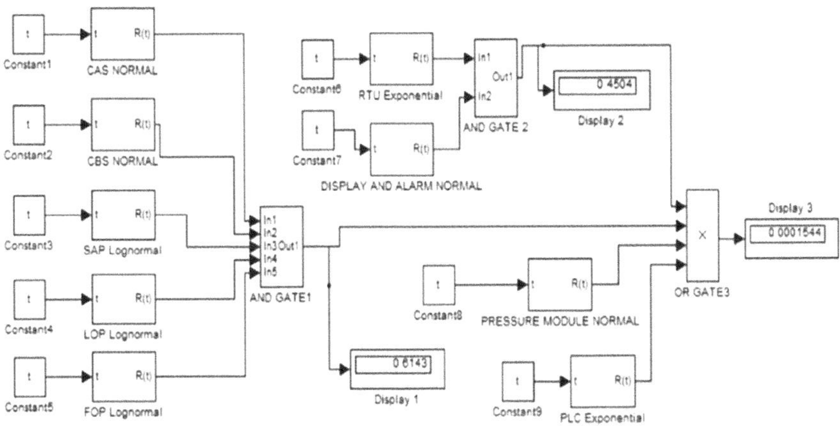

Figure 4.5 Model for pressure monitoring system.

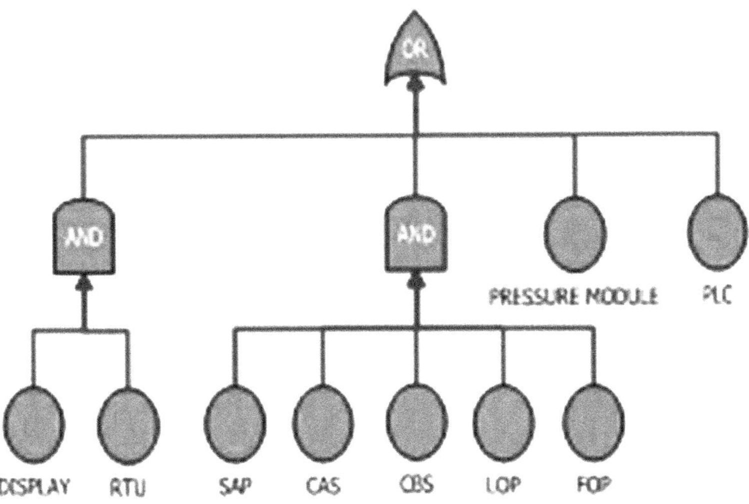

Figure 4.6 FTA analysis of pressure monitoring system.

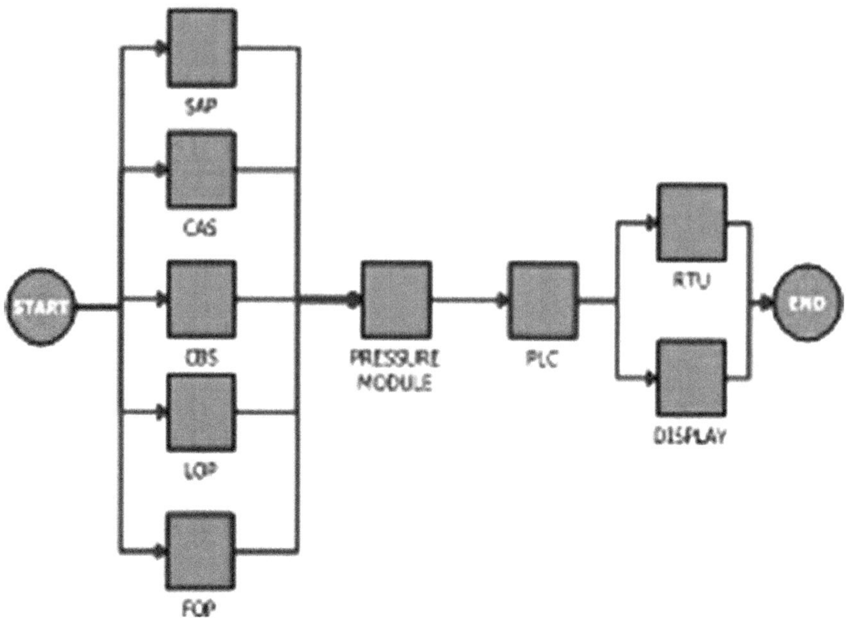

Figure 4.7 RBD analysis of pressure monitoring system.

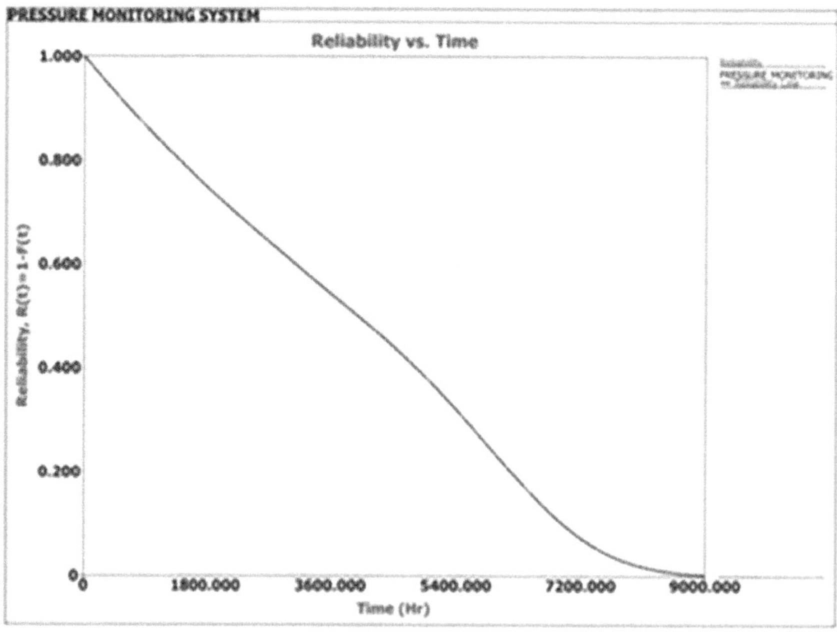

Figure 4.8 Graph of reliability vs Time for pressure monitoring system.

4.5 SIMULATION RESULTS OF PRESSURE MONITORING SYSTEM

Table 4.4 shows the reliability at various time intervals for the pressure monitoring system while Table 4.5 gives the reliability of individual components using FTA analysis.

Table 4.4 Simulation results of the pressure monitoring system of the PLC-based DG control system

		Time (hrs)					
		R(10)	*R(100)*	*R(1,000)*	*R(10,000)*	*R(100,000)*	*R(1,000,000)*
Pressure monitoring system	RBD (BlockSim®)	0.998434	0.984436	0.853763	0.000154	0	0
	FTA (BlockSim®)	0.998434	0.984436	0.853763	0.000154	0	0
	FTA (Simulink®)	0.9984	0.9844	0.8538	0.000154	0	0

Table 4.5 Reliability of individual component using FTA analysis

	Reliability hrs						
Component	*R(1)*	*R(10)*	*R(100)*	*R(1,000)*	*R(10,000)*	*R(10E5)*	*R(10E6)*
Air starter	1	1	0.999783	0.8867	0.135661	0.000326	5.38E–09
AVR	0.999823	0.998236	0.9825	0.838154	0.171095	2.15E–08	2.11E–77
Barring tool	1	0.999998	0.9998	0.977719	0.078871	1.00E–99	1.00E–99
CAS	0.953821	0.953617	0.95154	0.926451	0.25743	0	0
CBS	0.953821	0.953617	0.95154	0.926451	0.25743	0	0
display	0.989427	0.989349	0.988544	0.977129	0.219617	0	0
Elec. solenoid	0.902839	0.902474	0.898772	0.856031	0.14534	0	0
FOP	1	1	0.99999	0.935968	0.112333	0.000039	1.15E–11
Frequecny monitoring module	0.999792	0.999789	0.999753	0.998909	0.128475	0	0
Fuel pump	0.999855	0.998549	0.985587	0.864867	0.234148	4.95E–07	8.89E–64
Governor	1	1	1	0.986554	0.02851	8.75E–10	0
Load monitoring unit	0.999867	0.998669	0.986766	0.875267	0.263878	0.000002	1.38E–58
LOP	1	1	0.99999	0.935968	0.112333	0.000039	1.15E–11
PLC	0.999845	0.998446	0.984573	0.85601	0.211247	1.77E–07	3.01E–68
Pneu. solenoid	0.995243	0.995175	0.994447	0.979888	0.002344	0	0
Pressure module	1	1	1	1	0.002641	0	0
RTU	0.999878	0.998782	0.987891	0.885297	0.295726	0.000005	1.23E–53
SAP	1	1	0.99999	0.935968	0.112333	0.000039	1.15E–11
Speed control module	0.95636	0.95614	0.953894	0.926015	0.177673	0	0
Speed sensor 1	1	0.999999	0.999963	0.998829	0.96364	0.30999	8.22E–17

(Continued)

Table 4.5 (Continued)

	Reliability hrs						
Component	*R(1)*	*R(10)*	*R(100)*	*R(1,000)*	*R(10,000)*	*R(10E5)*	*R(10E6)*
Speed sensor 2	0.997635	0.997612	0.99737	0.993416	0.262705	0	0
Start stop module	0.938993	0.938675	0.935422	0.895152	0.084413	0	0
TAS1–TAS6	0.914486	0.914159	0.910845	0.872337	0.171776	0	0
TBS1–TAS2	0.914486	0.914159	0.910845	0.872337	0.171776	0	0
TM1	0.997405	0.997372	0.997018	0.990315	0.037862	0	0
TM2	0.997405	0.997372	0.997018	0.990315	0.037862	0	0
TSA	0.914486	0.914159	0.910845	0.872337	0.171776	0	0
TSB	0.914486	0.914159	0.910845	0.872337	0.171776	0	0
Voltage isolator	0.999878	0.998782	0.987891	0.885297	0.295726	0.000005	1.23E–53
Voltage monitoring unit	1	1	1	1	0.982323	0	0
Winding temp module	0.970918	0.970732	0.968819	0.943426	0.112956	0	0

4.6 MULTI-STATE RELIABILITY MODELLING

In traditional reliability theory components, and systems are in only two states: the working state and the failure state, while in the multi-state system (MSS), components and systems are in multiple states. In the traditional reliability theory, systems and components in the fault tree are in only two states, a multi-state system cannot be analysed.

BN has been widely applied to the field of reliability analysis, risk analysis, and maintenance of complex systems. The BN model can express the multi-state of the event and the uncertainty of logical causality and information. The system can be predicted, analysed, and diagnosed by calculating probability. Therefore, a BN can better suit the requirements of MSS reliability analysis.

4.7 MULTI-STATE SYSTEM

In addition to the "working" state and "complete failure" state, the system can also be in a variety of working (or failure) states, or the system can be run at different performance levels, such a system is called a MSS.

The multi-state system is divided into two types: the multi-work (or failure) state system and the multi-performance level system.

4.7.1 Multiple works (or failure) state system

Multiple works (or failure) state system is that in addition to the two states of the working and complete failure, the system can be in a variety of work (or failure) states. Such as the k/n (G) system in the traditional reliability is a typical multi-work (or failure) state system.

4.7.2 Multiple performance level system

The multi-performance level system is that system can run in a variety of performance levels. Such as for a 300 MW generator set, when it is completely normal, their power level is 300 MW when a component such as a ventilator fails, the electricity generation levels will decrease to 225 MW, 200 MW, 150 MW, and so on.

4.8 MULTI-STATE SERIES SYSTEM

Let us consider a system consisting of two three-state components in series. In terms of system and component states as defined above, there are three states: state 0, state 1, and state 2. p represents the probability of the system or component (refer to Figure 4.9). The E_1 *and* E_2 represent the states of the two components, and S represents the state of the system. In the BN a priori probability of the three states is given to the components $E1$, $E2$, respectively, with the conditional probability distribution table we can analyse states of system nodes under the conditions of different states of the two components E_1, E_2 as shown in the Table 4.6. Thus, multi-state components relationship, usually being difficult to express, will be able to be clearly and easily described.

To calculate the reliability of the system we use the bucket elimination algorithm.

$$\begin{aligned} p(S) &= \sum_{\text{E1E2}} p(\text{E1,E2,S}) \\ &= \sum_{\text{E1E2}} p(\text{S} \mid \text{E1,E2}) * p(\text{E1}) * p(\text{E2}) \\ &= \sum_{\text{E1}} p(\text{E1}) \sum_{\text{E2}} p(\text{S} \mid \text{E1,E2}) * p(\text{E2}) \end{aligned} \tag{4.1}$$

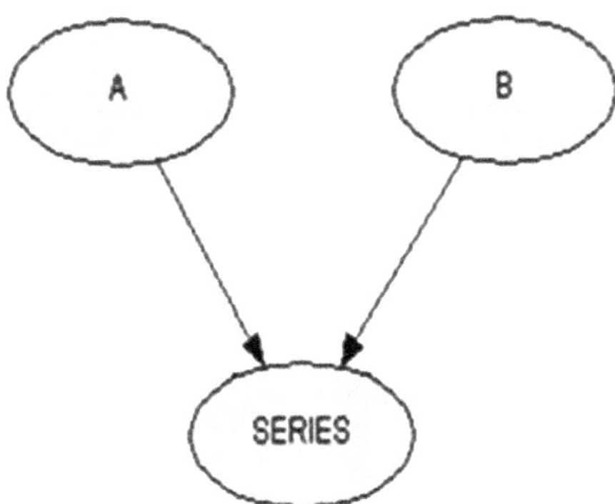

Figure 4.9 DAG of a two component three states in series.

Table 4.6 Truth table for two component three states in series

E1	*E2*	*S*	p(*S* = *1* | *E1*, *E2*)	p(*S* = 2 | *E1*, *E2*)
0	0	0	0	0
0	1	1	1	0
0	2	2	0	1
1	0	1	1	0
1	1	1	1	0
1	2	1	1	0
2	0	2	0	1
2	1	1	1	0
2	2	2	0	1

4.8.1 Solving a multi-state series system

Figure 4.10 indicates a MSS, and the truth logic is indicated in Table 4.7

$$\begin{aligned} I.\quad P(\text{series})_{\text{state0}} &= p(A_0) * p(B_0) \\ &= (.9 * .8) \\ &= 0.72 \end{aligned} \tag{4.2}$$

$$\begin{aligned} II.\quad P(\text{series})_{\text{state1}} &= p(B_0) * p(A_1) + p(B_1) * p(A_0) + p(B_1) * p(A_1) \\ &\quad + p(B_1) * p(A_2) + p(B_2) * p(A_1) \\ &= (0.8 * 0.08) + (0.12 * 0.9) + (0.12 * 0.08) \\ &\quad + (0.12 * 0.02) + (0.08 * 0.08) \\ &= 0.1904 \end{aligned} \tag{4.3}$$

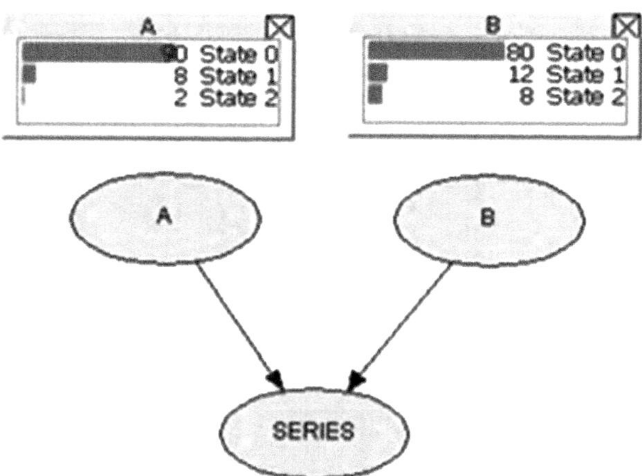

Figure 4.10 DAG of a multi-state (two-states) system in series.

Table 4.7 States for two component three-state system in series

Component	Probability of state 0	Probability of state 1	Probability of state 2
A	.90	.08	.02
B	.80	.12	.08

$$\begin{aligned} III.\quad P(\text{series})_{\text{state2}} &= p(B_0)*p(A_2)+p(B_2)*p(A_0)+p(B_2)*p(A_2) \\ &= (0.8*0.02)+(0.08*0.9)+(0.08*0.02) \\ &= 0.0896 \end{aligned} \tag{4.4}$$

4.8.2 Solving a multi-state series system using Hugin lite 7.8®

Figure 4.11 shows the solution of two component three-state series system using Hugin lite 7.8®

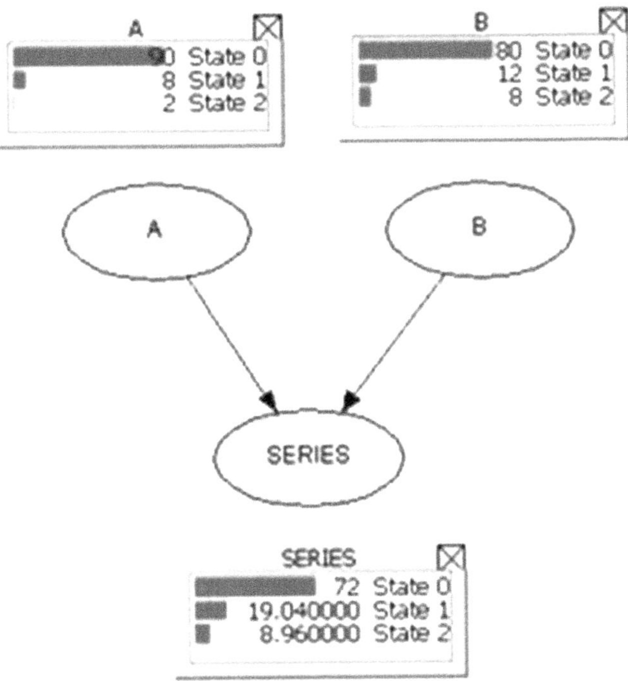

Figure 4.11 Solution of two component three-state series system using Hugin lite 7.8®.

4.9 MULTI-STATE PARALLEL SYSTEM

Let us now consider a system consisting of two three-state components in parallel. In terms of system and component states as defined above, there are three states: state 0, state 1, and state 2. p represents the probability of the system or component (refer to Figure 4.12). The E_1 *and* E_2 represent the states of the two components, and S represents the state of the system. In the BN a priori probability of the three states is given to the components $E1$, $E2$, respectively, with the conditional probability distribution table we can analyse states of system nodes under the conditions of different states of the two components E_1, E_2 as shown in Table 4.8.

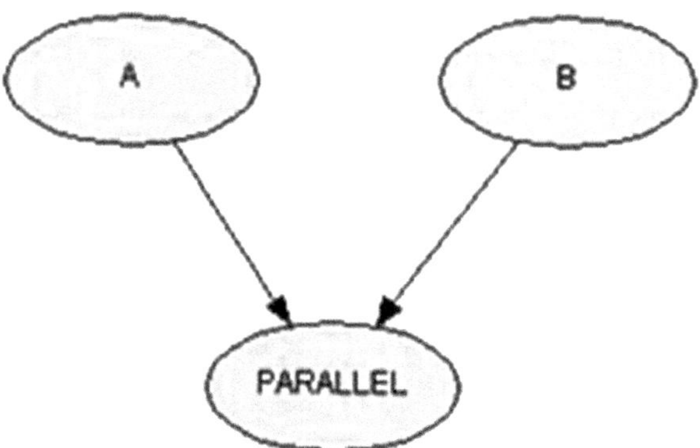

Figure 4.12 DAG of a two component three states in parallel.

Table 4.8 Truth table for two component three states in parallel

E1	E2	S	p(S = 1|E1, E2)	p(S = 2|E1, E2)
0	0	0	0	0
0	1	0	0	0
0	2	0	0	0
1	0	0	0	0
1	1	1	1	0
1	2	2	0	1
2	0	0	0	0
2	1	2	0	1
2	2	2	0	1

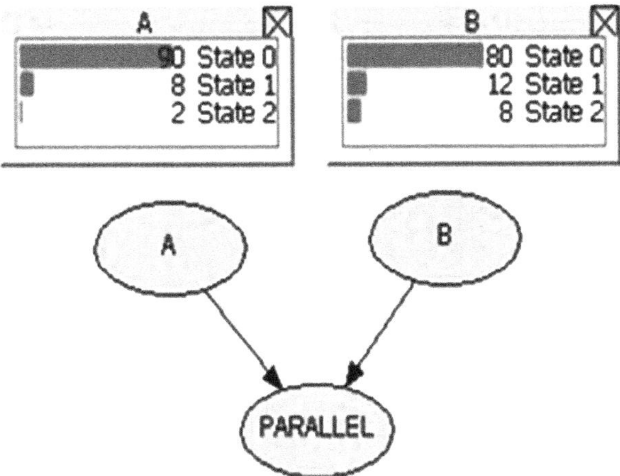

Figure 4.13 DAG of a multi-state (two-states) system in parallel.

4.9.1 Solving a multi-state parallel system

Figure 4.13 indicates a MSS in parallel, and the truth logic is indicated in Table 4.9.

$$\begin{aligned} I. \quad P(\text{parallel})_{\text{state0}} &= p(B_0)*p(A_0)+p(B_0)*p(A_1)+p(B_0)*p(A_2) \\ &\quad +p(B_1)*p(A_0)+p(B_2)*p(A_0) \\ &= (0.8*0.9)+(0.8*0.08)+(0.8*0.02)+(0.12*0.9) \\ &\quad +(0.08*0.9) \\ &= 0.98 \end{aligned} \tag{4.5}$$

$$\begin{aligned} II. \quad P(\text{parallel})_{\text{state1}} &= p(B_1)*p(A_1) \\ &= (0.12*0.08) \\ &= 0.0096 \end{aligned} \tag{4.6}$$

$$\begin{aligned} III. \quad P(\text{parallel})_{\text{state2}} &= p(B_1)*p(A_2)+p(B_2)*p(A_1)+p(B_2)*p(A_2) \\ &= (0.12*0.02)+(0.08*0.08)+(0.08*0.12) \\ &= 0.0184 \end{aligned} \tag{4.7}$$

Table 4.9 States for two component three-state system in parallel

Component	*Probability of state 0*	*Probability of state 1*	*Probability of state 2*
A	.90	.08	.02
B	.80	.12	.08

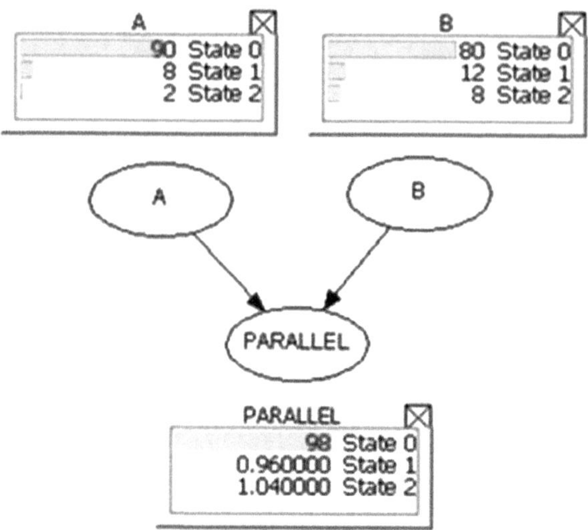

Figure 4.14 Solution of two component three state parallel system using Hugin lite 7.8®.

4.9.2 Solving a multi-state parallel system using Hugin lite 7.8®

Figure 4.14 shows the solution of two component three-state parallel system using Hugin lite 7.8®

4.10 MULTI-STATE SYSTEM OF THE PRESSURE MONITORING SYSTEM

We have already carried out the analysis of the pressure monitoring system of the main DG control system. In this section, a multi-state analysis of the system is carried out. The system has three states namely state 0, state 1, and state 2. State 0 is the normal working state and state 1 is the minor failure state (loose contact, etc.). State 2 is the state in which the component is rendered failed.

Using BNs we have carried out the analysis of the system for a multi-state nature. The prior probabilities in conjunction with the (conditional probability tables) CPT of the same can be looked upon as a strong basis for carrying out simulation studies of the system. Table 4.10 gives the prior probabilities of various components of the pressure monitoring system.

We have used the software package Hugin lite 7.8® to validate and calculate the total effect on the reliability of the system on account of the degradation of components.

Certainly, components in series will render the complete change of the system from one state to another but the components in parallel will only show the impact of its occurrence.

Table 4.10 Prior probability table of the pressure monitoring system

Component	*Probability of state 0*	*Probability of state 1*	*Probability of state 2*
CAS – Crankcase pressure sensor	0.95154	NA	0.04846
CBS – Crankcase pressure sensor	0.95154	NA	0.04846
LOP – Lube oil pressure sensor	0.999999	NA	0.000011
FOP – Fuel oil pressure sensor	0.999999	NA	0.000011
SAP – Starting air pressure sensor	0.999999	NA	0.000011
RTU – Remote transfer unit	0.98789	0.0090825	0.00302750
DISPLAY	0.98925136	0.0085891	0.00214973
PM – Pressure module	0.99999906	0.00000075	0.00000018
PLC – Programmable logic controller	0.98552229	0.01158217	0.00289554

4.11 SIMULATION OF THE MULTI-STATE SYSTEM USING HUGIN LITE 7.8®

The software package Hugin lite 7.8® has been used to model the multi-state model of the pressure monitoring system (refer to Figures 4.15 and 4.16).

Steps in modelling and simulation.

Step 1: Construct a model representing the system and define the number of states involved corresponding to each component. Tables 4.11 and 4.12 indicate operational logic for AND gate AND 1 and AND2, respectively.

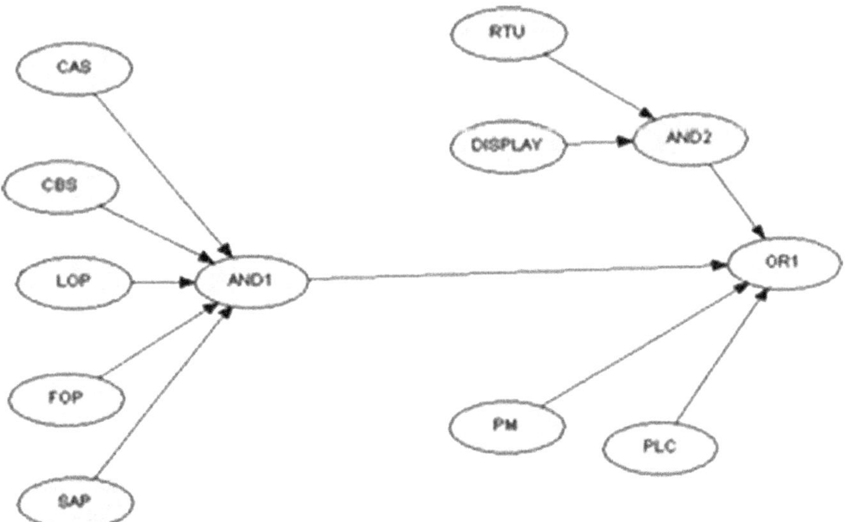

Figure 4.15 Multi-state model of the pressure monitoring system.

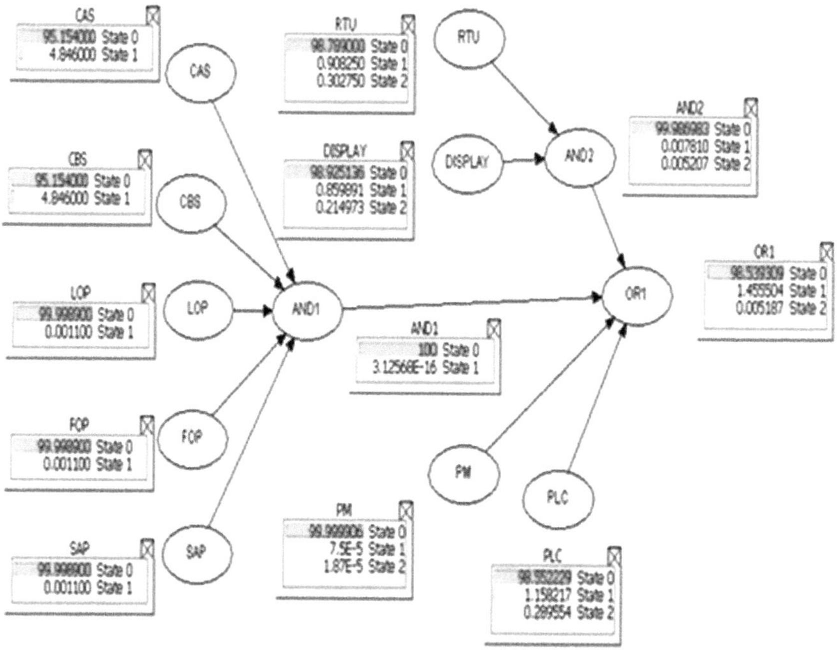

Figure 4.16 Multi-state model of the pressure monitoring system solved using Hugin lite 7.8®.

Step 2: We have to now define each connecting node to calculate the logic of AND & OR as applicable. AND logic for parallel systems and OR logic for series systems. The tables below represent the 'AND1' logic for five components, a two-state system in parallel, 'AND 2' logic for two components, three-states in parallel, and 'OR 1' for two component three-states, two subsystems in series.

Step 3: Run the system to simulate the model. Select show monitor window to display the probability at various nodes as required for carrying out the study.

Step 4: Now by choosing the evidence as RTU in state 1, simulate the system to obtain the reliability at the OR gate which indicates the reliability of the system. Continue the process by changing the states to calculate the system's reliability.

Figures 4.17–4.19 are results obtained using the Hugin lite software package. The effect of system reliability on the multi-states model is simulated. Table 4.13 indicates the reliability values of different states of the RTU, Display, PLC, and PM modelled to have a multi-state operation.

Table 4.11 AND logic for five component, two-states system in parallel (AND 1)

CAS	*S0*																*S1*															
CBS	*S0*								*S1*								*S0*								*S1*							
LOP	*S0*				*S1*				*S0*				*S1*				*S0*				*S1*				*S0*				*S1*			
FOP	*S0*		*S1*		*S0*		*S1*		*S0*		*S1*		*S0*		*S1*		*S0*		*S1*		*S0*		*S1*		*S0*		*S1*		*S0*		*S1*	
SAP	S_0	S_1	S_0	S_1	S_0	S_1	S_0	S_1	S_0	S_1	S_0	S_1	S_0	S_1	S_0	S_1	S_0	S_1	S_0	S_1	S_0	S_1	S_0	S_1	S_0	S_1	S_0	S_1	S_0	S_1	S_0	S_1
AND logic for 5 components, 2 states in parallel																																
State 0	1	1	1	1	1	1	1	1	1	1	1	1	1	1	1	1	1	1	1	1	1	1	1	1	1	1	1	1	1	1	1	0
State 1	0	0	0	0	0	0	0	0	0	0	0	0	0	0	0	0	0	0	0	0	0	0	0	0	0	0	0	0	0	0	0	1

Note: S0-state 0, S1-state 1

Table 4.12 AND logic for two component, three-states system in parallel (AND2)

Display	*State 0*			*State 1*			*State 2*		
RTU	*State 0*	*State 1*	*State 2*	*State 0*	*State 1*	*State 2*	*State 0*	*State 1*	*State 2*
State 0	1	1	1	1	0	0	1	0	0
State 1	0	0	0	0	1	0	0	0	0
State 2	0	0	0	0	0	1	0	1	1

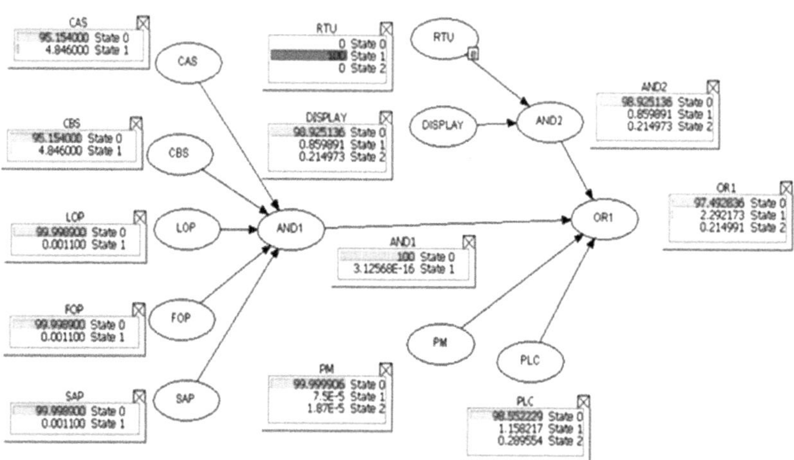

Figure 4.17 Multi-state model of the pressure monitoring system with the evidence of RTU in state 1 solved using Hugin lite 7.8®.

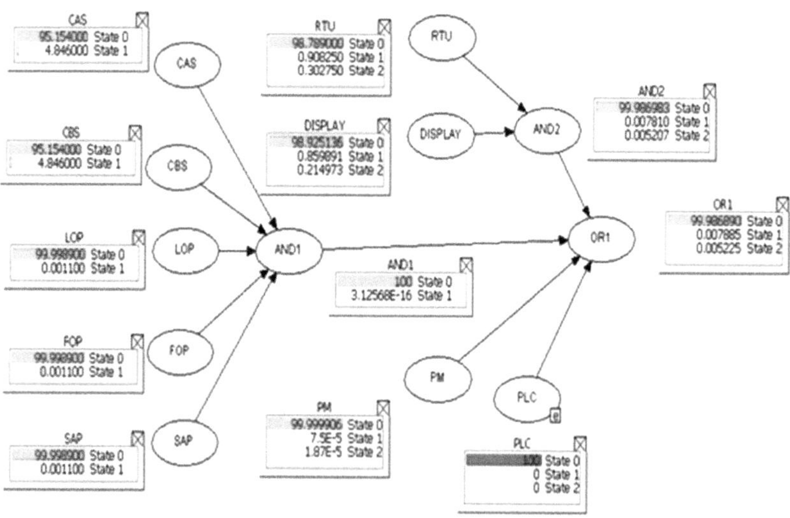

Figure 4.18 Multi-state model of the pressure monitoring system with the evidence of PLC in state 0 solved using Hugin lite 7.8®.

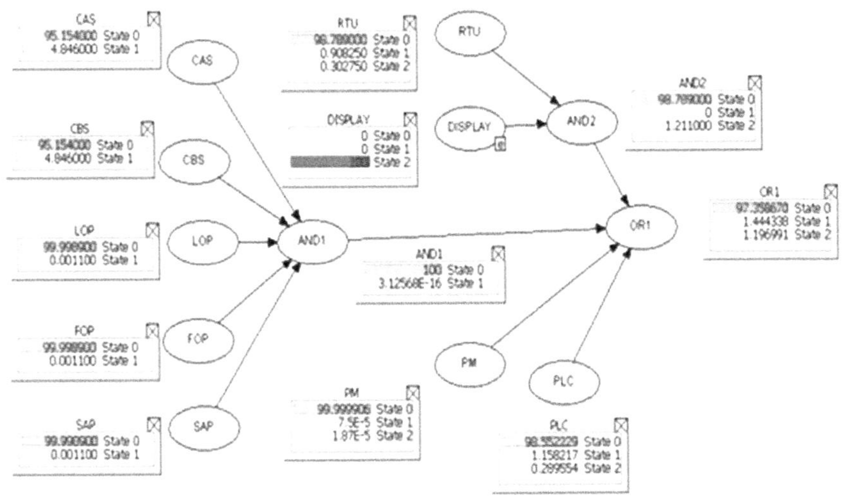

Figure 4.19 Multi-state model of the pressure monitoring system with the evidence of display in state 2 solved using Hugin lite 7.8®.

Table 4.13 Impact of states on the probability of system

		Probability of state (%)		
Component	*State*	*State 0*	*State 1*	*State 2*
RTU	0	98.552137	1.447845	1.84E-5
	1	97.492836	2.292173	0.214991
	2	97.492836	1.444733	1.062432
DISPLAY	0	98.552137	1.447845	1.84E-5
	1	97.358670	2.339438	0.301891
	2	97.358670	1.444338	1.196991
PLC	0	99.986890	0.007885	0.005225
	1	0	100	9.73E-10
	2	0	99.98696	0.013035
PM	0	98.539401	1.455430	0.005169
	1	0	99.99998	1.507E-5
	2	0	99.70525	0.294746

4.12 CONCLUSION

We have carried out a comprehensive study on the system simulation and developed a programme to calculate the reliability using the FTA analysis. A programme is developed in MATLAB Simulink to calculate the developed Cumulative Distribution Function (CDF) using standard distribution as blocks which can be assembled along with AND/OR logic to calculate the system reliability by defining the subsystems as series or parallel configurations. The blocks used for calculating the CDF are defined with their parameters based on the observed date of failure and can be changed to simulate a different failure scenario. Also, we can at any given time model the system with any distribution by merely changing the blocks to some other distribution. An FTA analysis was carried out successfully using this programme for calculating the system reliability for the simulation study of a PLC-based DG control system. We have further continued with a detailed analysis of one of the subsystems i.e. pressure monitoring system.

To take advantage of the multi-state nature of the BNs, we have used this approach to calculate the system reliability at a multi-state level of system reliability. Here, we have used the pressure monitoring system to extend the study of the effect of individual component degradation on the system. The study on multi-state systems has successfully worked out for the system. This is important since in carrying out system reliability we sometimes tend to ignore the effect of component degradation on the entire system. We have applied this approach to carry out a detailed analysis of systems in multi-states and have effectively proved it to be a factor worth considering. The FTA developed in Simulink can still be made a stronger tool as it is reinforced with the power of the BN multi-state model.

REFERENCES

1. Y. H. Lin, Y. F. Li, E. Zio, A reliability assessment framework for systems with degradation dependency by combining binary decision diagrams and Monte Carlo simulation. IEEE Transactions on Systems, Man, and Cybernetics: Systems, 2015; 46(11): 1–9. doi:10.1109/TSMC.2015.2500020.
2. Y. F. Li, E. Zio, A multi-state model for the reliability assessment of a distributed generation system via universal generating function. Reliability Engineering & System Safety, 2012; 106: 28–36. doi:10.1016/j.ress.2012.04.008.
3. T. Aven, U. Jensen, Stochastic models in reliability, Springer, NY, Berlin, New York, 1999.
4. J. Xue, K. Yang, Dynamic reliability analysis of coherent multistate systems. IEEE Transactions on Reliability, 1995; 44(4): 683–688.
5. R. D. Brunelle, K. C. Kapur, Review and classification of reliability measures for multi-state and continuum models. IEEE Transactions, 1999; 31(12): 1171–1180.
6. A. Lisnianski, Y. Ding, Redundancy analysis for repairable multistate system by using combined stochastic processes methods and universal generating function technique. Reliability Engineering & System Safety, 2009; 94(11): 1788–1795.
7. A. Lisnianski, G. Levitin, H. B. Haim, D. Elmakis, Power system structure optimization subject to reliability constraints. Electric Power Systems Research, 1996; 39(2): 145–152.
8. G. Levitin, D. X. Liu, H. B. Haim, Y. Dai, Multi-state systems with selective propagated failures and imperfect individual and group protections. Reliability Engineering & System Safety, 2011; 96(12): 1657–1666.
9. R. Billinton, W. Y. Li, Hybrid approach for reliability evaluation of composite generation and transmission systems using Monte Carlo simulation and enumeration technique. IEE Proceedings – C Generation, Transmission, and Distribution, 1991; 138(3): 233–241.

10. E. Zio, M. Marella, L. Podofillini, A Monte Carlo simulation approach to the availability assessment of multi-state systems with operational dependencies. Reliability Engineering & System Safety, 2007; 92: 871–882.
11. E. Jose, R. Marquez, D. W. Coit, A Monte Carlo simulation approach for approximating multi-state two-terminal reliability. Reliability Engineering & System Safety, 2005; 87(2): 253–264.
12. E. Zio, L. Podofillini, G. Levitin, Estimation of the importance measures of multi-state elements by Monte Carlo simulation. Reliability Engineering & System Safety, 2006; 55(2): 319–327.
13. S. Calik, Dynamic reliability evaluation for a multi-state component under stress-strength model. Journal of Nonlinear Sciences and Applications, 2017; 10: 377–385.
14. C. Zhang, A. Mostashari, Study on the influence of component uncertainty on reliability estimation of multi-state systems with continuous states. INCOSE International Symposium, 2009; 19(1): 457–465.
15. A. Salmasnia, E. Ameri, A. Ghorbanian, H. Mokhtari, A multi-objective multi-state degraded system to optimize maintenance/repair costs and system availability. Scientia Iranica, 2017; 24(1): 355–363.
16. L. Yong, L. Anxin, Z. Xiaonan, Complex multi-state system reliability modeling and assessment. International Journal of Engineering Science and Technology (IJEST), 2012; 4(09).

Chapter 5

Reliability analysis models for multi-state safety-critical systems

Pooja Singh, Lalit Kumar Singh, Jaishree Meena, and Gopika Vinod

5.1 INTRODUCTION: BACKGROUND AND DRIVING FORCES

The systems are created with the aim of executing their designated functions under the specified conditions. Certain systems exhibit different levels of efficiency in carrying out their tasks, often referred to as performance rates. A system capable of possessing a finite number of these performance rates is termed a multi-state system (MSS). The MSS comprises various elements, each of which may possess multi-states. An element represents the fundamental unit within the system that cannot be further subdivided. It may be possible that this element cannot be made from other parts; however, from reliability point-of-view, it is considered as a self-contained entity and the reliability of this element is evaluated without considering the reliability of its constituent parts.

The most basic form of MSS is a binary system, consisting of just two states: a successful state and a failed state. Here 'success state' indicates that the system is performing flawlessly, while 'failed state' means that the system is completely failed.

The basic illustration of MSS is the widely recognized k-out-of-n systems. These systems are composed of n identical binary units and can exhibit n+1 states, which vary depending upon the count of available units. The system reliability and performance depend on the number of units that are in 'success state'. In such systems, minimum number of units that must be in 'success state' is n, for acceptable performance rates of the entire system. In other words, MSS can execute their task with partial effectiveness. Failures of certain elements within the system result only in a decline in the overall system performance.

The traditional reliability models are based on binary states of the system. It assumes that the system and its components/elements have only two states. They are unable to characterize the multi-state nature of engineered systems. MSSs are able to expose deteriorating behavior by introducing many states in between 'success' and 'failed' states.

DOI: 10.1201/9781003546214-5

There are several engineering systems, which are of MSS, such as medical systems, aerospace systems, manufacturing systems, power systems, and computing systems. The assessment of MSS reliability relies on four distinct methodologies: an expansion of Boolean models to accommodate multi-valued scenarios, the stochastic processes (particularly Markov and semi-Markov) methodology, the universal generating function (UGF) method, and the Monte Carlo simulation technique. The primary challenges in MSS reliability analysis are the issue of 'dimension damnation', stemming from the fact that each system element can exhibit numerous states (as opposed to only two states in binary-state systems). This complexity often leads to excessive workload and time consumption when employing the structure function approach. However, the structure-based approaches are not able to capture the dynamic behavior of the MSS and hence may lead to inaccurate reliability prediction, which can be more critical for safety-critical systems (SCSs) because these systems do have a high reliability requirement.

A malfunctioning ceiling fan would not pose any deadly risks, but a failure of an aircraft control system could result in fatalities. A failure in a system occurs when it does not perform its function as intended, which can result in disastrous outcomes. The failure of some systems, therefore, may have unacceptable consequences. These systems are referred to as SCSs because if they malfunction, it could result in the endangerment of human lives, substantial property loss, harm to the environment, or the failure of mission objectives [1]. Thus, if the failure of a system could lead to outcomes that are deemed dangerously unacceptable, then those systems are classified as safety-critical [2]. To summarize, the malfunction of specific systems can lead to grave impacts on both human lives and the environment. Examples include nuclear reactors, airplane avionics systems, and medical infusion pumps. Nowadays, SCSs are utilized to facilitate a broad spectrum of crucial human activities and have witnessed increased prevalence in contemporary society as households increasingly depend on technology. Various facets of individuals' everyday routines are supported by computer-driven SCSs, encompassing healthcare systems, transportation services, nuclear power plants (NPPs), and defense apparatus.

SCSs present in various domains like automotive, NPP, space missions, and healthcare heavily depend on non-functional requirements and precise timing [3]. These systems are designed to provide high responsiveness and multitasking capabilities. Despite potential threats, these systems need to ensure safety and reliability. Furthermore, these systems must guarantee availability, even when faced with security breaches. The occurrence of any malicious cyberattack targeting these systems could result in significant consequences for safety and security, potentially resulting in serious and undesirable outcomes. Hence, researchers are actively working to improve the system's reliability. This examination of non-functional requirements plays a critical role in recognizing and addressing risks in SCS.

Several new reliability models for MSS are being introduced such as diagram-based method, the stochastic process, UGF, the recursive algorithm, and the simulation-based methods. The stochastic models are widely used due to its potential for capturing the system's dynamics.

In this chapter, we shall discuss stochastic-based reliability models for MSS. As the accurate reliability measurement for SCS is very essential, we shall focus our discussions on SCS; however, the models are well suited to all MSS. The outline of this chapter is as follows.

In the next section, we shall understand the need of SCS. The applications of SCS in different domains are given in Section 5.3. To derive the need of stringent reliability requirements for SCS, the past accidents due to failure of SCS are given in Section 5.4, along with their impacts. This supports the necessity of reliability analysis of SCS in Section 5.5. The stochastic modeling of SCS is discussed in Section 5.6. A widely known stochastic modeling technique, Petri nets (PNs), is given in Section 5.7. The working mechanism of PN and its properties are explained with examples. Another widely known stochastic technique, Markov chain (MC), is explained in Section 5.8. MCs and PNs can be used for reliability analysis. To demonstrate the method to evaluate reliability using PNs, a case study has been taken, which is described in Section 5.9. The reliability evaluation using PNs on the case study is described in Section 5.10.

5.2 NEED OF SAFETY-CRITICAL SYSTEMS (SCSs)

In the present time, several devices rely on computer systems with specific peripherals, and their functioning accuracy is heavily reliant on quality of software. However, due to the progressively complex nature of these computer-based systems, errors can occur frequently. Various reasons can cause this software failure, such as software errors, specification mistakes, errors in software or hardware design, and low-quality user interfaces [4]. If SCS does not function properly, it can cause users to feel insecure, frustrated, or upset. Many computer-based safety-critical systems (CBSCSs), including aerospace, aviation, NPPs, and healthcare, rely on the software to accurately carry out their critical functions.

The use of software-based technology has surpassed traditional hardware-based systems for ensuring safety and efficiency in various operations. Robotic tools driven by computer systems are now being employed within healthcare facilities to replace conventional surgical instruments [5]. Hardware components undergo routine inspection and maintenance to ensure their reliability, with failures typically attributed to manufacturing defects and aging. These measures are implemented to meet the stringent reliability standards for hardware components. As a result, hardware failures are often considered less significant compared to software failures.

However, this failure may endanger human lives, substantial property loss, harm to the environment, or the failure of mission objectives. Hence, it is essential to conduct thorough testing of SCS before deployment, alongside the precise and quantifiable prediction of failure probabilities. Therefore, it is essential to emphasize the estimate of performance and reliability of SCS [6].

5.3 APPLICATIONS OF SAFETY-CRITICAL SYSTEMS

SCSs are utilized in industries where their proper functioning is of utmost importance due to the significant potential risks they pose to human safety, the environment, or property in the event of failure. They are crucially important in various industries where safety is the top priority. Some applicable areas of SCS are shown in Figure 5.1 and are explained below:

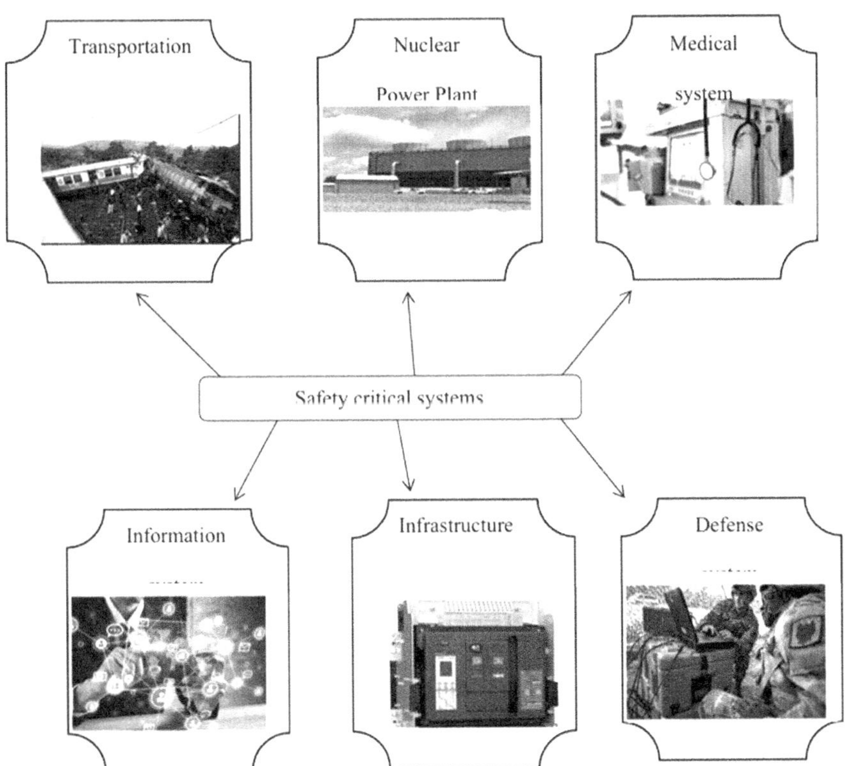

Figure 5.1 Application areas of safety-critical systems.

- **Transportation Systems:** These systems, which include railways, roadways, and aviation, are essential in facilitating the movement of people and goods. The intelligent transportation system utilizes advanced technologies to detect, analyze, control, and enable communication within ground transportation to enhance safety and reliability and increase overall efficiency in transportation networks. Any disruption or failure within these systems can result in significant consequences, potentially culminating in accidents that threaten human safety and potentially result in loss of life [6, 7].
- **NPPs:** NPPs are classified as SCSs because of inherent risks associated with nuclear energy and the heat generated from nuclear fission reactions. SCS of the NPP industry is crucial for the overall safe functioning of the plant. A failure in the SCS of NPP can potentially lead to radiation emission into the surroundings. As a result, it is imperative that these systems adhere to stringent dependability standards to ensure safety. Therefore, it is more advantageous to minimize or eliminate risks at the outset of design and development process rather than deferring action until the system becomes more intricate [3, 8, 9].
- **Medical Systems:** Medical devices play a vital role in the healthcare system as they are essential for diagnosing and treating patients. These devices incorporate software systems that facilitate communication between diagnostic instruments and the human body, allowing them to monitor heart rate, conduct blood glucose level tests, and administer necessary medications. Any potential malfunction in these systems can result in misdiagnosis, inappropriate treatment, or delayed reaction, all of which can jeopardize patients' lives [5, 10].
- **Information Systems:** These systems play a critical role in the effective task management, communication with customers and suppliers, and general functioning of modern enterprises and society. The functionality of these systems involves collecting, storing, processing, and distributing information. However, these information systems' possible security flaws could have disastrous repercussions. Due to the importance of information in many facets of society, such attacks can compromise sensitive data, necessitating stringent security and performance measures [11].
- **Defense and Military Systems:** These systems are described as safety-critical due to their vital contribution to safeguarding national security and the possible repercussions of operational failures. The primary objective of these systems is to protect a nation's inhabitants, territory, and interests from a range of potential dangers, encompassing armed wars, acts of terrorism, and other forms of hostile behavior. Therefore, the design and development of these systems undergo strict adherence to safety, reliability, and security standards [12].

Because of the severe consequences of potential failures, SCS undergoes thorough design, testing, and certification processes to comply with strict safety standards and regulations. These systems often employ redundancy, fault tolerance, and fail-safe mechanisms to improve reliability and reduce failure likelihood. High-quality software programs are crucial elements in the successful operations of SCS. Before implementation of SCS, these programs undergo a comprehensive design, testing, and official recognition process, ensuring they meet international standards. The objective is to achieve superior quality, high reliability, and minimal failures, thereby establishing SCS as a highly efficient and dependable system.

5.4 SOFTWARE-RELATED ACCIDENTS INVOLVING SCS

SCSs are crucial systems whose failure can result in catastrophic disasters. Due to their critical functions, these systems have received considerable attention regarding their reliability and performance [13, 14]. SCSs are susceptible to errors caused by flawed software or improper software utilization. Thus, even a minor software error can have devastating consequences, as shown in Figure 5.2.

Even with careful guidance and strict controls, the SCS industry remains susceptible to software failures. Table 5.1 highlights some notable accidents in the past four decades due to SCS software failures.

Numerous other software failures are discussed in [26, 27]. These include issues such as inaccurate or unclear requirement definitions, insufficient research, improper replication of outdated software, and the presence of software bugs. Thus, any effective strategy for quantifying software reliability must consider these critical factors to reduce catastrophic outcomes.

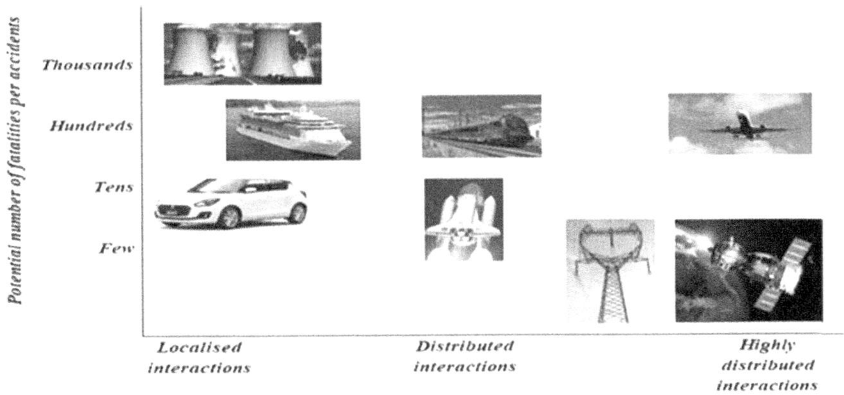

Figure 5.2 Failure effects of safety-critical systems.

Table 5.1 Catastrophic accidents of safety-critical systems

Accident name	*Year*	*Losses*	*Reason*
Odisha train collision [15]	2023	294 killed, 1175 injured	Error in electronic signal.
Flydubai 737–800, Russia [16]	2016	62 died	Outage in weather forecasting system.
Fukushima Daiichi nuclear disaster, Japan [17]	2011	37 injured, loss of $23.6 billion	Equipment failures because of earthquake and tsunami.
Railway signaling failure, Sydney [18]	2011	847 trains delayed, 240 canceled	Electrical fault in control signal system.
Germany, Saxony-Anhalt train accident [19]	2011	28 died	Abnormal working of traffic detecting software.
Deepwater Horizon oil spill [20]	2010	11 killed, 17 injured	Defective cement on the well.
Air France Flight 447 crash [21]	2009	228 died	23 error messages transmitted by on-board automotive report system.
Emergency shutdown of the Hatch NPP [22]	2008	loss of $5 million	S/W updates reset the data in I&C system.
Sri Lanka tsunami train wreck [23]	2004	30,000 died	The seismic monitoring station located in Pallekele, Sri Lanka, swiftly detected the earthquake, yet it did not assess the likelihood of a tsunami reaching the island.
Miscalculated radiation doses, Panama [24]	2001	8 died, 20 suffered	Excessive radiation therapy doses.
Bhopal gas tragedy, India [25]	1984	8000 died, 558, 125 injured	A release of methyl isocyanate occurred from the storage tank designated as E610 within the premises of the Union Carbide India Limited facility.

5.5 NECESSITY OF SAFETY AND RELIABILITY MODELING OF SCS

SCSs present in various domains like automotive, NPP, space missions, and healthcare heavily depend on non-functional requirements and precise timing. These systems are designed to provide high responsiveness and multitasking capabilities. Researchers have found that SCSs are highly susceptible to failures. Developing these systems is time-consuming and expensive, requiring a highly reliable and accurate mechanism to be added to avoid failure. However, this mechanism can impact the system's performance, making it necessary to study non-functional parameters based on software requirement specifications to ensure proper functioning [28]. Despite potential threats, these systems need to ensure safety and reliability. Furthermore, these systems must guarantee availability, even when faced with security breaches.

The reliability of an SCS is a vital aspect that significantly impacts its operational quality. It refers to the system's ability to function flawlessly within a specified environment for a specific duration [29]. The computer plays a crucial role in implementing the functionalities of SCS, which can introduce vulnerabilities to the system due to potential software faults. Reliability engineering aims to identify and correct failures based on analysis of past failure data and develop engineering techniques to reduce their failure rate [30]. While there are rapid strides in hardware advancements, the primary focus should be on software reliability and other metrics to enhance system safety. Fast and reliable software is crucial in developing SCS to ensure they are safe.

In short, the demand for SCSs is growing rapidly, resulting in a rise in the complexity of software. NPPs are designated as SCS due to the inherent risks associated with nuclear energy and the heat generated from nuclear fission reactions. The safety and reliability of these systems are paramount for the overall secure operation of the plant. Within NPPs, various crucial components like actuators, signal processing units, control valves, sensors, transformers, circuit-breakers, controllers, and heat exchangers are deemed critical [8]. These components are labeled safety-critical because any failure in their functioning could lead to catastrophic consequences in the event of an accident. Given that an infringement in the SCSs of an NPP could lead to the release of radiation into the environment, it is imperative that such systems adhere to stringent reliability standards.

5.6 STOCHASTIC MODELING OF SAFETY-CRITICAL SYSTEMS

The increasing complexities of SCS could affect its overall performance. Thus, when software components experience partial failure, conducting a non-functional analysis of SCSs becomes essential. Estimating reliability during the design phase can assist developers in selecting a more appropriate development approach. This awareness can minimize risk and ensure that the system meets the desired standards early. Several methods have been proposed to forecast reliability during the development process. However, most of these methods rely on estimating state transition probabilities, which are either assumed or determined based on operational profiles. Thus, the results are partially predictive. Thus, it is essential to create models of these systems to assess their reliability before implementing them in practice. Thus, employing a system model for dependability assessment aids in identifying potential issues and offering preemptive measures.

SCSs comprise several components and subsystems, and it is essential to validate the non-functional parameters to achieve a high level of system quality. For SCS, estimating reliability before implementation is preferable to reduce the risk of loss since changes during later stages can be costly and time-consuming. This evaluation provides additional insights into reliability and safety assessments and is commonly used in analyzing hardware and software systems that utilize fault tolerance mechanisms. Considerations like failure rate, duration of failure, and severity of software and hardware problems are crucial for determining a system's reliability. These failures, however, are considered stochastic events due to their random nature. One example of a modeling tool based on probabilistic and stochastic principles is PNs. Another tool rooted in probability is the MC, which can be derived from PN. PNs and MCs are powerful tools for analyzing the reliability and safety of SCS [31].

5.7 PETRI NETS

PNs are a directed weighted bipartite graphical and mathematical tool applicable for modeling various dynamic systems. Using them is very effective when analyzing and modeling concurrent, asynchronous, distributed, parallel, non-deterministic, and stochastic processes [32]. Being a visual tool, PN provides an intuitive means of facilitating communication, much like flowcharts, block diagrams, and networks. Furthermore, they allow the formulation of state, algebraic, and diverse mathematical expressions that describe a system's behavior. This makes PN accessible and beneficial to both practitioners and theoreticians alike. Consequently, they establish a potent bridge for communication between these two groups. Practitioners can gain insights from theoreticians on enhancing the methodological rigor of their models. Similarly, theoreticians can gather insights from practitioners to enrich their models with greater realistic detail.

A PN has two categories of nodes: places, represented by circles, and transitions, depicted as bars or rectangles. The place symbolizes a particular state, while a transition denotes an action or event that could transpire within the system. The place can hold a distinct number of tokens, depicted as black dots within them. These tokens serve to convey various information such as the condition of components, resource quantities, or the current state of the system.

A PN may contain two types of arcs: (1) Regular arcs, which are drawn with an arrow at the head, allow modifications of states by transferring tokens among input and output places of respective transition (denoted by $\rightarrow$). The positive weight-directed arcs establish connections between places and transitions. (2) Inhibitor arcs, drawn as an edge with a circle head, can disable the transitions when enough tokens are present in places (denoted by $\multimap$). The primary function of a place is to store tokens, whereas a transition is responsible for modifying the distribution of tokens [3].

Formally, a PN is described as a five-tuple, $PN = \{P, T, F, W, M_0\}$. Where $P = \{p_1, p_2, p_3, \ldots, p_m\}$ is a non-empty finite set of places (i.e., $P \neq \phi$) which describe the state of a system, $T = \{t_1, t_2, \ldots, t_n\}$ is a non-empty finite set of transitions (i.e., $T \neq \phi$) responsible for altering the system's state, $F \subseteq (P \times T) \cup (T \times P)$ denotes a collection of directed arcs connecting places to transitions or transitions to places, $W : F \rightarrow N$ signifies the weight function associated with every directed arc, $N = \{1, 2, \ldots,\}$ denotes the set of natural numbers, $M_0 : P \rightarrow \{0, 1, 2, \ldots\}$ stands for initial marking, i.e., an m-vector whose element represents the token present in each of the m places of PN. Label for unit weight is generally omitted. Arcs are drawn from place to transition or from transition to place. Thus, the following property holds good [8]: $P \cap T = \phi$ and $P \cup T \neq \phi$.

The tokens movement in PN illustrates the system's dynamic behavior, indicating alterations in the distribution of tokens among the various places. To modify the token distribution, it is essential that minimum one transition is enabled. The state/marking of PN evolves according to the enabling and firing rules outlined below [32]:

1. If each input place (p) of transition (t) holds the minimum number of tokens specified by the weight of the arc *(p, t)*, then transition *(t)* is considered enabled. In other words, t will be enable at a marking M if, for every places, p in the set of input places of transition t, denoted as $\bullet t$, the minimum token count at every place p (denoted by $M(p)$) is the weight of the directed arc from place p to transition t (represented as $W(p, t)$), i.e., if $\forall p \in \bullet t,\ M(p) \geq W(p, t)$.
2. An enabled transition is capable of firing.
3. When transition t is fired, it consumes a number of tokens equal to $W(p, t)$ from each of its input places and credits a quantity of tokens equal to $W(t, p)$ to each of its output places. Here, $W(p, t)$ represents the arc-weight from place p to transition t, and $W(t, p)$ represents the arc-weight from transition t to place p.

Example 1

Consider the model illustrated in Figure 5.3, which represents the functioning of a system using PN concepts. In this model, we have three places: X, Y, and Z, as well as two transitions: T_1 and T_2. Figure 5.3a presents the initial marking scenario, illustrating that place X and place Y contain a single token, whereas place Z is token-free. This configuration enables the transition T_1. Following the firing of T_1, the new arrangement of PN is shown in Figure 5.3b. When T_1 is fired, it removes tokens from both X and Y and places a single token into Z, as indicated by the weight of arc $W(T1, Z) = 1$. Consequently, place Z in Figure 5.3b contains one token, thereby enabling the transition T_2. Figure 5.3c shows the final arrangement of PN after firing of T_2.

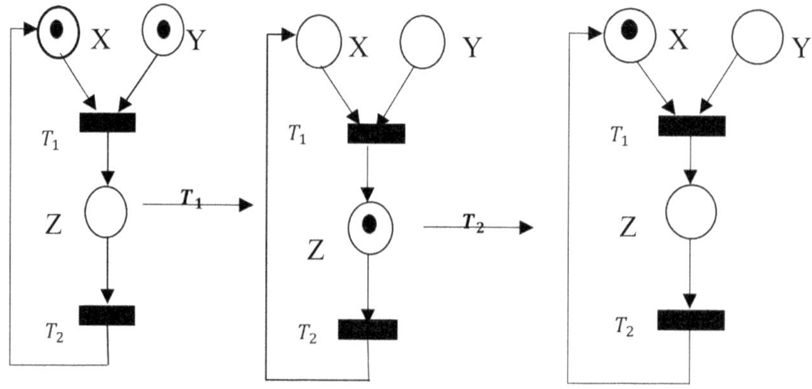

a. Initial Marking b. Marking after c. Marking after T_2 fires

Figure 5.3 Petri net execution.

The performance of any system depends on its reliability and safety. Hence, during performance evaluation, it is crucial to examine elements that could jeopardize the reliability and safety of SCS. Deadlock, boundedness, liveness, stability, reachability, and reversibility serve as vital metrics for assessing both safety and reliability. These metrics are commonly known as marking-dependent or behavioral properties of PN. The behavioral properties of the PN model are assessed through a reachability graph (RG).

RG: The RG of a PN visually represents all feasible states attainable after firing of transitions. In the RG, transitions are denoted by edges, while vertices symbolize system states or places within the PN. A state (marking) S_j of RG becomes accessible from another state (marking) Si if there exists a transition or sequence of transitions that, upon firing, facilitate the transformation from S_i to S_j.

Reachability Analysis: It provides the essential foundation for analyzing the dynamic characteristics of a system. The reachability analysis of a system is done by using an RG of PN. RG represents a directed graph illustrating the state transitions within the system. These states become accessible by executing a specific sequence of transitions from the initial marking or state. A marking M_n is considered reachable from an initial marking M_0, when a series of firing transitions occurs, starting from the initial state of the PN associated with M_0 and leading to a state associated with M_n. In other words, any marking within a PN can be reached from the initial marking if there exists a sequence of transitions being fired between them. This firing sequence is represented by $\sigma = M_0, t_1, M_1, t_2, \ldots t_n, M_n$ or simply $\sigma = t_1, t_2, \ldots, t_n$. The collection of every potential firing sequence originating from M_0 is represented as $R(M_0)$.

Boundedness and Stability Analysis: The boundedness property guarantees that there is no risk of overflow at any place within the PN model, regardless of the firing sequence. It ensures that the token count at any place in a bounded PN never exceeds a finite integer k for any reachable marking from the initial marking. In other words, if a marking M is reachable from the initial marking M_0, i.e., $M \in R(M_0)$, then the boundedness property should ensure that $\forall p,\ M(p) \leq k$, where k is a finite non-negative integer. If every potential firing sequence upholds the boundedness property, the PN is considered stable.

Mathematically, a PN is said to be bounded if $\exists k \in N,\ \forall M \in R(M_0),\ \forall p \in P,$ *then* $M(p) \leq k$. The reachability set of a bounded PN has a finite number of elements. If a PN is not bounded, then it will be unbounded.

Safeness: A bounded PN is deemed safe if it satisfies two conditions: (1) every marking of PN is attainable from the initial marking, and (2) $k = 1$. A PN with a bound of 1 is consistently safe. In mathematical terms, a PN is considered safe if $\forall M \in R(M_0),\ \forall p \in P,$ *then* $M(p) \leq 1$.

Siphon and Trap: The liveliness or deadlock presence in a PN is determined by a set of places known as a siphon. A non-empty set $S \subseteq P$ is called a siphon iff $^{\circ}S \subseteq S^{\circ}$, i.e., all transitions with an output place in S must also have an input place within S, and it is a trap iff $S^{\circ} \subseteq {}^{\circ}S$, i.e., all transitions with an input place in S must also have an output place within S. In notation, $^{\circ}S$ represents the collection of input transitions for the set of places denoted by S, while S° refers to the set of output transitions for the same set of places. When a siphon becomes empty of tokens in a particular marking, it stays empty in subsequent markings. Conversely, if a trap contains tokens, it remains marked indefinitely.

Example 2

Figure 5.4 depicts a siphon wherein the token count within the siphon remains unchanged after the firing of t_1 but decreases after the firing of t_2. Thus, once a siphon becomes token-free in a certain marking, it stays empty in subsequent markings. Conversely, a trap maintains its token count by firing t_1 but experiences an increase upon firing t_2. Therefore, if a trap contains any tokens, it remains marked indefinitely.

Deadlock and Liveness: Deadlock can lead to a system's processing being delayed or even to the system's state being indefinitely halted. The term 'liveness' refers to a set of characteristics essential for a system to continue progressing despite its components running concurrently. A live PN always ensures that the system operates without deadlocks, regardless of the firing sequence of transitions. The behavioral properties of a PN, namely siphons and marked traps, are utilized to assess deadlock and liveness. Deadlock or liveness is examined in two ways: (1) If a PN model lacks siphons and traps, then the model is always free of deadlocks and is therefore live; or (2) If any PN has a marked trap, there is no risk of potential deadlock, making the PN both live and deadlock-free. Thus, the presence of a marked trap in the PN guarantees the absence of potential deadlock in any siphon, ensuring that the PN is both deadlock-free and live.

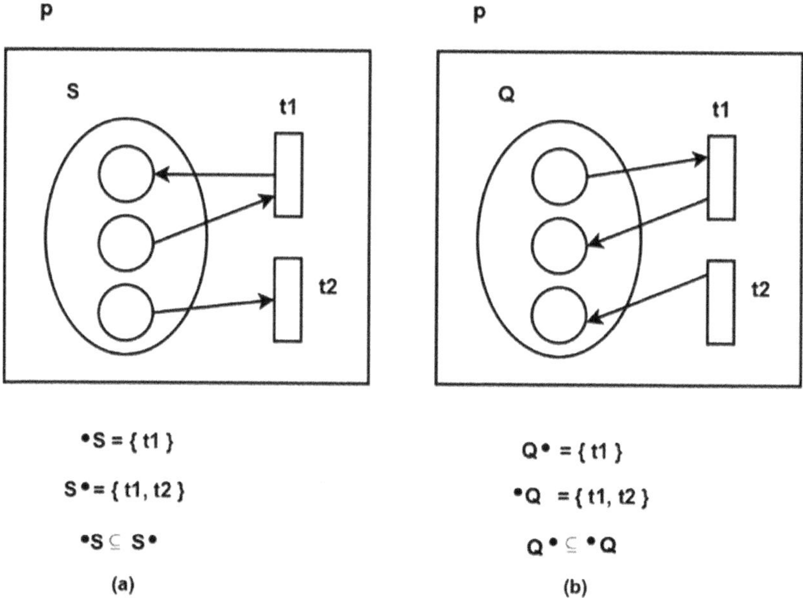

Figure 5.4 An example of (a) siphon, and (b) trap.

Consistency and Reversibility: This process ensures that the initial marking state of the RG can always be reached from any other marking state. To uphold the consistency property, all transitions of the PN should fire at least once. Additionally, consistency also indicates the reversibility property, which allows us to return to the initial marking M_0 from any reachable marking $M \in R(M_0)$ upon the firing of transitions.

Controllability: A PN is considered as fully controllable when it is possible to reach any marking from any other marking within the system.

Ordinary Petri nets (OPNs) are effective for modeling small systems, but their effectiveness goes down while dealing with larger systems. They are not appropriate for systems involving data and time, i.e., OPN lacks the capability to quantitatively analyze the system behavior. To address this limitation, OPN has been expanded to include various other elements such as time, color, stochastic elements, fuzzy logic, and hierarchy so that it can cover the diverse features of a real-world system. In recent years, many PN-based models have been developed in the literature. Timed PN (TPN), Stochastic PN (SPN), General SPN (GSPN), Deterministic & SN (DSPN), and Continuous PN (CPN) are some of the methods to model the system.

TPNs: The original definition of PN does not explicitly incorporate the concept of time. However, when assessing the performance of dynamic systems, it becomes essential and advantageous to incorporate time delays to the transitions and/or places within their PN model. This modified PN, which includes time delays, is referred to as a TPN. If the delays within this model are provided in a deterministic manner, it is termed deterministic TPN, or stochastic if the delays are defined probabilistically.

The transitions in PN can fire either instantly or with a delay once they receive tokens from their input places upon becoming enabled. A time delay of a minimum Y_i seconds means that each input place of transition t_i will retain tokens for at least Y_i seconds before they are removed upon firing of t_i. Formally, a TPN can be described as $TPN = (PN, H)$. Where PN is an OPN, and the additional element, H, is a function that associates timing information with each transition. Depending on the time delay, the transition set T is divided into three categories as follows: (a) immediate transitions set – the set which transitions have zero firing delay, i.e., the transitions fire immediately after receiving the token; (b) deterministic timed transitions set – the set in which transitions have constant firing delay; and (c) stochastic transitions set – the set in which firing delay is decided by some probability time distribution.

SPNs: Each transition of PN is associated to a time delay that follows an exponential distribution. Formally, an SPN can be described as $SPN = (PN, \Lambda)$. Where PN is an ordinary PN, and $\Lambda = (\lambda_1, \lambda_2, \ldots, \lambda_n)$ represents the array containing the firing rates corresponding to transitions $t_1, t_2, \ldots, t_n$, respectively.

GSPNs: It extends SPN by including the immediate transition having zero firing delay. GSPNs contain two categories of transitions, timed and immediate. Timed transitions are having exponentially distributed firing delays, whereas immediate transitions are having zero firing delays.

DSPNs: The DSPNs are extensions of GSPN in which the allowable transitions are immediate, exponentially distributed, and deterministic firing timed transitions.

5.8 MARKOV CHAINS

An MC is a mathematical idea used for the purpose of representing a series of events or states of a system, where the future state is solely dependent on the current state and not on the sequence of events that led to the current state. As a result, it possesses the property of being memoryless. Because of the memoryless characteristic exhibited by the exponential distribution of firing delays, it has been demonstrated that the RG of a bounded SPN is equivalent to a finite MC [3]. The MC associated with an SPN can be derived from the RG using the following method:

- The state space of the MC will be the reachability set $R(M_0)$, i.e., the number of states will be identical both in the RG and in the corresponding MC.

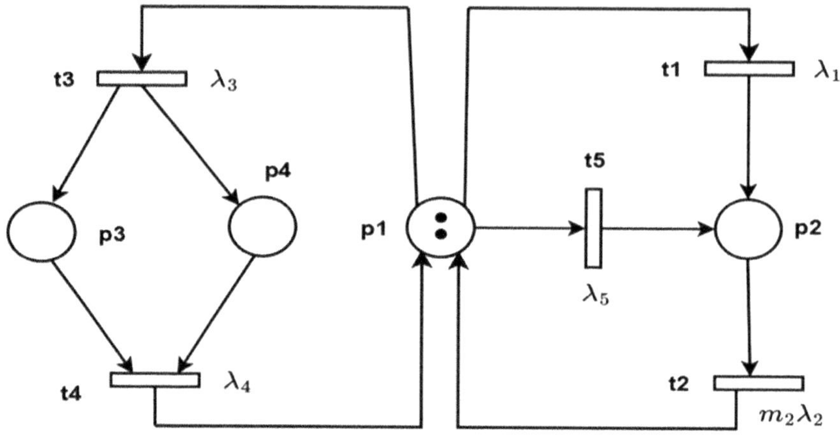

Figure 5.5 The Stochastic Petri net.

- The transition firing rate q_{ij} from state M_i to state M_j is evaluated as follows:

$$q_{ij} = \begin{cases} \lambda_{i,} \text{ if transition } t_i \text{ transforms } M_i \text{ to } M_j \text{ in RG} \\ \lambda_i + \lambda_j + \ldots, \text{ if two or more transitions } t_i \text{ and } t_j \\ \quad \text{transform } M_i \text{ to } M_j \text{ in RG} \\ 0, \text{ if no transition exist between Mi and Mj in RG} \end{cases} \tag{5.1}$$

where λi and λj denote the firing rate of transitions t_i and t_j, respectively. The SPN is shown in Figure 5.5 and its RG and corresponding MC are shown in Figures 5.6 and 5.7 respectively. The TimeNET tool [33] is used to generate the RG from any given PN.

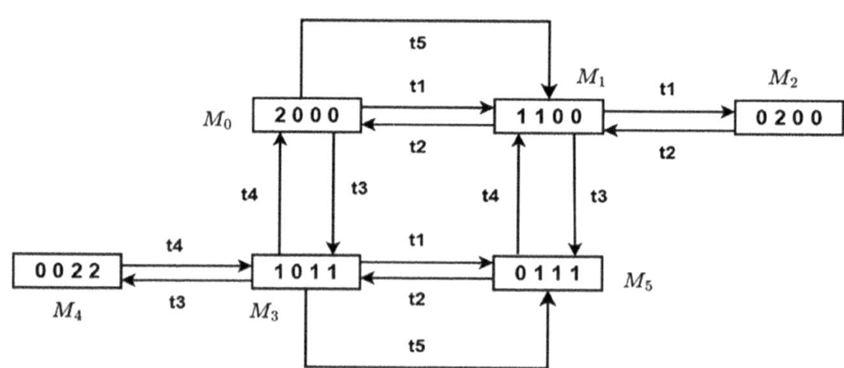

Figure 5.6 Reachability Graph for Stochastic Petri net of Figure 5.5.

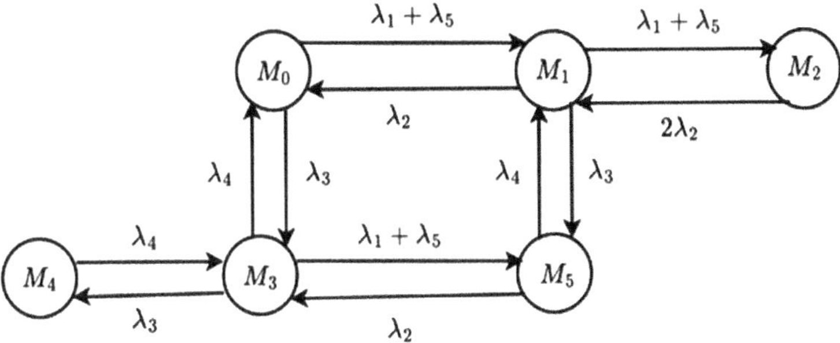

Figure 5.7 Markov chain for Stochastic Petri net of Figure 5.5.

Given that the transition throughput q_{ij} represents the movement from one state M_i to M_j , then the diagonal elements of Q can be computed by using below equation:

$$Q_{ij} = -\sum_{j=1}^{j=n,\, j\neq i} Q_{ij} \tag{5.2}$$

$Q = [q_{ij}]$ is the transition rate matrix (or infinitesimal generator matrix) such that (i ≠ j) and q_{ij} denotes the transition rate from state M_i to M_j . For no transition, $q_{ij} = 0$. In the transition rate matrix, the row-wise summation of the transition rate of each state is 0. The transition rate matrix Q of Figure 5.7 is shown in Eq. (5.3), where $\lambda_{15} = \lambda_1 + \lambda_5$.

$$Q = \begin{bmatrix} & M_0 & M_1 & M_2 & M_3 & M_4 & M_5 \\ M_0 & -\lambda_{15} - \lambda_3 & \lambda_{15} & 0 & \lambda_3 & 0 & 0 \\ M_1 & \lambda_2 & -\lambda_{15} - \lambda_2 - \lambda_3 & \lambda_{15} & 0 & 0 & \lambda_3 \\ M_2 & 0 & 2\lambda_2 & 2\lambda_2 & 0 & 0 & 0 \\ M_3 & \lambda_4 & 0 & 0 & -\lambda_{15} - \lambda_3 - \lambda_4 & \lambda_3 & \lambda_{15} \\ M_4 & 0 & 0 & 0 & \lambda_4 & -\lambda_4 & 0 \\ M_5 & 0 & \lambda_4 & 0 & \lambda_2 & 0 & -\lambda_2 - \lambda_4 \end{bmatrix} \tag{5.3}$$

The PN model's steady-state probability distribution is computed after creating an equivalent MC from its RG and solving the following linear system:

$$\begin{cases} \prod \times Q = 0 \\ \sum_{i=0}^{n} \pi_i = 1 \end{cases} \tag{5.4}$$

where, $\Pi = (\pi_1, \pi_2, \pi_3, \cdots, \pi_n)$ is the steady-state probability and π_i denotes the probability of being in state M_i. Let $\lambda_1 = \lambda_5 = 1/2$ and $\lambda_2 = \lambda_3 = \lambda_4 = \lambda_5 = 1$. Then, using Eq. (5.4), the steady-state probability values are computed as follows: $\pi_2 = 1/11$, $\pi_0 = \pi_1 = \pi_3 = \pi_4 = \pi_5 = 2/11$. The steady-state probability values can be used to find various performance metrics. For example, the mean token count at any place can be calculated. Since p_2 holds single token at M_1 and M_5, and two tokens at M_2, therefore, mean token count at place p_2 is evaluated as follows:

$$E[p_2] = \pi_1 + \pi_5 + 2\pi_2 = 6/11$$

5.9 A CASE STUDY OF NPP: SHUT DOWN SYSTEM-2 (SDS-2)

The SDS is a safety system that allows the reactor to shut down in any unfavorable plant conditions to avoid potentially dangerous situations. Safety systems of NPP are deployed to ensure the safety of the plant and public in all normal operating conditions, anticipated operational occurrences, and emergency conditions. The regulatory board of each country sets and imposes the guidelines/standards for the robust design of these safety systems. The pressurized heavy water reactor has two independent, fast-acting, and diverse SDS to ensure safe shutdown. Each of these SDSs, SDS1, and SDS2 operates on a distinct concept and they have the capability to entirely shut down the reactor in an emergency.

Both the systems are fully automated, however, can be activated manually also, for increased reliability. SDS1 stops the reactor operation and keeps it safe by dropping mechanical rods into the reactor core. SDS2 is intended to function at a greater 'trip' set-point as compared to SDS1 to ensure the reactor shutdown in case of unavailability or failure of SDS1. It rapidly injects the poison into the NPP reactor, which absorbs neutrons and terminates the fission reaction. We have taken SDS2 as a case study to illustrate our approach to measuring reliability and performance [8].

The considered SCS of the NPP is a complete system, representing a combined hardware-software cyber-physical system (CPS). The CPS of the NPP involves integrating digital computing elements with physical processes, sensors, and actuators to monitor, control, and ensure the safety and reliability of the overall system. CPS includes the use of various sensors to monitor physical parameters such as temperature, pressure, radiation levels, and coolant flow in the NPP. However, it is assumed that the signals are continuously received by the sensors to collect data, which is then processed by digital systems to make real-time decisions. The integration of cyber and physical components in an SCS through a well-designed CPS enhances overall efficiency, safety, reliability, and performability while addressing the unique challenges posed by the complex nature of nuclear facilities. The software portion of an SCS is a concurrent computing system in which the data is captured in parallel with computation to produce the output for the actuator. The human intervention is very important for any safety system in the following two contexts: (1) The human operator monitors the process parameters in the graphical user interface unit, which can be any computer or alarm window. (2) In case of any abnormalities in automatic mode, operators may need to take control, assess the situation, and take manual actions regarding the shutdown or isolation process.

5.9.1 SDS-2 and its functional requirements

To meet shutdown requirements, monitoring specific critical factors called 'trip parameters' is crucial. These trip parameters fall into two categories: absolute and conditional. Absolute trip parameters remain applicable across all power levels of the reactor, whereas conditional parameters are applicable when the reactor's power level reaches or exceeds 2% of its total capacity, as detailed in [34]. The SDS-2 responds to activate in automatic mode whenever any of the nine parameters, as outlined in Table 5.2, deviate from their normal operating ranges. Figure 5.8 shows the simplified diagram of SDS-2 and is explained as follows.

The SDS-2 system includes a poison tank, which injects poison into the calandria, where the nuclear chain reaction takes place, to terminate the reaction. The cylindrical poison tank is connected to the outer gate of the reactor chamber [8]. A nozzle facilitates the release of poison from the tank to the moderator. Inside the tank, there is a plastic ball that floats on the poison's surface. Poison injection occurs when any of the trip parameters deviates from their normal range, triggering the instrumentation logic. The poison tank is linked to the helium supply tank through four quick opening valves (QOVs), arranged in series and parallel combinations. These QOVs normally remain closed during reactor operation and ensure reliable and prompt response when opening, operating based on air closure and spring-driven mechanisms. Each line has two parallel QOVs, with one vent valve (VV) in each line to release helium

Table 5.2 Trip parameters and detectors used

S. no.	*Trip parameter*	*Detector*
1	Neutron power	Vertical in-core detector
2	Rate log neutron power	Ion chamber
3	Heat transport system flow	Differential pressure transmitter (DPT)
4	Heat transport system pressure	Pressure transmitters
5	Reactor building pressure	DPT
6	Steam generator Level	DPT on each steam generator
7	Steam generator feedline pressure	Pressure transmitter individual feedline
8	Moderator level	DPT
9	Low pressurizer level	DPT

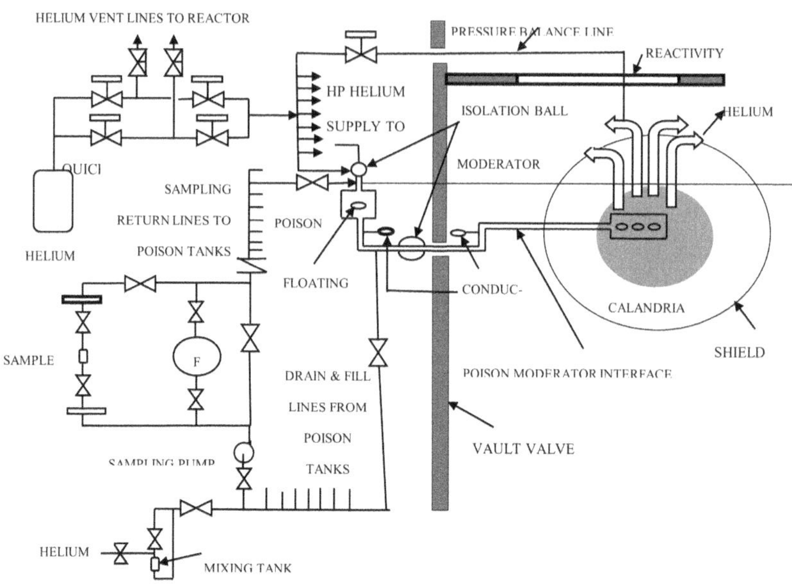

Figure 5.8 Schematic representation for liquid poison injection system of SDS-2.

pressure during reactor operation. These VVs typically remain open, but if any trip parameters exceed their normal range, relays are activated to close them. Subsequently, the QOV opens, allowing helium to pressurize and inject poison into the calandria. This action moves the poison ball to the tank's bottom exit, preventing helium gas from causing overpressure in the calandria.

Following reactor shutdown, maintenance is conducted, and to restart the reactor, the VVs are opened first, followed by the closure of the QOV.

The functional requirements of SDS2 are realized through a CBS, comprising various hardware and software elements. These include sensors, actuators, digital I/O cards, relay output modules, data processing software, and a graphical user interface, among others. The injection of liquid poison into the calandria is achieved using a two-out-of-two-trip circuit that utilizes control valves [3].

5.10 RELIABILITY EVALUATION OF SDS-2 USING PN AND MARKOV CHAINS

This segment introduces a framework for assessing the performability of SCS. The framework comprises eight sequential steps, illustrated in Figure 5.9, and is detailed below.

Step 1: PN model creation of SDS-2

In this step, we picked the target system SDS-2 and thoroughly studied it to capture the functional requirements. The functional requirements of

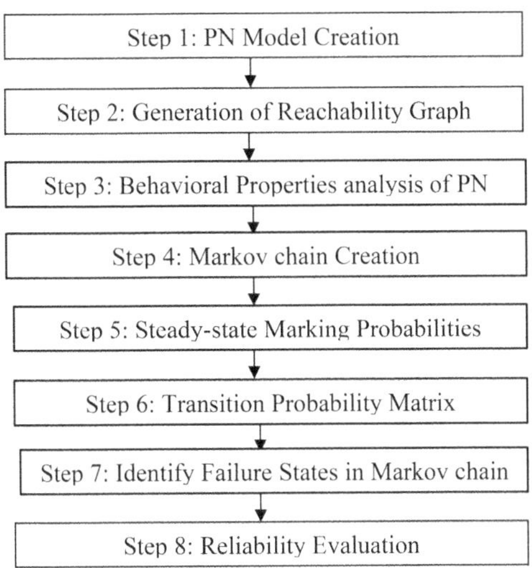

Figure 5.9 Reliability evaluation framework.

SDS-2 are captured and discussed in Section 5.1. The PN model of SDS-2 is shown in Figure 5.10. The potential failure of SDS-2 carries the risk of a significant increase in power levels, which could lead to the design parameters exceeding their safe limits. Such a situation could threaten the integrity of the mechanical components within SDS2, potentially exposing

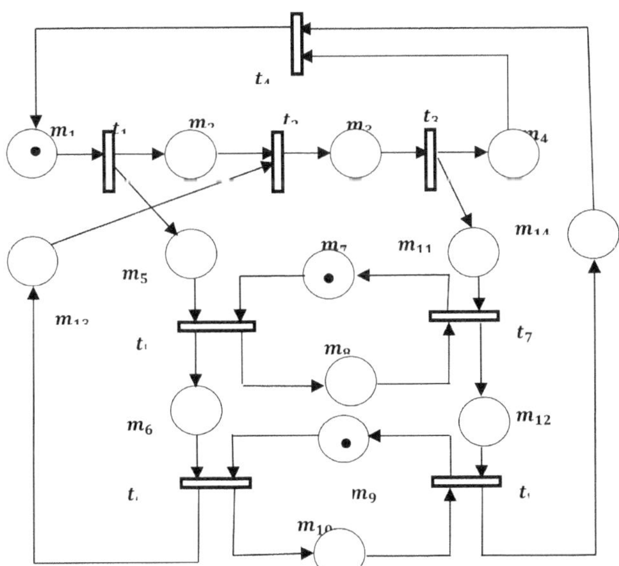

Figure 5.10 Petri net model for liquid poison injection system of SDS-2.

radioactivity to the public. SDS-2 is a complex system comprising various elements, such as sensors, logic systems, actuators, and a specialized human-machine interface, all working together to accomplish its intended purpose. In normal operating conditions, both QOV lines are equipped with VVs that remain open to release pressure within the line as needed, thus preventing any inadvertent poison injection. Figure 5.10 illustrates the PN model of SDS-2, which is further detailed below.

A token in m_1 place signifies any deviation of trip parameters from their designated limits. Within m_2, a token signifies the creation of a logic condition (LC), while m_3 denotes the LC being in a hold state. The activation of a relay to close VVs is depicted by a token in m_5. Injection of poison into the moderator occurs upon opening of the QOV, indicated by a token in m_{10}. Enhanced reliability is achieved by monitoring duplicate information regarding the QOV state from redundant sensors, represented by a token in m_8. m_{13} is utilized to prioritize t_6 over t_2 in instances of race conditions, ensuring the QOV opens even if false information (closed state) arises due to a security threat. Tables 5.3 and 5.4 provide detailed descriptions of the transitions and placeholders depicted in Figure 5.10, respectively.

During this phase, the model's delay for each transition is incorporated based on specifications, expert input, and insights drawn from analogous projects. Subsequently, the TimeNET tool was employed to execute the model and determine the firing rate (throughput) for the transitions, as outlined in Table 5.5. Here, λ_i signifies the throughput of transition t_i (with i ranging from 1 to 8).

Step 2: Generation of RG

The RG serves to establish the boundary conditions of the system, offering insights into the potential number of states achievable throughout its operational lifespan. It delineates the entirety of possible markings, representing the diverse states the system may traverse. Utilizing the PN model

Table 5.3 SDS-2 process transitions

Transitions	*Description*
t_1	Sending signal to create LC and for energizing relays to close the vent valves
t_2	Triggering signal to hold LC in created state
t_3	Triggering signal to restore LC and relay de-energizes
t_4	Resending signals for opening QOV if it fails to open
t_5	Triggering for closing the vent valves (VVs)
t_6	Triggering for opening of every QOV
t_7	Triggering for opening the VV
t_8	Triggering for closing each QOV

Table 5.4 SDS-2 process places

Places	*Description*
m_1	Trip parameters deviation
m_2	LC creation
m_3	Holding state of LC
m_4	Restore LC
m_5	Energizing relays for closing the VV
m_6	Closing of VV
m_7	Duplicate information of QOV at closing state
m_8	Duplicate information of QOV at opening state
m_9	Closing of QOV
m_{10}	Opening of QOV
m_{11}	De-energizing relays for opening the VV
m_{12}	Opening of VV
m_{13}	Ensuring priority of t_6 over t_2
m_{14}	Ensures reversibility properties

Table 5.5 Transition's firing rate (in per ms)

λ_1	λ_2	λ_3	λ_4
0.01256	0.820	0.532	0.650
λ_5	λ_6	λ_7	λ_8
0.148	0.020	0.0185	0.150

illustrated in Figure 5.10, a corresponding RG is constructed and depicted in Figure 5.11, as detailed in references [32, 35]. In Section 4.1, we discussed reachability.

Step 3: Behavioral properties analysis of PN

Analyzing the behavioral properties of PN is crucial for evaluating various factors that could affect the system's reliability. These assessments utilize RGs to get essential insights into reachability analysis, boundedness, liveness, and deadlock. While evaluating the system's reliability, it is essential to consider other performance metrics as well. For instance, deadlock situations can significantly slow down processes or cause them to remain indefinitely stuck in a particular state. Liveness refers to the system's capability to progress regardless of concurrent components. Boundedness is employed to prevent overflow issues within the system, ensuring that all elements remain within a finite space. Stability is another vital property aimed at maintaining control over the system's output.

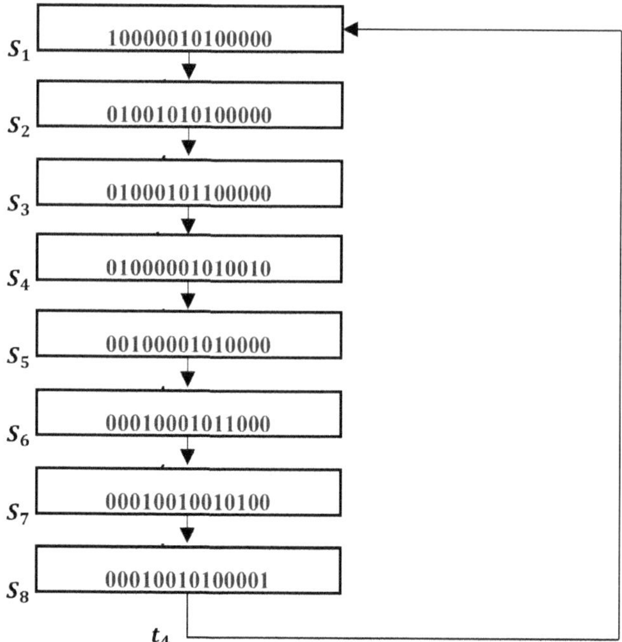

Figure 5.11 Reachability Graph of the Petri net model of SDS-2.

Steady-state analysis confirms the system's consistent behavior. Hence, these metrics serve as valuable performance indicators and, hence, must be analyzed.

Reacha`bility, Stability, Boundedness, Consistency, Reversibility, and Safety Analysis:

From SDS-2 PN model, shown in Figure 5.10, each place is reachable from every other place and contains either zero or one token for each marking, which is reachable from the initial marking M_0, i.e., $M_0 \leq 1$. This observation confirms the system's reachability and stability. Furthermore, since the PN model is one-bounded, it indicates their safety. Hence, the analysis of the PN model of SDS-2 affirms that it meets all performance criteria. In the PN model depicted in Figure 5.10, each transition must have been fired at least once to uphold the consistency requirement. This consistency condition also unveils the reversibility aspect, ensuring that the system can revert to its initial marking, S_1 from any reachable marking in the RG illustrated in Figure 5.11, following the firing of transitions.

Deadlock and liveness analysis:

The modeling of SDS-2 was carried out using a TPN, as shown in Figure 5.10. The analysis of deadlock and liveness, employing siphons and traps, is elaborated in Section 4.1. To determine the count of siphons and traps within the SDS-2 PN model, we utilized the TimeNET tool [32], revealing 12 minimal siphons and 12 marked traps. These siphons include the following:

$$s_1 = \{m_6, m_{13}, m_3, m_{11}, m_7\}, s_2 = \{m_3, m_{11}, m_{12}, m_{14}, m_1, m_5, m_6, m_{13}\},$$
$$s_3 = \{m_1, m_5, m_6, m_{13}, m_3, m_4\}, s_4 = \{m_1, m_2, m_3, m_4\},$$
$$s_5 = \{m_1, m_2, m_3, m_{11}, m_{12}, m_{14}\}, s_6 = \{m_3, m_{11}, m_{12}, m_9, m_{13}\},$$
$$s_7 = \{m_3, m_4, m_1, m_5, m_8, m_{12}, m_9, m_{13}\},$$
$$s_8 = \{m_1, m_2, m_3, m_{11}, m_7, m_6, m_{10}, m_{14}\}, s_9 = \{m_7, m_8\},$$
$$s_{10} = \{m_5, m_8, m_{12}, m_{14}, m_1\}, s_{11} = \{m_9, m_{10}\}, \text{ and}$$
$$s_{12} = \{m_5, m_6, m_{10}, m_{14}, m_1\}.$$

The traps are as follows: $T_1 = \{m_1, m_5, m_8, m_{12}, m_{14}\}$, $T_2 = \{m_1, m_5, m_6, m_{13}, m_3, m_{11}, m_{12}, m_{14}\}$, $T_3 = \{m_1, m_2, m_3, m_{11}, m_{12}, m_{14}\}$, $T_4 = \{m_3, m_{11}, m_7, m_6, m_{13}\}$, $T_5 = \{m_3, m_{11}, m_{12}, m_9, m_{13}\}$, $T_6 = \{m_1, m_2, m_3, m_4\}$, $T_7 = \{m_1, m_5, m_8, m_{12}, m_9, m_{13}, m_3, m_4\}$, $T_8 = \{m_7, m_8\}$, $T_9 = \{m_1, m_5, m_6, m_{10}, m_{14}\}$, $T_{10} = \{m_9, m_{10}\}$, $T_{11} = \{m_3, m_4, m_1, m_5, m_6, m_{13}\}$, and $T_{12} = \{m_1, m_2, m_3, m_{11}, m_7, m_6, m_{10}, m_{14}\}$.

We notice that states s_1, s_2, s_3, s_4, s_5, s_6, s_7, s_9, s_{10}, s_{11}, and s_{12} exhibit trap markings, indicating that they can potentially lead to deadlock situations. However, state s_8 does not possess any trap markings, indicating that our PN model is free from deadlocks. Furthermore, PN fulfills the liveness conditions outlined in Section 4.1, as it does not have any potential deadlocks.

Step 4: MC creation:
The MC is derived from the RG of the PN model, [3, 32]. Figure 5.12 provides a visualization of the MC corresponding to the timed PN model depicted in Figure 5.10.

Step 5: Steady-state marking probability calculation
Eqs. (5.2) and (5.4) offer a means to compute the probabilities of steady-state markings. Eq. (5.5) displays the transition rate matrix Q. Eq. (5.6)

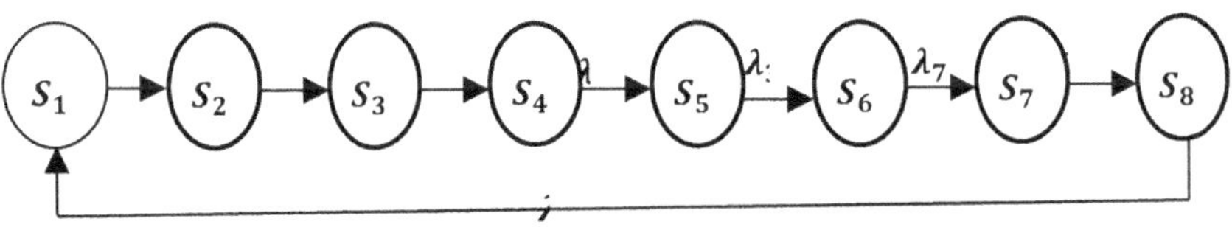

Figure 5.12 Markov chain for the Petri net model of SDS-2.

Table 5.6 Steady-state probabilities

π_1	π_2	π_3	π_4
$1.4988 \times e^{-15}$	$1.7347 \times e^{-18}$	$2.2204 \times e^{-16}$	0.0191
π_5	π_6	π_7	π_8
0.0295	0.8470	0.1045	$2.896 \times e^{-15}$

presents the resulting equation derived from these calculations. Utilizing Eq. (5.6) along with the firing rate values of the transitions outlined in Table 5.6, we determine the steady-state marking probabilities, as demonstrated in Table 5.7.

$$Q = \begin{bmatrix} & S_1 & S_2 & S_3 & S_4 & S_5 & S_6 & S_7 & S_8 \\ S_1 & -\lambda_1 & \lambda_1 & 0 & 0 & 0 & 0 & 0 & 0 \\ S_2 & 0 & -\lambda_5 & \lambda_5 & 0 & 0 & 0 & 0 & 0 \\ S_3 & 0 & 0 & -\lambda_6 & \lambda_6 & 0 & 0 & 0 & 0 \\ S_4 & 0 & 0 & 0 & -\lambda_2 & \lambda_2 & 0 & 0 & 0 \\ S_5 & 0 & 0 & 0 & 0 & -\lambda_3 & \lambda_3 & 0 & 0 \\ S_6 & 0 & 0 & 0 & 0 & 0 & -\lambda_7 & \lambda_7 & 0 \\ S_7 & 0 & 0 & 0 & 0 & 0 & 0 & -\lambda_8 & \lambda_8 \\ S_8 & \lambda_4 & 0 & 0 & 0 & 0 & 0 & 0 & -\lambda_4 \end{bmatrix} \cdots\cdots\cdots\cdots \quad (5.5)$$

$$\begin{aligned} \pi_1 \bullet \lambda_1 &= \pi_8 \bullet \lambda_4 \\ \pi_2 \bullet \lambda_5 &= \pi_1 \bullet \lambda_1 \\ \pi_3 \bullet \lambda_6 &= \pi_2 \bullet \lambda_5 \\ \pi_4 \bullet \lambda_2 &= \pi_3 \bullet \lambda_6 \\ \pi_5 \bullet \lambda_3 &= \pi_4 \bullet \lambda_2 \\ \pi_6 \bullet \lambda_7 &= \pi_5 \bullet \lambda_3 \\ \pi_7 \bullet \lambda_8 &= \pi_6 \bullet \lambda_7 \\ \pi_8 \bullet \lambda_4 &= \pi_7 \bullet \lambda_8 \end{aligned} \quad (5.6)$$

Step 6: Transition probability matrix generation

The transition probability matrix P is a square matrix where each element represents the likelihood of transitioning from one state to another in unit amount of time in the MC. The probability P_{ij} of transitioning from state S_i to S_j is determined by dividing Q_{ij} by the total throughput of all transitions, excluding self-loop transitions. If a transition returns to its initial state indefinitely, rendering it non-ergodic, thus their transition probability will be zero [8]. Consequently, the elements of R can be computed using the equation provided below:

$$R_{ij} = \begin{cases} \dfrac{Q_{ij}}{\Sigma_{k \neq i} Q_{ik}}, \text{ if } i \neq j \\ 0, \text{ otherwise} \end{cases} \tag{5.7}$$

Using Eqs. (5.6) and (5.7) and MC of the RG of Figure 5.11, we compute the transition rate matrix R. The matrix R is shown in Eq. (5.8). The matrix R = $[R_1 R_2 R_3 \ldots R_8] \times [R_1 R_2 R_3 \ldots R_8]$ is of order 8×8, as there are eight places in MC of the RG.

$$R = \begin{bmatrix} & S_1 & S_2 & S_3 & S_4 & S_5 & S_6 & S_7 & S_8 \\ S_1 & 0 & 1 & 0 & 0 & 0 & 0 & 0 & 0 \\ S_2 & 0 & 0 & 1 & 0 & 0 & 0 & 0 & 0 \\ S_3 & 0 & 0 & 0 & 1 & 0 & 0 & 0 & 0 \\ S_4 & 0 & 0 & 0 & 0 & 1 & 0 & 0 & 0 \\ S_5 & 0 & 0 & 0 & 0 & 0 & 1 & 0 & 0 \\ S_6 & 0 & 0 & 0 & 0 & 0 & 0 & 1 & 0 \\ S_7 & 0 & 0 & 0 & 0 & 0 & 0 & 0 & 1 \\ S_8 & 1 & 0 & 0 & 0 & 0 & 0 & 0 & 0 \end{bmatrix} \tag{5.8}$$

Step 7: Identify failure states in MC

In the context of safety-critical functions, there are scenarios where the reactor fails to trip due to component failure. Consequently, all such failure states within the Continuous-time Markov Chain (CTMC) are taken into account for reliability assessment. These failure components might necessitate a certain duration for repair before the system can resume functioning. The subsequent state following the failure state in the CTMC is considered a repairable state. In case, where the reactor fails to trip upon deviation from normal design parameters, the transition t_4 fires to retry the reactor trip process.

Step 8: Reliability evaluation

In this stage, the system's reliability is calculated using the matrix P and the failure states identified in CTMC. Subsequently, the reliability of the SDS-2 can be assessed by removing all probabilities associated with the failure states of the CTMC. The probability value of each state is determined as follows.

Let $p_i(t)$ denote the probability that a component remains in the state i at time t. As a component operates for an extended duration $(t \rightarrow \infty)$, the probability tends to stabilize into a steady-state distribution [36], i.e.,

$$\vec{p} = \left[p(S_1), p(S_2), \ldots, p(S_8)\right] \tag{5.9}$$

$$\vec{p} = \vec{p}\ R \tag{5.10}$$

and from the total law of probability,

$$\sum_i p(i) = 1 \tag{5.11}$$

From Eqs. (5.10) and (5.11), we get Eqs. (5.12) and (5.13)

$$\sum_{i=1}^{8} \mu_i = 1 \tag{5.12}$$

$$[P_1 P_2 P_3, \ldots P_8] = [P_1 P_2 P_3 \ldots P_8] \times R \tag{5.13}$$

On solving Eqs. (5.12) and (5.13), the state's probabilities value is computed as follows:

$$p_1 = 0.00512,\ p_2 = 0.1659,\ p_3 = 0.0818,\ p_4 = 0.1946,\ p_5 = 0.0595,$$
$$p_5 = 0.0595,\ p_6 = 0.0376,\ p_7 = 0.2454,\ p_8 = 0.2100.$$

The reliability of SDS-2 system can be measured after identification of the failure states. S_1 is the only failure state, as identified in step 7, because S_1 is generated after the firing of transition t_4. Therefore, the reliability can be computed as follows:

$Reliability = 1 - sum\ of\ probability\ of\ failure\ states = 1 - P((S_1)) = 1 - 0.00512 =$ 0.99488 i.e., our model gives a reliability of 99.48%.

REFERENCES

1. A. Avizienis, J.-C. Laprie, and B. Randell, "Fundamental concepts of dependability," Department of Computing Science Technical Report Series, 2001, Accessed: Mar. 02, 2024.
2. T. Zhang, Y. Wang, and M. Xie, "Analysis of the performance of safety-critical systems with diagnosis and periodic inspection," in 2008 Annual Reliability and Maintainability Symposium. Jan. 2008, pp. 143–148. doi: 10.1109/RAMS.2008.4925785.
3. N. K. Jyotish, L. K. Singh, C. Kumar, and P. Singh, "Reliability and performance measurement of safety-critical systems based on petri nets: a case study of nuclear power plant," IEEE Transactions on Reliability, pp. 1–17, 2023, doi: 10.1109/TR.2023.3244365.

4. H. Thimbleby, P. Oladimeji, and P. Cairns, "Unreliable numbers: error and harm induced by bad design can be reduced by better design," Journal of The Royal Society Interface, vol. 12, no. 110, p. 20150685, 2015.
5. H. Alemzadeh, R. K. Iyer, Z. Kalbarczyk, and J. Raman, "Analysis of safety-critical computer failures in medical devices," IEEE Security & Privacy, vol. 11, no. 4, pp. 14–26, 2013.
6. S. Biswas, R. Tatchikou, and F. Dion, "Vehicle-to-vehicle wireless communication protocols for enhancing highway traffic safety," IEEE Communications Magazine, vol. 44, no. 1, pp. 74–82, 2006.
7. P. Papadimitratos, A. De La Fortelle, K. Evenssen, R. Brignolo, and S. Cosenza, "Vehicular communication systems: enabling technologies, applications, and future outlook on intelligent transportation," IEEE Communications Magazine, vol. 47, no. 11, pp. 84–95, 2009.
8. L. K. Singh and H. Rajput, "Dependability analysis of safety critical real-time systems by using petri nets," IEEE Transactions on Control Systems Technology, vol. 26, no. 2, pp. 415–426, 2017.
9. N. K. Jyotish, L. K. Singh, C. Kumar, and P. Singh, "Reliability and performance evaluation of safety-critical instrumentation and control systems of nuclear power plant," IEEE Transactions on Reliability, vol. 72, pp. 1523–1539, 2023.
10. A. Shaban, A. Abdelwahed, G. Di Gravio, I. H. Afefy, and R. Patriarca, "A systems-theoretic hazard analysis for safety-critical medical gas pipeline and oxygen supply systems," Journal of Loss Prevention in the Process Industries, vol. 77, p. 104782, 2022.
11. R. Baldoni, L. Montanari, and M. Rizzuto, "On-line failure prediction in safety-critical systems," Future Generation Computer Systems, vol. 45, pp. 123–132, 2015.
12. P. Fraga-Lamas, T. M. Fernández-Caramés, M. Suárez-Albela, L. Castedo, and M. GonzálezLópez, "A review on internet of things for defense and public safety," Sensors, vol. 16, no. 10, p. 1644, 2016.
13. L. K. Singh, G. Vinod, and A. K. Tripathi, "Approach for parameter estimation in markov model of software reliability for early prediction: a case study," IET Software, vol. 9, no. 3, pp. 65–75, 2015.
14. N. K. Jyotish, L. K. Singh, C. Kumar, and P. Singh, "Batch deterministic and stochastic petri nets modeling for reliability quantification for safety critical systems of nuclear power plants," Nuclear Engineering and Design, vol. 404, p. 112191, 2023.
15. Wikipedia, "2023 Odisha train collision, India Train Disaster," 2023, https://en.wikipedia.org/wiki/2023 Odisha train collision.
16. Wikipedia, "Flydubai flight 981," 2016, https://en.wikipedia.org/wiki/Flydubai Flight 981.
17. B. B. Wittneben, "The impact of the fukushima nuclear accident on european energy policy," Environmental Science & Policy, vol. 15, no. 1, pp. 1–3, 2012.
18. The Guardian, "Railway signaling failure, Sydney," 2011, https://www.theguardian.com/australia-news/2023/mar/08/sydney-train-network-trains-down-rail-delays-halt-communications-fault-update, Last accessed on 2023-07-25.
19. Wikipedia, "Rhordorf train collision," 2011, https://en.wikipedia.org/wiki/Hordorf train collision, Hordorf train collision.
20. J. L. Ramseur, "Deepwater Horizon oil spill: the fate of the oil." Congressional Research Service, Library of Congress, Washington, DC, 2010.
21. L. D. Stone, C. M. Keller, T. M. Kratzke, and J. P. Strumpfer, "Search for the wreckage of air France Flight AF 447," 2014.
22. B. Krebs, "Cyber incident blamed for nuclear power plant shutdown," Washington Post, June, vol. 5, no. 2008, p. 5, 2008.

23. T. Rossetto, N. Peiris, A. Pomonis, S. Wilkinson, D. Del Re, R. Koo, and S. Gallocher, "The Indian ocean tsunami of December 26, 2004: observations in Sri Lanka and Thailand," Natural Hazards, vol. 42, pp. 105–124, 2007.
24. C. Borrás, "Overexposure of radiation therapy patients in panama: problem recognition and follow-up measures," Revista Panamericana de Salud P´ublica, vol. 20, no. 2–3, pp. 173–187, 2006.
25. P. K. Mishra, R. M. Samarth, N. Pathak, S. K. Jain, S. Banerjee, and K. K. Maudar, "Bhopal gas tragedy: review of clinical and experimental findings after 25 years," International Journal of Occupational Medicine and Environmental Health, vol. 22, no. 3, p. 193, 2009.
26. S. Collins and S. McCombie, "Stuxnet: the emergence of a new cyber weapon and its implications," Journal of Policing, Intelligence and Counter Terrorism, vol. 7, no. 1, pp. 80–91, 2012.
27. W. E. Wong, V. Debroy, A. Surampudi, H. Kim, and M. F. Siok, "Recent catastrophic accidents: Investigating how software was responsible," in 2010 Fourth International Conference on Secure Software Integration and Reliability Improvement. IEEE, 2010, pp. 14–22.
28. M. Bashiri, and S. G. Miremadi, "Performability guarantee for periodic tasks in real-time systems," Scientia Iranica, vol. 21, no. 6, pp. 2127–2137, 2014.
29. N. K. Jyotish, L. K. Singh, and C. Kumar, "A state-of-the-art review on performance measurement petri net models for safety critical systems of NPP," Annals of Nuclear Energy, vol. 165, p. 108635, 2022.
30. M. R. Lyu, "Software reliability engineering: a roadmap, future of software engineering (fose'07), Minneapolis, 23–25 may 200, proceedings. Minneapolis," 2007.
31. N. K. Jyotish, L. K. Singh, and C. Kumar, "Availability analysis of safety-critical systems of nuclear power plant using ordinary differential equations and reachability graph," Progress in Nuclear Energy, vol. 159, p. 104624, 2023.
32. T. Murata, "Petri nets: properties, analysis and applications," Proceedings of the IEEE, vol. 77, no. 4, pp. 541–580, 1989.
33. A. Zimmermann, "Modeling and evaluation of stochastic Petri nets with timenet 4.1," in 6th International ICST Conference on Performance Evaluation Methodologies and Tools. IEEE, 2012, pp. 54–63.
34. D. Torgerson, B. A. Shalaby, and S. Pang, "Candu technology for generation iii+ and iv reactors," Nuclear Engineering and Design, vol. 236, no. 14–16, pp. 1565–1572, 2006.
35. N. K. Jyotish, L. K. Singh, and C. Kumar, "Reliability assessment of safety-critical systems of nuclear power plant using ordinary differential equations and reachability graph," Nuclear Engineering and Design, vol. 412, p. 112469, 2023.
36. M. Asadi, R. P. Torghabeh, and N. P. Santhanam, "Stationary and transition probabilities in slow mixing, long memory Markov processes," IEEE Transactions on Information Theory, vol. 60, no. 9, pp. 5682–5701, 2014.

Chapter 6

Study on improving the lifetime of mechanical systems during transit established on quantum/transport life-stress prototype and sample size

Seong-woo Woo, Dennis L. O'Neal, Yimer Mohammed Hassen, and Gezae Mebrahtu

6.1 INTRODUCTION

Because of competitive forces in the marketplace, manufacturers need to innovate and enhance the design of their products on a regular basis. Such innovations require new product features to be rapidly incorporated into the product and introduced into the marketplace. Companies often incorporate new features into new products' specifications during each design cycle. Without enough reliability testing or expectations of how modifications may be utilized in the field, the introduction of these new attributes may cause unanticipated failures in the field and influence the manufactured product's brand in a negative way [1, 2]. Since the 1970s, it has also been acknowledged that there exists a big gap between reliability theory and its application to industrial fields. Prior to release, any product imperfections should be identified and rectified by reliability testing such as parametric accelerated life testing (ALT) [3–6].

Mechanical fatigue accounts for over 70% of structural failures in machinery parts and load-bearing components. These failures often start as cracks from stress raisers, such as sharp edges, notches, grooves, and thin surfaces, and grow when subjected to repetitive (impact) loads [7]. Fatigue affects the lifetime of mechanical devices, such as cars, aircraft wings and fuselages, ships, nuclear reactors, motors, and ground turbines, so that structural fatigue failures may have significant negative outcomes with regard to loss of human lives and environmental catastrophes. Significant effort is being directed to low-cycle fatigue in the field of turbine-engine designs. However, it is difficult to collect test data for product lifetime for low-cycle fatigue because of the lack of a structured method such as a (generalized) life-stress (LS) model and sample size. Reliability testing such as ALT is also affected by the cyclic stress amplitude, mean stress or stress ratio, R ($=\sigma_{min}/\sigma_{max}$), which can be stated as the proportion of the minimum cyclic stress to the maximum cyclic stress as an inverse of accelerated factor (AF) [8]. Quantum mechanics provides designers with a way to better understand how fatigue starts from the smaller nanometer-sized material

DOI: 10.1201/9781003546214-6

defects that have not been explained by the traditional continuum theory used in the strength of materials [9]. Field failure often stochastically occurs in areas of locally high stress concentrations.

The Boeing 737 MAX adopted the more efficient CFM International LEAP-1B engines using the optimized 68-in. fan design. These engines consumed 12% less fuel and were 7% lighter than other engines [10]. If an airplane engine had insufficient stiffness due to decreased engine weight and increased power when subjected to the adjusted (or abruptly increased) random loads from take-offs, cruising, and landings, it could unexpectedly fail due to fatigue [11, 12]. The requirements for stiffness, which is the resistance against reversible deformation, rely on the (transit) product application. Large oscillations usually occur near the natural frequency of the component or part. An engine should be designed to allow it to survive random vibration loads that are generated in a multimodule product internally (or externally). These random loads impact the dynamic behavior and cannot easily be predicted with a deterministic model [13]. To approximate the fatigue damage of a system in the field, the equivalent damage method, represented by the time-dependent random vibrational spectra transmitted from the base supporting structures, needs to be considered for developing an accelerated test [14, 15]. This method allows the conversion of these complicated signals into a straightforward Gaussian distribution, represented by the power spectral density (PSD) in the frequency domain. Based on classical vibration theory, it is possible to obtain identical field spectral (or PSD) damage of mechanical systems that can be applied to systems for accelerated tests. However, it is necessary to have numerous additional fundamental concepts for ALT – fatigue damage spectrum (FDS), LS prototype, and sample size equation.

As an alternative, the ALT based on reliability block diagrams could be utilized [16–18]. It involves test schemes for the product, failure mechanism, an established elevated testing method, sample size, etc. Elsayed [19] grouped physics/statistics, statistical, and physics/test-based prototypes for assessment. Meeker [20] proposed numerous feasible recommendations to organize an ALT. Modern experimental methods may fail to reproduce field failure(s) produced by design imperfections because they may not include adequate sample sizes or proper testing periods to identify the most fragile components in multimodule products. Carrying out an ALT also requires numerous concepts to be developed, such as fracture mechanics [21], the BX life idea, LS model, and sample size formulation.

With a statistical methodology [22] or design of experiments (DOE) [23], design faults of a product might be identified and altered during the product development process. DOE is a structured method to determine the relationship between factors affecting a process and its production. Its purpose is to assure that the components are designed in the optimum way for the functioning (or environmental) condition. DOE is performed for some factors that influence the product designs, and its effectiveness is quantified using analysis of variance (ANOVA). Taguchi's design method expresses the system's robustness for design and assessment. It places the optimum design where the "noise" factor does not create failure and determines the design factor. However, many test cases may be needed to complete Taguchi's array. Thus, this approach may necessitate significant computations to arrive at a best answer but may not provide optimal results because it does not include a failure model such as fatigue.

An additional engineering perspective for fatigue failures incorporates strength of materials and the finite element method (FEM) [24, 25]. Uncommon product failures may be evaluated using (1) rigorous formulation employing Newtonian or Lagrangian skills, (2) evaluating the product response for (dynamic) loads and identifying its stress, (3) employing the rainflow counting procedure through von Mises stress, and (4) roughly calculating product damage by applying the Palmgren-Miner rule [26, 27]. Employing a systematic method that may give a closed-form solution frequently necessitates invoking numerous assumptions that may not identify multimodule product failures, such as small cracks or preexisting defects.

Random vibration testing in the laboratory can be conducted by using FDS-measured environmental data. Some of the most typical test guidelines are found in MIL-STD-810 [28, 29]. However, there are inherent limitations in this approach because the actual environmental stresses (singularly or in combination) in the field may not be exactly matched in the laboratory. Consequently, an engineer should not assume that a successful item tested in the laboratory will necessarily survive in the field. As an alternative approach, companies can carry out highly ALT (HALT) to find product problems in transit [30]. However, because no systematic concepts are used, such as a generalized stress model, accelerated factor, sample size, and lifetime target, the tests will be performed without the reliability quantitative (RQ) specification – mission cycle, sample size, etc.

The purpose of this chapter is to offer ALT as a general structured method that may create RQ specifications, such as the assigned time for finding the design defects of mechanical systems in transit, and provide an approach for changing and improving the imperfection designs of the product. This established process for identifying the causes of failure mode and improving product designs examines appropriate design responses to address (low-cycle) fatigue failures in mechanical products, such as airplanes, automobiles, appliances, and construction machinery. The process incorporates: (1) an ALT strategy created on a newly defined system BX life, (2) a PSD load examination, (3) a customization of ALTs with alternations, and (4) an assessment of whether the latest redesign(s) of the system fulfills the objective BX life. To secure the effectiveness of the ALT, it would be necessary to assess the introduction of the new design in the marketplace to ensure that it passed the target reliability. A quantum-transported time-to-failure prototype and sample size formulation are also proposed. A case study of a refrigerator subjected to repeated random vibrations in railway transit is provided as an example of the application of this structured method.

6.2 PARAMETRIC ALT FOR MECHANICAL SYSTEMS IN TRANSIT

6.2.1 Product life and its targeting

In a mechanical product, power transfer is used to produce mechanical advantages via forces and motion by implementing proper mechanisms [31]. Most mechanical products are composed of multimodule structures. If the modules are assembled, a mechanical product can work properly and perform its intended functions. If there is a design imperfection in a product, when the structure is subjected to stress, it may stochastically fail before its expected life. Products should have a defined proper lifetime index for design such as BX life (or "bearing life"). BX life is the period (or time) when X% of the components in a collection of items under consideration shall have been unsuccessful.

If a random variable, T, represents the system life, the Weibull cumulative distribution function (CDF), indicated by $F(t)$, can be expressed as all items having $T \le t$, split by the whole number of items. It also takes the chance P of arbitrarily selecting an item having a value of T equal to or less than t, and then we obtain

$$P(T \le t) = F(t) \tag{6.1}$$

where $F(t) = 1 - e^{-\left(\frac{t}{\eta}\right)^{\beta}}$ for the Weibull distribution.

The failure rate may be expressed as

$$\lambda = \frac{f}{R} = \frac{dF/dt}{R} = \frac{(1-R)'}{R} = \frac{-R'}{R} \tag{6.2}$$

where f is the density function, and λ is the failure rate.

If Eq. (6.2) is integrated over the period, the $X\%$ accumulative failure, $F(L_B)$, at $T = L_B$ (BX lifetime) might be attained:

$$\int \lambda \, dt = -lnR \tag{6.3}$$

In other words,

$$A(=X) = (\lambda) \cdot L_B = \int_0^{T=L_B} \lambda(t) \cdot dt = -lnR(L_B) = -ln(1-F) \cong F(L_B) \tag{6.4}$$

Therefore, the sum of $F(t)$ and $R(t)$ is 1 or $F(t) + R(t) = 1$ for $t \ge 0$. Assuming that T_1 is the period of the first failure, reliability, $R(t)$, can be stated as follows:

$$R(t) = P(T_1 > t) = P(\text{no failure in}\,(0, t]) = \frac{(m)^0 e^{-m}}{0!} = e^{-m} = e^{-\lambda t} \tag{6.5}$$

where m is the Poisson parameter that can be expressed as λt (mean).

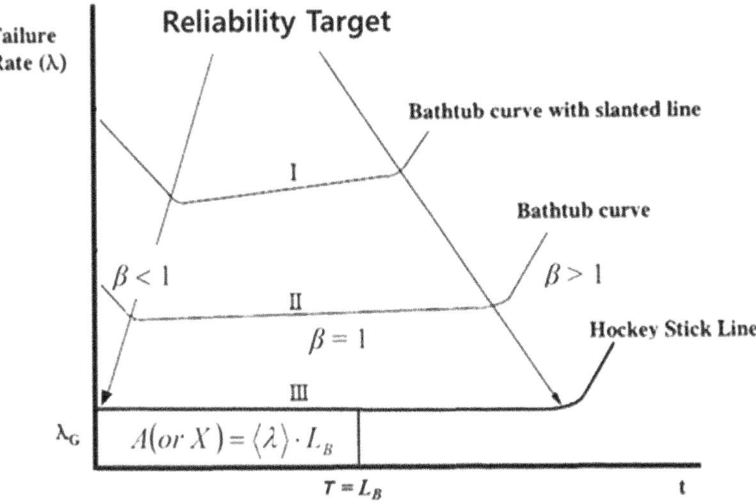

Figure 6.1 A lifetime index – BX life (L_B) – on the bathtub curve (or hockey stick line).

Reliability is interpreted as the product's ability to handle specified circumstances for a stated period of time or total distance [32]. It can often be explained by the bathtub curve shown as line II. If a product's reliability follows the traditional bathtub curve, it may have difficulties achieving success in the market because of design flaws in the product that create large failures early in the introduction of the product. As the design of a mechanical system is improved, the normal (slanted) bathtub might be changed to a more linear line (or hockey stick line) as shown in line III (Figure 6.1).

Thus, the reliability function of a mechanical system in Eq. (6.5) can be obtained from the multiplication of the system life L_B and failure rate λ:

$$R(L_B) - 1 - F(L_B) - e^{-\lambda L_B} \cong 1 - \lambda L_B \tag{6.6}$$

Practically, Eq. (6.6) is relevant to less than 20% of cumulative failure rates [33].

To keep food fresh, a refrigerator, operating on the vapor-compression refrigeration cycle, is designed to supply cooled air from a heat exchanger (evaporator) to the refrigerator and freezer compartments. The refrigerant is circulated through the evaporator, compressor, condenser, and a capillary tube. If there are some design defects in a multimodule (A–E) refrigerator, it will not achieve its expected design lifetime. A design engineer often does not know how a customer will use a product under specific environmental conditions in the field. By robustly designing problematic parts in product modules through parametric ALT, the engineer should be able to achieve the lifetime goal for the product. A refrigerator can be divided into five modules (A–E) as shown in Figure 6.2.

To fulfill the target of a (new) product lifetime using ALT when there are no (expected) field reliability data (expressed as product of failure rate and

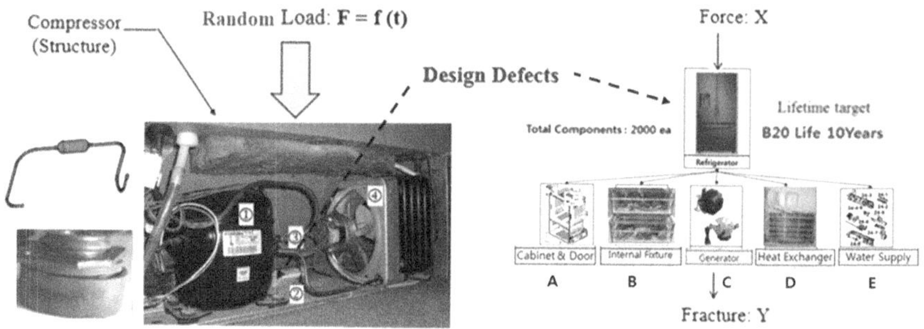

Figure 6.2 Refrigerator mechanical compartment: (1) compressor, (2) rubber, (3) connecting tubes, and (4) fan and condenser.

Table 6.1 Entire ALT strategy of a mechanical system such as a refrigerator

	Market data		*Predicted reliability*				*Intended reliability*	
Modules	*Failure rate per year, λ (%/year)*	*BX life, L_B (year)*	*Failure rate per year, λ (%/year)*			*BX life, L_B (year)*	*Failure rate per year, λ (%/year)*	*BX life, L_B (year)*
A	0.25	7.2	Modified	×1	0.50	3.6	0.10	10(BX = 1.0)
B	0.30	6.0	Given	×1	0.30	6.0	0.10	10(BX = 1.0)
C	0.40	4.5	New	×5	2.00	0.9	0.10	10(BX = 1.0)
D	0.20	9.0	Given	×1	0.20	9.0	0.15	10(BX = 1.0)
E	0.20	9.0	Modified	×2	0.40	4.5	0.10	10(BX = 1.0)
Others	0.50	12.0	Given	×1	0.50	12.0	0.45	10(BX = 4.5)
System	1.85	4.5	–	–	3.90	0.9	1.00	10(BX = 10)

lifetime), three system modules can be specified to capture product market reliability data: (1) an altered module, (2) a newly altered module, and (3) a similar module. For example, module A – mechanical compartment, including compressor – which is discussed as a case study in Section 6.2.2 is considered an altered module because the compressor rubber mount was redesigned to dampen the transmitted impact in rail transportation. It had yearly failure rates of 0.25% per year and a B1.8 life 7.2 years in the marketplace. After modification, the module was expected to have yearly failure rates of 0.50% per year and a B1.8 life 3.6 years. To achieve the targeted product lifetime, a new lifetime of module A was set to a B1 life 10 years with 0.1% per year (Table 6.1).

6.2.2 Quantum/Transport-based (generalized) life-stress prototype and sample size formulation

Fatigue cracks can start to appear because of the existence of minute cracks or preexisting faults on the microscale structural surface of a component. If there are large stresses in a part, such as a groove, notch, and thin area, a crack can grow from microscale voids and propagate to the surface of the part. Thus, a quantum/transport-based LS model might be needed.

Consider noninteracting electrons moving (freely) in a constant potential wall that is assumed to be zero. Time-independent Schrodinger's equation in one dimension can be expressed as follows:

$$-\frac{h^2}{8\pi^2 m}\frac{d^2\psi_n(x)}{dx^2} = E_n\psi_n \tag{6.7}$$

where ψ_n is the wave function, E_n is the (electron) energy, h is the Planck constant, and m is the electron mass.

The boundary conditions are as follows: (1) ψ_n finite inside the metal but decaying exponentially. That is, $\psi_n \to 0\ as\ x \to \infty$, (2) $\psi_n = 0$ at walls. Therefore, we can find the solution as follows:

$$\psi_n(x) = \sqrt{\frac{2}{a}} sin\left(\frac{n\pi}{a}\right)x;\ E_n = \frac{n^2h^2}{8ma^2}\quad n > 0 \tag{6.8}$$

where $\psi_n(x+a) = \psi_n(x)$, a is the (periodic) distance, and n is the principal quantum number.

Linear transport such as solid-state diffusion of impurities of silicon takes place in a material when an electric magnetic field, ξ (driving force), is applied. It can be summarized as follows: (1) electromigration-induced voiding, (2) buildup of chloride ions, and (3) trapping of electrons or holes [34–36]:

$$J = [aC(x-a)] \cdot exp\left[-\frac{q}{kT}\left(w - \frac{1}{2}a\xi\right)\right] \cdot v$$

[Density/Area]·[Jump Chance]·[Jump Rate]

$$\begin{aligned} &= -\left[a^2ve^{-qw/kT}\right] \cdot cosh\frac{qa\xi}{2kT}\frac{\partial C}{\partial x} + \left[2ave^{-qw/kT}\right]C\ sinh\frac{qa\xi}{2kT} \\ &= \Phi(x,t,T)\ sinh\ (a\xi)\ exp\left(-\frac{Q}{kT}\right) = Bsinh(a\xi)\ exp\left(-\frac{Q}{kT}\right) \end{aligned} \tag{6.9}$$

where C is the concentration, q is the extent of electric charge, ν is the frequency, a is the distance between successive potential barriers, ξ is the exerted electromagnetic field, k is the Boltzmann's quantity, T is the absolute temperature, Q is the energy, and Φ() and B are constants.

The inverse of Eq. (6.9) can be used to express the (generalized) LS prototype:

$$TF = A[sinh(aS)]^{-1} exp\left(\frac{E_a}{kT}\right) \tag{6.10}$$

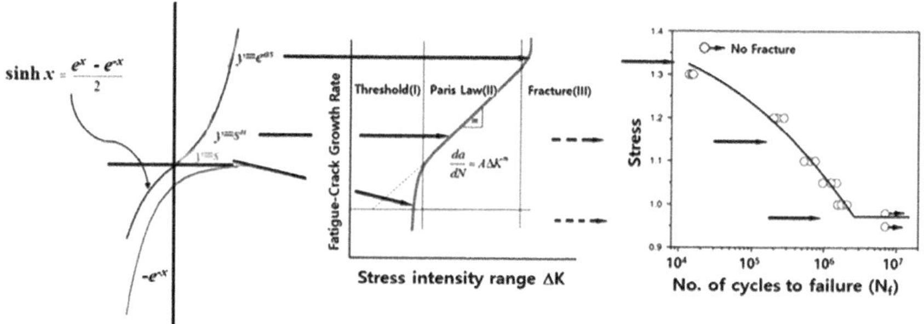

Figure 6.3 A meaning of hyperbolic sine stress expression in Paris law or S–N curve.

The sine hyperbolic term $\left[\sinh(aS)\right]^{-1}$ in Eq. (6.10) might be represented as follows: (1) $(S)^{-1}$ initially at low stresses is nearly linear, (2) $(S)^{-n}$ has what is considered a middle stress region, and (3) $\left(e^{aS}\right)^{-1}$ at final is a large stress region (Figure 6.3).

Stress is a physical quantity that designates the inner forces that neighboring subatomic particles of a continuous matter mutually apply. As the power is specified as the product of effort and flows, stresses can start from effort in a multiport system (Table 6.2) [37].

Because an ALT is performed in the middle stress region (Figure 6.3) and the stresses in transit come from the transmitted vibration loads (F_T) that can be expressed as the PSD level of acceleration for a certain frequency band. Eq. (6.10) can be redefined as

$$TF = A(S)^{-n} exp\left(\frac{E_a}{kT}\right) = A(e)^{-\lambda} exp\left(\frac{E_a}{kT}\right) = B(F_T)^{-\lambda} exp\left(\frac{E_a}{kT}\right) \quad (6.11)$$

So, the force transmissibility, Q, due to the base load, Y, can be modeled as [38]

$$Q = \frac{X(t)}{F_T(t)} = \frac{X(t)}{Y(t)} = \left[\frac{1+(2\zeta r)^2}{\left(1-r^2\right)^2+(2\zeta r)^2}\right]^{1/2} = |H(f)| \quad (6.12)$$

Table 6.2 Power idea in a multiport product indicated as effort and flow

System	*Effort, e(t)*	*Flow, f(t)*
Translation system	Force, $F(t)$	Velocity, $V(t)$
Rotation system	Torque, $\tau(t)$	Angular velocity, $\omega(t)$
Pump, compressor	Pressure difference, $\Delta P(t)$	Volume flow rate, $Q(t)$
Electric system	Voltage, $V(t)$	Current, $i(t)$
Magnetic	Magneto-motive force, e_m	Magnetic flux, φ

where r is the frequency ratio $(=\omega_n/\omega = f_n/f)$, ζ is the damping ratio $(c/c_c = c/2\omega_n)$, k is the spring constant, and $Y(t)$ is the range of sinusoidal base excitations.

We also know that the PSD level for the given frequency band, a, is calculated from Eqs. (A10) and (A11). Thus, the acceleration factor (AF) is specified as the relationship between the typical circumstance and the elevated circumstance. It is defined as the multiplication of the amplitude proportion of gravitational acceleration, R, and force transmissibility, Q [39]. In other words,

$$\begin{aligned} AF &= \left(\frac{S_1}{S_0}\right)^n \left[\frac{E_a}{k}\left(\frac{1}{T_0}-\frac{1}{T_1}\right)\right] = \left(\frac{a_1}{a_0}\frac{X}{Y}\right)^\lambda \left[\frac{E_a}{k}\left(\frac{1}{T_0}-\frac{1}{T_1}\right)\right] \\ &= (R\times Q)^\lambda \left[\frac{E_a}{k}\left(\frac{1}{T_0}-\frac{1}{T_1}\right)\right] \end{aligned} \tag{6.13}$$

where a_1 is the accelerated PSD level for the determined frequency band, a_0 is the normal PSD level for the determined frequency band, T_0 is the normal temperature, and T_1 is the accelerated temperature.

Because elevated testing is conducted at room temperature, Eq. (6.13) is rewritten as

$$AF = \left(\frac{S_1}{S_0}\right)^n = \left(\frac{F_1}{F_0}\right)^\lambda = \left(\frac{a_1}{a_0}\frac{X}{Y}\right)^\lambda = (R\times Q)^\lambda \tag{6.14}$$

To attain the assignment cycle – RQ specification – of parametric ALTs from the objective BX life on the test strategy, the sample size combined with the AF in Eq. (6.14) needs to be estimated. Some procedures have been presented to resolve problems with sample size.

Product failures can be categorized as infant mortality, random failure, and wear-out failure, which can be described in accordance with the shape parameter in the Weibull distribution. For estimating lifetime, the shape parameter is greater than one. The Weibayes method is defined as Weibull analysis with a given shape parameter that is developed from the real test data. Using the Weibayes method, the characteristic life can be derived from employing maximum likelihood estimation (MLE) in statistics, which is a common method used to estimate the parameters of a prototype. The characteristic lifetime η_{MLE} might be determined as [40]

$$\eta_{MLE}^{\beta} = \sum_{i=1}^{n} \frac{t_i^{\beta}}{r} \tag{6.15}$$

where η_{MLE} is the maximum likelihood estimate, n is the sample size, t_i is the test cycle for each sample, and r is the failure number.

The Weibull assessment is one procedure used to examine reliability data. However, due to the complications of the mathematical formulation, it is not easy to directly apply this method. Thus, it is recommended that the occasions of failures ($r \geq 1$) and no failures ($r = 0$) in reliability tests might be separated. As an outcome, it is practicable to derive an expected sample size formulation.

When the failure number, r, is greater than or equal to 1 and the confidence level is 100 $(1-\alpha)$, the characteristic lifetime, η_α, may be assessed from Eq. (6.15):

$$\eta_\alpha^\beta = \frac{2r}{\chi_\alpha^2(2r+2)} \times \eta_{MLE}^\beta = \frac{2}{\chi_\alpha^2(2r+2)} \times \sum_{i=1}^{n} t_i^\beta \text{ for r} \geq 1 \tag{6.16}$$

where $\chi_\alpha^2()$ is the chi-square function in case the p-value is α.

Supposing there is no failure, ln $(1/\alpha)$ is similar to the chi-square amount, $\frac{\chi_\alpha^2(2)}{2}$:

$$p-value\text{: } \alpha = \int_{\chi_\alpha^2(2)}^{\infty} \left(\frac{e^{-\frac{x}{2}} \chi^{\frac{v}{2}-1}}{2^{\frac{v}{2}} \Gamma\left(\frac{v}{2}\right)} \right) dx = \int_{2 ln \alpha^{-1}}^{\infty} \left(\frac{e^{-\frac{x}{2}} \chi^{\frac{v}{2}-1}}{2^{\frac{v}{2}} \Gamma\left(\frac{v}{2}\right)} \right) dx \tag{6.17}$$

where Γ is the gamma function and ν is the shape parameter.

As all occasions where $r \geq 0$ from Eqs. (6.16) and (6.17) are valid, the characteristic lifetime, η_α, can be defined as

$$\eta_\alpha^\beta = \frac{2}{\chi_\alpha^2(2r+2)} \times \sum_{i=1}^{n} t_i^\beta \text{ for r} \geq 0 \tag{6.18}$$

where t_i is the period for the tested samples.

If the Weibull distribution in Eq. (6.1) takes a logarithm formulation, the connection between characteristic life, η, and BX life, L_B, could be expressed as

$$L_B^\beta = \left(ln \frac{1}{1-x} \right) \times \eta_\alpha^\beta \tag{6.19}$$

where x is $x = 0.01 \cdot X$, on the condition that $x \leq 0.2$.

If the estimated calculated characteristic lifetime of the p-value α, η_α, in Eq. (6.18), is exchanged into Eq. (6.19), the BX lifetime, L_B, can be attained:

$$L_B^\beta = \left(ln \frac{1}{1-x} \right) \times \frac{2}{\chi_\alpha^2(2r+2)} \times \sum_{i=1}^{n} t_i^\beta \tag{6.20}$$

Because the entire reliability test has a limited sample number to estimate the lifetime for the allocated failure numbers, which could be less than that of the sample size, the test scheme might be expressed as

$$nh^{\beta} \geq \sum t_i^{\beta} \geq (n-r)h^{\beta} \tag{6.21}$$

where h is the test time (or cycles).

If Eq. (6.21) is exchanged into Eq. (6.20), the BX lifetime shall be restated as

$$L_B^{\beta} = \left(ln\frac{1}{1-x}\right) \times \frac{2}{\chi_{\alpha}^2(2r+2)} \times nh^{\beta} \geq \left(ln\frac{1}{1-x}\right) \times \frac{2}{\chi_{\alpha}^2(2r+2)} \times (n-r)h^{\beta} \tag{6.22}$$

where h is the test time (or cycles).

If Eq. (6.22) is reorganized, the sample size might be explained as follows:

$$n \geq \frac{\chi_{\alpha}^2(2r+2)}{2} \times \frac{1}{ln(1-x)^{-1}} \times \left(\frac{L_B^*}{h}\right)^{\beta} + r \tag{6.23}$$

Because, for $\alpha = 0.6$, $\frac{\chi_{\alpha}^2(2r+2)}{2} \cong (r+1)$ and $ln(1-x)^{-1} = x + \frac{x^2}{2} + \frac{x^3}{3} + \cdots \cong x$, the sample size formulation (6.23) shall be straightforwardly adjacent to

$$n \geq (r+1) \times \frac{1}{x} \times \left(\frac{L_B^*}{h}\right)^{\beta} + r \tag{6.24}$$

where the sample size formulation may be expressed as n ~ (failure numbers + 1)·(1/accumulative failure rate)·((objective life/(planned testing period)) ^ $\beta + r$.

Finally, we can obtain the sample size combined with the AF in Eqs. (6.13) or (6.14):

$$n \geq (r+1) \times \frac{1}{x} \times \left(\frac{L_B^*}{AF \cdot h_a}\right)^{\beta} + r \tag{6.25}$$

If the lifetime of a mechanical product, such as a compressor, is targeted to have a B1 life 10 years, the assigned cycles shall be obtained for a set of sample sizes subjected to (impact) loading. In parametric ALTs, the design imperfections of a newly designed product shall be found to help achieve the life objective.

For $n \gg r$, the sample size is expressed as [41]

$$\begin{aligned} n &= -\frac{\chi_{\alpha}^2(2r+2)}{2m^{\beta} lnR_L} = \frac{\chi_{\alpha}^2(2r+2)}{2m^{\beta} lnR_L^{-1}} = \frac{\chi_{\alpha}^2(2r+2)}{2m^{\beta} ln(1-F_L)^{-1}} \\ &= \frac{\chi_{\alpha}^2(2r+2)}{2} \times \frac{1}{ln(1-F_L)^{-1}} \times \left(\frac{L_B}{h}\right)^{\beta} \end{aligned} \tag{6.26}$$

where m $\cong h/L_B$, R_L is reliability, and F_L is unreliability

If $r = 0$, the sample size can be expressed as

$$n = \frac{ln(1-CL)}{m^{\beta} ln R_L} = \frac{-ln(1-CL)}{-m^{\beta} ln R_L} = \frac{ln(1-CL)^{-1}}{m^{\beta} ln R_L^{-1}} = \frac{ln\alpha^{-1}}{m^{\beta} ln R_L^{-1}}$$
$$= \frac{\chi_{\alpha}^{2}(2)}{2} \times \frac{1}{ln(1-F_L)^{-1}} \times \left(\frac{L_B}{h}\right)^{\beta} \tag{6.27}$$

where $2ln\alpha^{-1} = \chi_{\alpha}^{2}(2)$ and CL is confidence level.

Eqs. (6.26) and (6.27) as well as Eq. (B6) have similar formulations to Eq. (6.25).

6.2.3 Case study: Improving the fatigue of a domestic refrigerator subjected to (random) vibrations during rail transit

Rail transportation is a leading way of moving products from cargo ports to the product's eventual destination – which could be a distributor, warehouse, or end-user. Two cases are discussed: (1) imported refrigerators were shipped from London in the United Kingdom over two days to customers who lived in Skopje, the capital and largest city of North Macedonia in southeastern Europe – a total distance of 2,520 km, and (2) refrigerators were delivered from the West Coast of the United States to customers (or end-users) who lived on the East Coast. This trip was 7,200 km for a total travel time of seven days. According to market data, after shipping units in both cases, the compressor rubber mounts in the refrigerator were torn and the connecting tubes to the compressor had fractured under unknown stress conditions during railway transport. For Europe, the distance at which failure first occurred during rail transportation was approximately 2,400 km in Nis, Serbia. Over two days, when the refrigerators traveled the 2,520-km distance from London to Skopje, 10% of the products failed. On the other hand, in the United States, the distance at which failure first occurred during rail transportation was approximately 2,500 km over two days. In Chicago, 27% of the transported products failed. Over seven days, when the refrigerators traveled the 7,200-km distance from Los Angeles to Boston, 67% of the products failed (Figure 6.4).

Because of the large failure rates, it was evident that the refrigerators had design flaws. After identifying the refrigerator design flaws in laboratory tests, the manufacturer had to modify the problematic designs.

The rail routes, shown in Figure 6.4(b), included mainline, side-line, and industrial lines in the United States. To obtain the PSD along these rail routes, it was measured in both the vertical and horizontal directions. A SAVER 3X90 shock and vibration field data recorder (Lansmont Corp., CA, USA) was utilized to gather the spectral data for the whole trip. The SAVER was attached to the ground situated in the middle of the storage area. After analyzing the data, it was possible to obtain the vibration environment spectra,

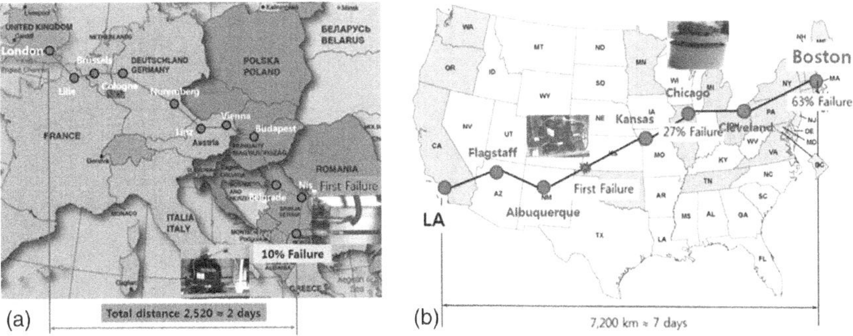

Figure 6.4 Failed places from the marketplace: (a) mechanical compartment and (b) refrigerator with modules A–E.

which showed a graphic chart of the PSD quantities versus frequency. In this investigation, the PSD spectra included frequencies from 0.3 to 400 Hz. These were used to assess the impact of the vibrations on the design of the refrigerator.

After analyzing the measured vibration spectra, it was possible to calculate the acceleration factor that multiplies the amplitude proportion of acceleration R and force transmissibility Q from Eq. (6.14). The damping ratio, natural frequency, the normal PSD level, and accelerated PSD level were known. A programmed shaker table was used to apply the elevated external random vibrations, expressed by PSD, to each refrigerator sample. Tables were set up with refrigerator test samples that could be driven by actuators. Random waves and periodic waves, such as sine, rectangular, and triangular waves, could be selected as the input motion. Random shaking using observed strong PSD motions could help improve the robustness of the product and help identify design weaknesses for RQ specification. The amplitude of the powerful random vibrations, expressed as G_{rms}, could be adjusted according to the capacity of the specimen for accelerated loads.

In applying the harmonic waves with various frequencies on the refrigerator, the natural frequency of the vibration (left ↔ right) parallel to the plane of the horizon was found to be 5 Hz. The natural frequency of the vibration (up ↔ down) at right angles to a horizontal plane was 9 Hz in the vibration test. To obtain the AF for horizontal or vertical directions, based on random vibrations observed in the field, amplified PSD loads for each orientation were applied to the refrigerator on a shaker table (Figure 6.5).

To determine the stress quantity for the parametric ALT and assess the lifetime, the step-stress life test was conducted under constant use-conditions for various amplitudes of random vibrations. These vibrations included 0.40 G_{rms}, 0.60 G_{rms}, 0.80 G_{rms}, and 1.00 G_{rms} at the natural frequency ($r = 1.0$, $\zeta \approx 0.1$) [41]. The damping ratio ($\zeta = 0.096 \approx 0.1$) had a settling time of 2 sec and approximately a 5% overshoot due to direct contact between the train and refrigerator. The frequency ratio was expected, r ($=\omega/\omega_n$) = 1 at the

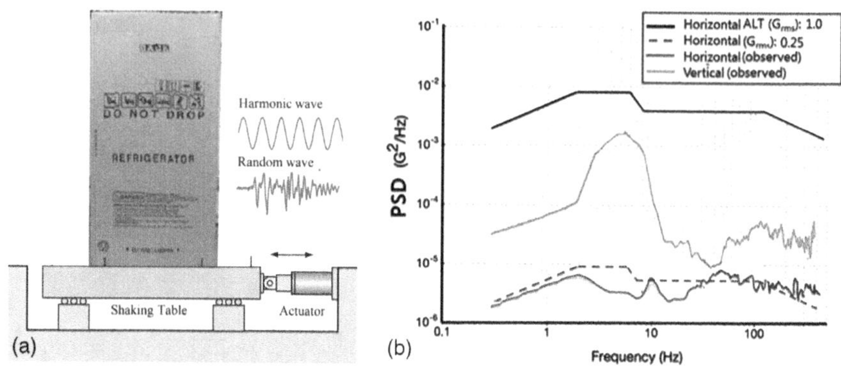

Figure 6.5 Vibration test for the parametric ALT: (a) programmed shaking table and (b) applied PSD loads.

natural frequency ω_n that was applied to the refrigerator by rail, and the force transmissibility, Q, had a magnitude of approximately 5.3 from Eq. (6.12). The AF due to gravitational acceleration for 1 G_{rms} was 4.0 because the refrigerator reached an acceleration of 1 G_{rms} on a shaker table, compared to that of worst-case 0.25 G_{rms}. Using a cumulative damage quantity, λ, of 2.0, the entire AF in Eq. (6.14) was determined to be 450.0 (Table 6.3).

Based on the calculated AF, the needed testing time for a given sample size was estimated when the lifetime target was assigned. That is, if it is assumed that the initial shape parameter in the Weibull plot was 2.0 because the market and ALT, β, were not obtainable at the period of test design (see Figure 6.1), the lifetime target was set to be B1 life for seven days. The test time acquired from Eq. (6.25) was approximately 130 minutes for three sample pieces. If the refrigerator failed less than once in 130 minutes, the refrigerator design was suitable for the whole travel distance of 7,200 km (seven days) to endure the fatigue damage by random vibration and have a B1 lifetime with an approximately 60% level of confidence.

The assessed life L_B in each ALT can be expressed as

$$L_B^{\beta} \cong x \cdot \frac{n \cdot (h_a \cdot AF)^{\beta}}{r+1} \tag{6.28}$$

Table 6.3 Conditions for parametric ALT of the refrigerator

System states	*Worst case*	*ALT*	*AF*
Transmissibility, Q (r = 1.0, ζ = 0.096 ≈ 0.1)	-	5.30 (from Eq. 6.17)	28.1 ①
Amplitude proportion of acceleration, R (a_1/a_0)	0.25 G_{rms}	1 G_{rms}	16.0 ②
Entire AF (= ① × ②)			450

If $x = \lambda \cdot L_B$, the assessed failure rate, λ, may be stated as

$$\lambda \cong \frac{1}{L_B} \cdot (r+1) \cdot \frac{L_B^{\beta}}{n \cdot (h_a \cdot AF)^{\beta}} \tag{6.29}$$

In every ALT stage, if the measured cycles, h_a, the lifetime target, x, sample size, n, accelerated factor, AF, shape parameter, β, and failure number, r, are known, the assessed life, L_B, and failure rate, λ, can be assessed from Eqs. (6.28) and (6.29).

6.3 RESULTS AND DISCUSSION

In the first ALT, the failure cycles of the succeeding stress levels were 0.40 G_{rms}, 0.60 G_{rms}, 0.80 G_{rms}, and 1.00 G_{rms} at the natural frequency ($r = 1.0$, $\zeta \approx 0.1$). For these conditions, the refrigerator samples had torn rubber mounts and broken tubes. For 0.40 G_{rms}, they fractured at 70 minutes. For 0.60 G_{rms}, they fractured at 60 minutes. For 0.80 G_{rms}, they fractured at 45 minutes. For 1.00 G_{rms}, they fractured at 20 minutes. Therefore, a 1.00 G_{rms} was used because it had a good result with accelerated testing and was nearly linear on the Weibull plot.

As 1.00 G_{rms} at the natural frequency ($r = 1.0$, $\zeta \approx 0.1$) was applied, three refrigerators in the first parametric ALT fractured as follows: one sample at 20 minutes and two samples at 40 minutes from the horizontal vibration (left ↔ right). As the rubber mounts tore, the connecting refrigerant tubing in the machine compartment broke, and the refrigerant gas leaked out of the system, and the three samples no longer worked. Figure 6.6 shows the fractured products returned from the field and the unsuccessful samples from the parametric ALT, individually. The forms and places of the failures in the parametric ALT were similar to those shown in the field. To explore the fracture surfaces made of ethylene propylene diene monomer (EPDM) rubber, they were also captured by an SEM. Some irregular/coarser voids were discovered because of the small rubber particles that fell out of the mounts during the tests.

Figure 6.7 shows a plot of the market data and parametric ALT outcomes on a Weibull chart. The shape parameters of the ALT (β_1) and field data (β_2) were similar. Under the close repeated stresses near the natural frequency, the failure patterns displayed in the first ALT were similar to the failure patterns in the failed refrigerators returned from the field. Therefore, the shape parameter, β, was confirmed to be 6.13. Based on both test results and the Weibull chart, the ALT pinpointed the design defects in the refrigerator.

The rubber tearing of the mount in the first ALT occurred because there was no support for the rubber mount to withstand horizontal vibrations (left ↔ right). Due to the design imperfections – such as no rubber support in the high-stress regions – the repeated random loads produced the rubber

Figure 6.6 Failed refrigerator tubes and rubber from the marketplace and first ALT: (a) field and (b) first ALT results.

tearing and broken connection pipes. The refrigerator was redesigned as follows: (1) a reshaped compressor rubber mount, C1 (Figure 6.8a); and (2) a reshaped connecting tube, C2 (Figure 6.8b). Second ALTs were then conducted using these modified designs.

The quantities of AF and β in Table 6.3 and Figure 6.7 were affirmed to be 452.0 and 6.41, respectively. Based on the test data, because the lifetime objectives of a new sample were less than a B1 lifetime for the whole travel interval (seven days), the recomputed test cycles in Eq. (6.25) for the three refrigerators were 40 minutes, which would be the specification of the

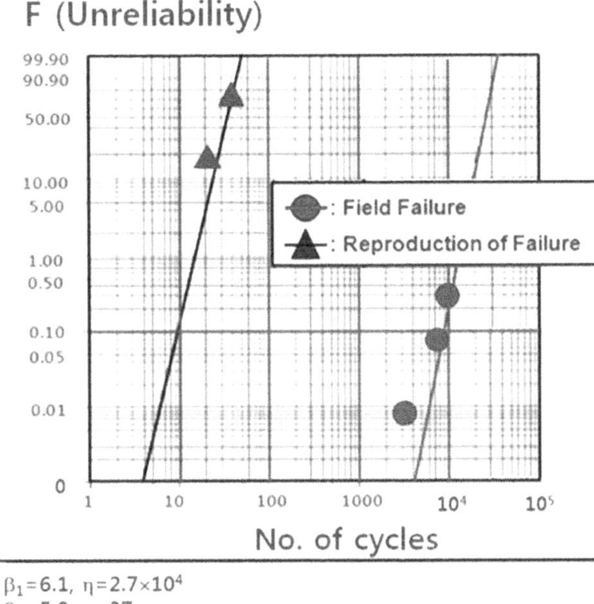

Figure 6.7 Market data and outcomes of ALT on the Weibull plot.

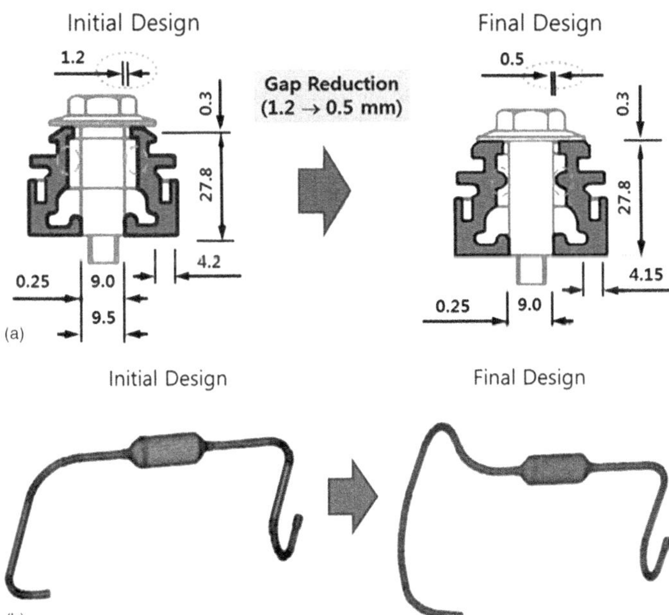

Figure 6.8 Design modifications: (a) problematic rubber shape in refrigerator and (b) problematic connecting tube shape in refrigerator.

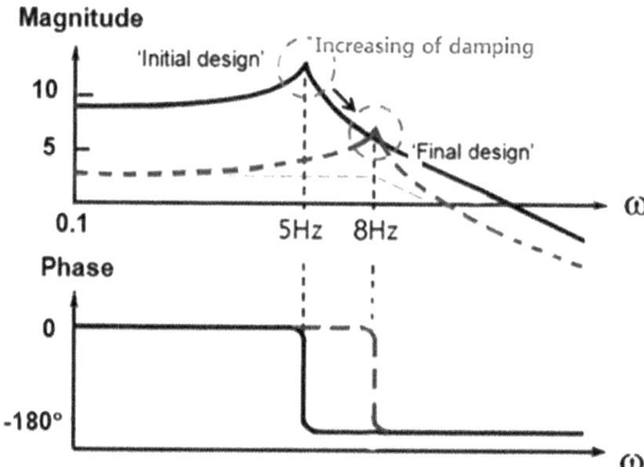

Figure 6.9 Change of the natural frequency of horizontal vibration due to the modifications of product design.

parametric ALT. During the second ALT, the altered designs were successful in protecting the refrigerator from the random-induced vibrations. As a result, the refrigerator tubing or the compressor mounts did not fracture until 60 minutes. When a refrigerator reached an acceleration of 1.00 G_{rms} on a shaking table, the natural frequency of the vibration (left ↔ right) parallel to the plane of the horizon was moved from 5 to 8 Hz due to the increase of damping in the system (Figure 6.9).

Through two rounds of ALTs, the new refrigerator was assured to have a B1 lifetime for the whole travel distance of 7,200 km (seven days) from Eqs. (6.28) and (6.29) when the actual running time, $h_a = 60$ minutes, in plugging in lifetime target, $x = 0.01$, sample size, $n = 3$, accelerated factor, $AF = 452.0$, shape parameter, $\beta = 6.4$, and failure number, $r = 0.0$.

6.4 CONCLUSIONS

To extend the life of a mechanical product in transit, parametric ALT with RQ specification was developed. The quantum/transport-based (generalized) LS equation and sample size were suggested. As a case study, the lifetime design of a refrigerator fatigued by random vibrations during rail transportation was investigated.

- By inspecting problematic domestic refrigerators returned from the marketplace, it was found that fracturing occurred in compressor mounts made of EPDM rubber and connecting refrigerant tubes. Because of the fracturing of the refrigerant tubing, refrigerant leaked from the refrigerators, and the refrigerators did not work.
- To reproduce the problems in the refrigerators returned from the field, amplified PSD loads – 1 G_{rms} at resonance in first parametric ALT were performed for RQ specification – 40 minutes. Like that of field

failure, we found the rubber mounts and refrigerant tubing fractured at 20 minutes and 40 minutes. After the compressor mounts and the tubes were redesigned, in the second ALT, there were no failures for RQ specification – 40 minutes. Thus, the refrigerator was assured to survive the fatigue damage induced by random vibrations and fulfill the product lifetime requirements – B1 life for the whole transit distance of 7,200 km (seven days).

- Examination of the unsuccessful product, load examination, and two sequences of ALTs for RQ specification revealed that the lifetime of the redesigned refrigerator in transit was dramatically increased. This structured reliability method is applicable to other engineered mechanical systems, such as cars, airplanes, and machine tools in transit.

LIST OF ABBREVIATIONS

AF	acceleration factor
BX	period that shall be X% of the cumulated failures D (thermodynamic or propelling) force
e	effort
E_a	activation energy, eV
$F(t)$	unreliability
f	flow
h	test time (or cycles)
h_a	actual test time (or cycles)
h^*	nondimensional test time, $h^* = h/L_B \geq 1$
k	Boltzmann constant, 8.62×10^{-5} eV/°C
K	reaction speed
L	Phenomenological transport quantity
L_B	objective BX life
n	number of tested samples
r	unsuccessful numbers
r	frequency ratio ($= \omega_n/\omega = f_n/f$)
Q	amount of energy absorbed or released during the reaction.
R	amplitude proportion of gravitational acceleration
S	stress, Pa
t_i	period for tested samples
T	absolute temperature, K
TF	time to failure, h
X	required reliability target or cumulated failure, %
Y	range of sinusoidal (base) excitations
x	$x = 0.01 \cdot X$, on condition that $x \leq 0.2$

Greek symbols

Ξ	Electrical field applied
α	confidence level
η	characteristic lifetime

λ failure rate
λ accumulative damage quantity in Palmgren-Miner's rule
ζ damping ratio ($c/c_c = c/2\omega_n$)
Γ gamma function
$\chi^2_\alpha()$ chi-square function in case the p-value is α.

Superscripts

β shape parameter in a Weibull function
n stress dependence, $n = -\left[\frac{\partial ln(T_f)}{\partial ln(S)}\right]_T$

Subscripts

0 usual stress circumstance
1 elevated stress circumstance

REFERENCES

1. Magaziner IC, Patinkin M. Cold competition: GE wages the refrigerator war. Harvard Business Review. 1989; 89(2): 114–124.
2. WRDA 2020 Updates. The Final Report of the US House Committee on Transportation and Infrastructure on the Boeing 737 Max, 2020. Available online: https://democrats-transportation.house.gov/committee-activity/boeing-737-max-investigation (accessed date September 2020).
3. Woo S, Pecht M, O'Neal D. Reliability design and case study of the domestic compressor subjected to repetitive internal stresses. Reliability Engineering & System Safety. 2020; 193: 106604.
4. CMMI Product Team. Capability Maturity Model Integration (CMMI) Version 2.0, Continuous Representation; Report CMU/SEI-2002-TR-011; Software Engineering Institute: Pittsburgh, PA, USA, 2018.
5. NASA. System Engineering Handbook; NASA/SP-2020-6105 Rev 2, NASA Headquarters: Washington, DC, USA, 2020; p. 92.
6. Kong Y, Bennett CJ, Hyde CJ. A review of nondestructive testing techniques for the in-situ investigation of fretting fatigue cracks. Materials & Design. 2020; 196: 109093.
7. Duga JJ, Fisher WH, Buxaum RW, Rosenfield AR, Buhr AR, Honton EJ, McMillan SC. The Economic Effects of Fracture in the United States, Final Report, September 30, 1982, Battelle Laboratories, Columbus, OH. Available as NBS Special Publication 647-2.
8. Fatigue. In: Campbell, F.C., editors. Elements of Metallurgy and Engineering Alloys, Materials Park, OH: ASM International; 2008.
9. Weingart RG, Stephen P. Timoshenko: Father of Engineering Mechanics in the U.S. Structure Magazine, 1 August 2007.
10. DVB Bank SE Aviation Research (AR). An Overview of Commercial Jet Aircraft 2013; 2014, p. 20.
11. National Transportation Safety Board. April 2018. Left Engine Failure and Subsequent Depressurization, Southwest Airlines Flight 1380, Boeing 737-7H4, N772SW, Philadelphia, Pennsylvania, Aircraft Accident Report, NTSB/AAR-19/03.
12. Wang T, Wang Q. Cyclic responses and microstructure sensitivity of Cr-based turbine steel under different strain ratios in low cycle fatigue regime. Materials & Design. 2021; 201: 109529.
13. Inman DJ. Engineering Vibration. 4th ed. Upper Saddle River (NJ): Prentice-Hall; 2013.

14. Lalanne C. Mechanical Vibration and Shock Analysis, Volume 5, Specification Development. 3rd ed. Hoboken (NJ): John Wiley & Sons, Inc.; 2014.
15. Khalij L, Gautrelet C, Guilletb A. Fatigue curves of a low carbon steel obtained from vibration experiments with an electrodynamic shaker. Materials & Design. 2015; 86(5): 640–648.
16. Modarres M, Kaminskiy M, Krivtsov V. Reliability Engineering and Risk Analysis: A Practical Guide. 3rd ed. Boca Raton (FL): CRC Press; 2016.
17. Özsoy S, Çelik M, Kadıoğlu FS. An accelerated life test approach for aerospace structural components. Engineering Failure Analysis. 2008; 15(7): 946–957.
18. Lu Y, Zheng H, Zeng J, Chen T, Wu P. Fatigue life reliability evaluation in a high-speed train bogie frame using accelerated life and numerical test. Reliability Engineering & System Safety. 2019; 188: 221–232.
19. Elsayed EA. Reliability Engineering. 3rd ed. Hoboken (NJ): John Wiley & Sons, Inc.; 2021.
20. Hahn GJ, Meeker WQ. How to Plan an Accelerated Life Test (E-Book). Milwaukee (WI): ASQ Quality Press; 2004.
21. Ritchie RO, Liu D. Introduction to Fracture Mechanics. 1st ed. Amsterdam: Elsevier; 2021.
22. Taguchi G. Off-line and on-line quality control systems. Proceedings of the international conference on quality control. 1978.
23. Montgomery D. Design and Analysis of Experiments. 10th ed. Hoboken (NJ): John Wiley & Sons, Inc.; 2020.
24. Reddy JN. An Introduction to the Finite Element Method. 4th ed. New York: McGraw-Hill; 2020.
25. Wang Z, Wang S, Niu Y, Zhao H. A study of effect factors on thermal drift rates during cryogenic indentation via Taguchi design and finite element method. Materials & Design. 2020; 219: 110764.
26. Miner MA. Cumulative damage in fatigue. Journal of Applied Mechanics. 1945; 12: 149–164.
27. Sun YS. Revised Miner's rule and its application in calculating equivalent loads for components. Reliability Engineering & System Safety. 1994; 43(3): 319–324.
28. MIL-STD-810H. January 2019. Department of Defense Test Method Standard for Environmental Engineering Considerations and Laboratory Tests. United States Department of Defense.
29. MIL-STD-810B. June 1967. Environmental Test Methods. Ohio: United States Department of Defense.
30. McLean HW. HALT, HASS and HASA Explained: Accelerated Reliability Techniques. Revised ed. Milwaukee (WI): ASQ Quality; 2009.
31. Cengel YA, Boles MA. Thermodynamics: An Engineering Approach. 9th ed. New York: McGraw-Hill; 2019.
32. IEEE Standard Glossary of Software Engineering Terminology. IEEE STD 610.12-1990. Standards Coordinating Committee of the Computer Society of IEEE (Reaffirmed September 2002).
33. Kreyszig E. Advanced Engineering Mathematics. 10th ed. Hoboken, NJ: John Wiley & Son; 2020, p. 683.
34. Plawsky JL. Transport Phenomena Fundamentals. 4th ed. Boca Raton (FL): CRC Press; 2020.
35. Woo S, O'Neal D. Reliability design and case study of mechanical system like a hinge kit system in refrigerator subjected to repetitive stresses. Engineering Failure Analysis. 2019; 99: 319–329.
36. Grove A. Physics and Technology of Semiconductor Device. 1st ed. Hoboken: Wiley International Edition; 1967, p. 37.
37. Karnopp DC, Margolis DL, Rosenberg RC. System Dynamics: Modeling, Simulation, and Control of Mechatronic Systems. 5th ed. New York: John Wiley & Sons; 2000.

38. Rao SS. Mechanical Vibration. 6th ed. New York: Pearson Prentice Hall; 2017.
39. Woo S. Design of Mechanical Systems Based on Statistics. Boca Raton (FL): CRC Press; 2021.
40. Abernethy RB. The New Weibull Handbook. Florida (FL): Reliability Analysis Center; 2000.
41. Wasserman G. Reliability Verification, Testing, and Analysis in Engineering Design; New York: Marcel Dekker; 2003, p. 228.

APPENDIX A. THE SYSTEM RESPONSE FROM A SIMPLIFIED MODELING OF MECHANICAL SYSTEMS IN TRANSIT

To find out the critical mode due to design imperfection in a transit mechanical system, the motion of the equation for N degrees of freedom (DOFs) is expressed as follows:

$$M\ddot{X} + C\dot{X} + KX = F_T(t) \tag{A1}$$

where $X(t)$ is the steady-state solution of the system, $F_T(t)$ is the excited force due to sinusoidal base excitations.

The solution of Eq. (A1) is assumed as follows:

$$X = uq(t) \tag{A2}$$

where X is generalized coordinate, and $q(t)$ is modal coordinate (or natural coordinate).

By differentiating Eq. (A2), it is

$$\dot{X} = u\dot{q}(t) \text{ and } \ddot{X} = u\ddot{q}(t) \tag{A3}$$

Inserting them into Eq. (A1) for X, $\dot{X}$, $\ddot{X}$ and multiplying by u^T, it produces

$$u^T Mu\ddot{q}(t) + u^T Cu\dot{q}(t) + u^T Kuq(t) = u^T F \text{ for ideal conditions.} \tag{A4}$$

Because the mode shapes are orthogonal to one another, Eq. (A4) is

$$[M]\ddot{q} + [C]\dot{q} + [K]q = u^T F = Q(t) \text{ (N independent DOF systems)} \tag{A5}$$

where $[M] = u^T Mu = \begin{bmatrix} \ddots & & \\ & M_i & \\ & & \ddots \end{bmatrix}$, $[K] = u^T Ku = \begin{bmatrix} \ddots & & \\ & K_i & \\ & & \ddots \end{bmatrix}$, $[C] = u^T Cu$, $Q(t) = u^T F$

From the characteristic equations with response to the initial conditions, the steady-steady response for each critical mode, $\{u\}^i$, is

$$X = \sum_i \{u\}^i q_i(t) \tag{A6}$$

To assess the ride attribute of a system installed on a vehicle, one of the most insightful mathematical representations of a vehicle suspension system is a quarter car prototype that can be simply analyzed for vibration modes. The quarter car is used to study wheel-hop as well as the bounce mode. It is set up using mutual connections of masses, springs, and dampers. Although a quarter car prototype has two DOFs and four state variables, it fulfills the aim of identifying the vehicle movement in transit. In these diagrams, the presumed prototype of the vehicle includes the unsprung mass and the sprung mass, individually. The sprung mass, m_s, denotes one-fourth of the body of the vehicle; the unsprung mass, m_u, denotes one wheel of the vehicle. The principal suspension is formulated as a spring k_s and a damper b_s in parallel, which attach the unsprung mass to the sprung mass. The tire (or rail) is prototyped as a spring constant of the tire (or rail), k_t, and denotes the transfer of the road force to the unsprung mass (Figure 6.A1).

After the quarter car prototype is decomposed, the differential equations of movement can be represented as

$$m_s\ddot{x}_3 = -k_s\left(x_3 - x_2\right) - b_s\left(\dot{x}_3 - \dot{x}_2\right) \tag{A7}$$

$$m_u\ddot{x}_2 = k_s\left(x_3 - x_2\right) + b_s\left(\dot{x}_3 - \dot{x}_2\right) - k_t\left(x_2 - y\right) \tag{A8}$$

So, Eqs. (A7) and (A8) might be simply defined as

$$\begin{bmatrix} m_s & 0 \\ 0 & m_u \end{bmatrix}\begin{bmatrix} \ddot{x}_3 \\ \ddot{x}_2 \end{bmatrix} + \begin{bmatrix} b_s & -b_s \\ -b_s & b_s \end{bmatrix}\begin{bmatrix} \dot{x}_3 \\ \dot{x}_2 \end{bmatrix} + \begin{bmatrix} k_s & -k_s \\ -k_s & k_t + k_s \end{bmatrix}\begin{bmatrix} x_3 \\ x_2 \end{bmatrix} = \begin{bmatrix} 0 \\ k_t y \end{bmatrix} \tag{A9}$$

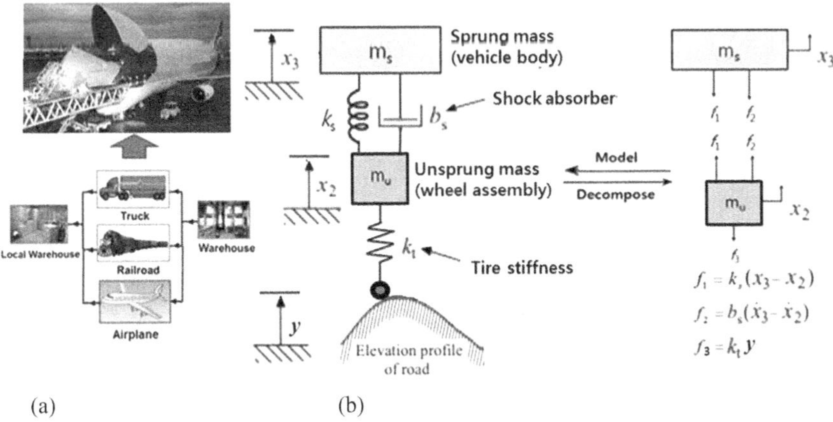

Figure 6.A1 A quarter car model subjected to random loads from the base (or road): (a) a transportation mode and (b) a dynamic model.

As a consequence, Eq. (A9) might be expressed as follows:

$$[M]\ddot{X}+[C]\dot{X}+[K]X=F_T e^{i\omega t} \tag{A10}$$

For Eq. (A10), the (random) system response of an equivalent single DOF (SDOF) is described (see the left side picture of Figure 6.A2(a)). That is,

$$m_e\ddot{X}(t)+c_e\dot{X}(t)+k_eX(t)=c_e\dot{Y}(t)+kY(t) \tag{A11}$$

As the variables in the time domain are changed into the frequency domain, we can obtain a frequency domain function called FDSs. Therefore, Eq. (A11) can be stated as follows:

$$\ddot{X}(f)+2j\varsigma\omega_n\dot{X}(f)+\omega_n^2X(f)=\left[-2j\varsigma\frac{f_n}{f}-\left(\frac{f_n}{f}\right)^2\right]\ddot{Y}(f) \tag{A12}$$

where ζ is the damping proportion, ω_n ($=2\pi f_n$) is the natural frequency, and $\omega(=2\pi f)$ is the excitation frequency.

The acceleration $\ddot{X}(f)$ can be expressed in the enforced acceleration $\ddot{Y}(f)$. That is,

$$\ddot{X}(f)=\left[\frac{2j\varsigma\left(\frac{f_n}{f}\right)+1}{1-\left(\frac{f_n}{f}\right)^2+2j\varsigma\left(\frac{f_n}{f}\right)}\right]\ddot{Y}(f)=H(jf)\ddot{Y}(f) \tag{A13}$$

where $H(jf)$ is the frequency transfer function (FRF).

From Eq. (A13), the PSD as the strength of the variations (energy) in the frequency domain can be attained. PSD is used extensively to analyze the mechanical system characteristics for random loads such as forces and moments. The PSD of the acceleration $W_{\ddot{X}}$ can be stated in the PSD of the transmitted acceleration $W_{\ddot{Y}}$. That is,

$$W_{\ddot{X}}(f)=|H(f)|^2\,W_{\ddot{Y}}(f) \tag{A14}$$

where

$$|H(f)|=\left[\frac{\left(2\varsigma\frac{f_n}{f}\right)^2+1}{\left(1-\left(\frac{f_n}{f}\right)^2\right)^2+\left(2\varsigma\frac{f_n}{f}\right)^2}\right]^{1/2} \tag{A15}$$

The root mean square (RMS) of the acceleration, $\ddot{X}$, can be computed as

$$a(=G_{rms}) = \ddot{X}_{rms} = \int_0^{\infty} W_{\ddot{X}}(f)df \tag{A16}$$

The integral in Eq. (A16) can also be approximated (numerical quadrature) by the trapezoidal method to calculate $\ddot{X}_{rms}$ numerically:

$$\ddot{X}_{rms}(f_n) = \sqrt{\sum_{k=1}^{N} \left|H(f_n, f_k)\right|^2 W_{\ddot{X}}(f_k)\Delta f_k} \tag{A17}$$

If the input from the base is a random vibration characterized as (1) Gaussian distribution, (2) constant PSD, and (3) the damping ratio is less than 0.1 ($\zeta < 0.1$), it is possible to assume that the amplitude versus occurrence in the system response pursues the normal distribution (i.e., Gaussian distribution) and the peak versus occurrence manifests a Rayleigh distribution (Figure A2(a)). However, in a number of cases, the field data may not follow a Gaussian distribution because of the base random excitation. If the response of the equivalent SDOF in Eq. (A11) is displayed in the frequency domain, a plot (Figure 6.A2(b)) of the PSD can be obtained.

If the (accelerated) PSD response of the product is applied, it is possible to evaluate the design robustness due to the amplitude of the product at its natural frequency during its transit. That is, the resonance-induced impact of weak material parts due to repeated large displacements, strains, and stresses near natural frequency can be assessed on the potential for fracture, which starts at voids and then propagates to the end.

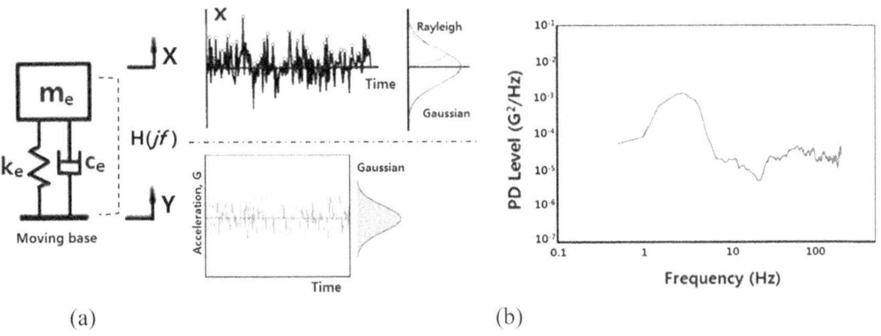

Figure 6.A2 (Intermodal) vehicle responses for base random vibrations: (a) time domain typical rail random and (b) vibration in frequency domain.

APPENDIX B. DERIVATION OF ANOTHER SAMPLE SIZE EQUATION

The cumulative probability of each Bernoulli trial on the condition that n is not small and p is tiny is approached as a Poisson distribution:

$$L(p)=\sum_{r=0}^{c}\binom{n}{r}p^{r}\cdot(1-p)^{n-r}=\sum_{r=0}^{c}\frac{1}{r!}m^{r}\cdot e^{-m}\leq\alpha \tag{B1}$$

where n is the number of test samples and c is the assumed number of failures, $m=\text{parameter}=n\cdot p$.

In the case of following the Weibull distribution, Eq. (B1) is expressed as

$$L(p)=\sum_{r=0}^{c}\binom{n}{r}\left(1-e^{-\left(\frac{t}{\eta}\right)^{\beta}}\right)^{r}\cdot\left(e^{-\left(\frac{t}{\eta}\right)^{\beta}}\right)^{n-r}\leq\alpha \tag{B2}$$

Because $e^{-\left(\frac{t}{\eta}\right)^{\beta}}\cong 1-\left(\frac{t}{\eta}\right)^{\beta}$, Eq. (B2) can be approximated as

$$L(p)\cong\sum_{r=0}^{c}\frac{1}{r!}\left(\frac{t}{\eta}\right)^{\beta r}\cdot\left(1-\left(\frac{t}{\eta}\right)^{\beta}\right)^{n-r}\leq\alpha \tag{B3}$$

Because parameter m follows the chi-square distribution, $\chi_{\alpha}^{2}()$ and Eqs. (B1) and (B3) have a similar form, the characteristic life with a confidence level of 100 $(1-\alpha)$ can be stated as

$$m=n\cdot\left(\frac{t}{\eta}\right)^{\beta}\sim\frac{\chi_{\alpha}^{2}(2r+2)}{2}\quad or\quad \eta_{\alpha}^{\beta}=\frac{2}{\chi_{\alpha}^{2}(2r+2)}\cdot n\cdot t^{\beta} \tag{B4}$$

At BX life, L_B, in Eq. (1), testing time, t, is h:

$$L_{B}^{\beta}\cong x\cdot\eta_{\alpha}^{\beta}=x\cdot\frac{2}{\chi_{\alpha}^{2}(2r+2)}\cdot n\cdot h^{\beta}\text{ or }n\geq\frac{\chi_{\alpha}^{2}(2r+2)}{2}\times\frac{1}{x}\times\left(\frac{L_{B}^{*}}{h}\right)^{\beta}\quad\text{for }x\leq 0.2 \tag{B5}$$

where $x=0.01F(t)$

As the term $\frac{\chi_{\alpha}^{2}(2r+2)}{2}$ in Eq. (B5) is close to $(r+1)$ in a 60% confidence level, Eq. (B5) is

$$n\geq(r+1)\times\frac{1}{x}\times\left(\frac{L_{B}^{*}}{h}\right)^{\beta} \tag{B6}$$

Chapter 7

Topology optimization of binary-state networks with the use of cumulative updating of network reliability bounds

Denis A. Migov, Kseniya A. Volzhankina, and Alexey S. Rodionov

7.1 INTRODUCTION

In the chapter, a problem of reliability-based optimization of a binary-state network topology is studied when we need to maximize the network reliability within given certain constraints. For modeling a binary-state network, the elements of which are subject to random failures, we use random graphs with unreliable edges [1]. However, such problems of graph optimization in conditions of different constraints are mostly "NP"-hard. Nevertheless, these problems can be effectively solved by optimization methods based on biological processes [2, 3], such as genetic algorithms (GAs), clonal selection algorithms (CSAs), and others.

The main advantage of such a bio-inspired approach is that these techniques can provide an applicable solution for network topology optimization within an acceptable time. In the related problems of network reliability-based optimization, a fitness function of algorithms for describing biological processes is a network reliability measure. In the case of such complex fitness functions, whose calculation demands exponential or even more complexity, their calculation or evaluation is the most time-consuming part of a topology optimization process. Instead of exhaustive reliability calculation, we propose to use the following approach.

In [4], the problem of determination whether the network reliability exceeds a given value (threshold) is considered. In other words, we have to find out if the network is reliable enough. The idea is to consistently (cumulatively) refine the upper and lower bounds of the network reliability, which can be done using the factorization method (see Section 7.3). Each time the factorization process produces a subgraph, whose reliability can be calculated directly, we refine the upper and lower bounds of the reliability of the original network. If one of the obtained values crosses the threshold line, we make a decision about its reliability. Thus, it's possible to draw a valid conclusion about the reliability of the network without a completed calculation. This approach has been further developed in our previous work that used cut-point decomposition and two-vertex cuts [5] and for other reliability indices [6–8].

In order to speed up fitness function calculation, we improve operators of a GA and a CSA by using the method of cumulative updating of lower and upper bounds of network reliability. Based on this approach, we propose the

DOI: 10.1201/9781003546214-7

GA and the CSA for network topology optimization. As reliability measures, we consider the network probabilistic connectivity (so-called all-terminal network reliability) and the diameter constrained network reliability (DCNR). The results of the numerical experiments are also presented for demonstration of the applicability of the proposed approach.

7.2 BASIC DEFINITIONS AND NOTATIONS

For modeling a network, we use an undirected random graph $G = (V, E)p$ with given edge presence probabilities $0 \leq r_e \leq 1$. It is assumed that the network has perfectly reliable nodes and unreliable links. $K \subseteq V$ is a given set of terminals, i.e., selected nodes that have to be connected.

Let us assume $Q = (V, E_Q)$ is a subgraph, where E_Q is defined by the existence or absence of each edge $e \in E_Q$. An edge e is called *operational* if it exists in E_Q; otherwise, we call it *faulty*.

The probability of an elementary event Q is defined as

$$P(Q) = \prod_{e \in E_Q} r_e \prod_{e \notin E_Q} (1 - r_e). \tag{7.1}$$

The reliability of network G, $R_K(G)$, is defined as the sum of probabilities of elementary events in which every pair of terminals $u, v \in K$ can be connected by a path. This reliability measure is well-known as K-terminal network reliability. If $K = V$, it turns to the all-terminal network reliability, or just probabilistic connectivity. We use notation $R(G)$ instead of $R_V(G)$

The reliability with diameter constraint d of network G, $R_K^d(G)$, is defined as the sum of probabilities of elementary events in which every pair of terminals $u, v \in K$ can be connected by a path p of length at most d, where the length of the path p is the number of edges belonging to this path.

Below, we consider two indices: the all-terminal network reliability (all-terminal reliability (ATR)) and the DCNR.

Let us consider the problem of obtaining the most reliable network structure on a given set of nodes V under certain constraints. Suppose we are given a set of vertices V and a redundant set of edges S. For each edge e, c_e is the cost of its installation, and r_e is the reliability value. $Weight(G)$ is defined as the sum of the edge weights of the graph G on the sets V, S. A natural number d and a real positive number C^* are also given. Therefore, a problem of constructing the most reliable network on a given set of vertices and redundant set of edges with a budget constraint can be formulated as follows:

Problem 1 is to find a graph G with a diameter not greater than d, $E \subseteq S$ with the following restrictions:

$$\begin{cases} R(G) \to \max; \\ Weight(G) < C^*. \end{cases} \tag{7.2}$$

Preference is given to the cheapest solution in the case of equal probability values. If $R_K^d(G)$ is the reliability measure, then a statement of Problem 1 should be slightly changed:

Problem 2 is to find a graph G, $E \subseteq S$ with the following restrictions:

$$\begin{cases} R_k^d(G) \to \max; \\ Weight(G) < C^*. \end{cases} \tag{7.3}$$

7.3 METHODS FOR NETWORK RELIABILITY CALCULATION

The problem of network reliability calculation is NP-hard, as are the the ATR computation problem [9] and the DCNR computation problem [10].

Nevertheless, exact methods are widely studied for reliability calculation. The best-known method is the factoring method (branching method or the Moore-Shannon method) [11, 12]. Inclusion-exclusion methods and methods based on binary decision diagrams are also used for network reliability calculation [13–15]. Important approaches to speed up the reliability calculation (by various methods) are the reduction and decomposition methods [16–18]. One of the most effective reduction methods is the parallel-series transformation, which removes chains and multiple edges from a graph. For ATR [19, 20] and for DCNR [21], we can apply this reduction in every call of the factoring procedure.

The factoring method is based on the partition of a probability space into two subsets. For this purpose, an arbitrary unreliable network element e is selected, and two new networks are reviewed, in which one of the elements is absolutely reliable G/e; in the other, it is absent or deleted $G \backslash e$. The resulting networks are recursively factorized, and the full probability formula expresses the chosen reliability index Rel of the network under factorization:

$$Rel(G) = r_e Rel(G/e) + (1 - r_e) Rel(G \backslash e). \tag{7.4}$$

We continue the recursions until either the resulting network is unreliable ($Rel = 0$) or it is absolutely reliable ($Rel = 1$). For ATR, we can stop the recursion process if the resulting graph has not more than five vertices, so the formulas for four- and five-vertex graph reliability can be used [5].

In [10], authors have proposed the modification of factoring method for the DCNR calculation. Below, we call this method the CPFM (Cancela & Petingi factoring method). As experiments show, this method is much faster than the original factorization (7.4). The method proposed in [10] deals with the list of relevant paths $P_{st}(d)$ between all pairs of terminals s and t with limited length $\leq d$.

Now, the factorization can be described in terms of $P = \cup_{s,t \in T} P_{st}(d)$ instead of graphs. For example, the failure of the pivot factored element disconnects each path it belongs to, so we just remove such paths from consideration.

The CPFM uses the following parameters for description of graphs during factorization:

- np_{st} is the number of $s-t$ paths, the length of which does not exceed d;
- $links_p$ is the number of edges not yet considered in path p;
- $feasible_p$ is a Boolean variable that is equal to False if the path contains a failed edge and equal to True otherwise;
- $connected_{st}$ is a Boolean variable that is equal to True if s and t are connected by a path consists of absolutely reliable edges, whose number does not exceed d, and equal to False otherwise;
- *connectedPairs* is the number of pairs of terminals that are connected by a path consists of absolutely reliable edges, whose number does not exceed d.

7.4 THE CUMULATIVE UPDATING OF LOWER AND UPPER BOUNDS OF NETWORK RELIABILITY

In [4], authors try to determine if the ATR of a given network with unreliable edges is not less than a predefined value $0 \le R_0 \le 1$. In general words, this can be described as a reliability exact calculation with simultaneous calculation of lower and upper ATR bounds (RL and RU, respectively), but the calculation stops if a corresponding bound becomes lesser or greater than R_0.

For this purpose, a number of exact reliability calculations may be used. The main method is the factorization. However, the binary decision diagrams or the spanning tree decomposition can be used as well. For all these methods, their modifications are provided for obtaining the lower and upper ATR bounds in the cumulative way [4]. Therefore, the bounds are updated at the next step according to the rule $RL_i \ge RL_{i-1}$, $RU_i \le RU_{i-1}$. For ATR, $RL = 0$ and $RU = 1$. If RL_i becomes greater than R_0, then the initial network is reliable. If RU_i becomes lesser than R_0, then the network is obviously unreliable. In both cases, the problem is solved and the calculation process stops. So, instead of the whole calculation, which is an NP-hard problem, we do the partial calculation.

Let us describe this procedure for the factoring method, in an intermediate step of which we obtain graphs $G_1, G_2, \ldots, G_L$ with associated probabilities $P_1, P_2, \ldots, P_L$. Obviously, $\sum_{i=1}^{L} P_i = 1$. Alo, we have an inequality [4]:

$$\sum_{i=1}^{k} P_i R(G_i) \le R(G) \le 1 - \sum_{i=1}^{k} P_i (1 - R(G_i)). \tag{7.5}$$

Based on it, we obtain the algorithm for cumulative updating of the ATR bounds. If we can calculate the ATR of a next graph G_i (for example, because it contains less than six vertices), then the bounds RL_i and RU_i are updated in accordance with the rule:

$$RL_i = RL_{i-1} + P_i R(G_i)$$
$$RU_i = RU_{i-1} - P_i(1 - R(G_i)). \tag{7.6}$$

Since this process is cumulative, the bounds tend to $R(G)$, and sooner or later one of them will cross the threshold R_0. In the worst case, after the exhaustive calculation, if $R_0 = R(G)$.

Later, we adapted this technique for cumulative updating DCNR-based CPFM [6]. If a pair of terminals becomes disconnected, or all the pairs of terminals become reliable connected, then the bounds are updated. It can be done for and only for the final graphs $G_1, G_2, \ldots, G_L$ obtained with the corresponding probabilities $P_1, P_2, \ldots, P_L$. After factorization by an edge e, we multiply P_i by r_e or $1 - r_e$.

Example of how this method works is given in the diagram (Figure 7.1). This diagram shows how lower and upper DCNR bounds are updated. The analyzed graph is the graph of Internet2 network (Figure 7.2). The following parameters were used for testing: $p_e = 0.9$ for each edge e, $d = 25$, $R_0 = R_K^d(G)$. The whole time of calculation is about 1s. Here the X axis corresponds to the recursion number (in scale 10^4). The Y axis represents the calculated values of DCNR bounds.

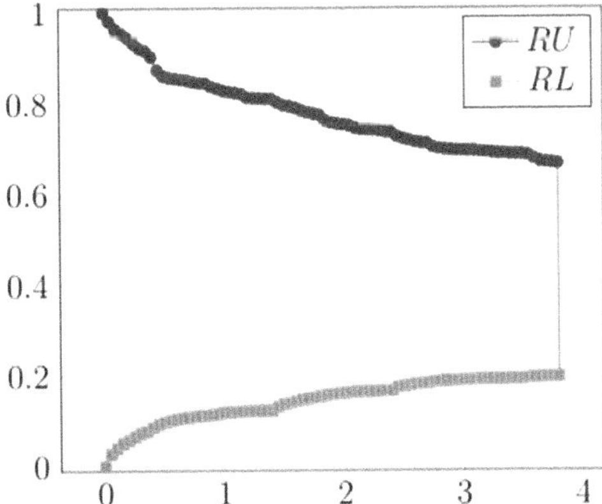

Figure 7.1 Behavior of reliability bounds.

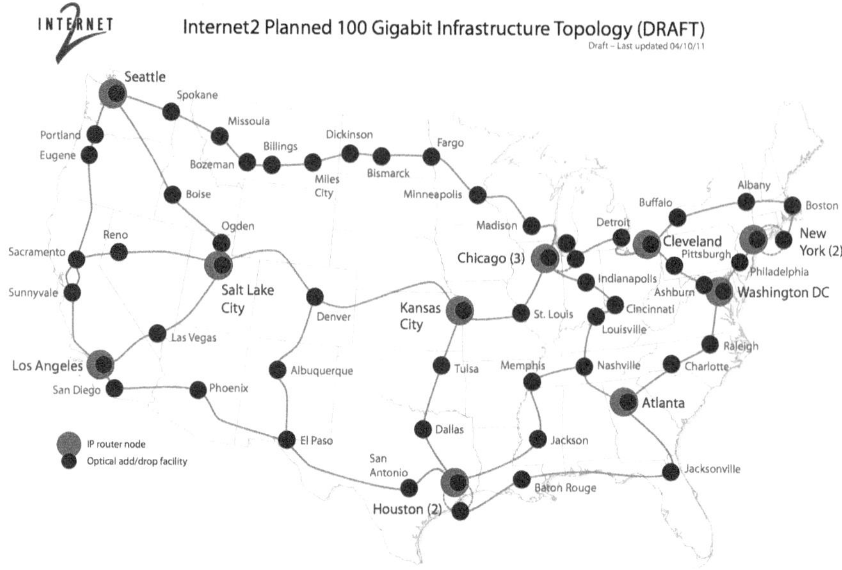

Figure 7.2 Tested network.

7.5 BIO-INSPIRED ALGORITHMS FOR NETWORK TOPOLOGY OPTIMIZATION

The most popular approach among evolutionary algorithms (optimization techniques) based on the principles of natural selection is GAs. Further, we denote them by the abbreviation GAs. The theory presented by Charles Darwin in [22] in 1859 provided the basis for such algorithms.

Let us describe in general terms the optimization process by a GA. First, individuals (or chromosomes) are defined; they are represented, for example, by bit strings or an ordered sequence of elements. Each individual represents a solution. The population is randomly formed from the set of individuals. Next, the strongest and fittest individuals are ranged by a fitness function. Pairs of parents are selected using the selection operator (for example, tournament selection). Next comes the crossing (the "crossover" operator). After crossing, the mutation operator applies to the new generation. The next generation formed from the fittest individuals of the current population and newly formed offspring. The most unsuitable individuals are eliminated. New individuals are added up to population size. This prevents the algorithm from falling into the local optimum. The stopping criterion for the algorithm can be the number of generations, a time limit, or stagnation (several generations pass without improvement). Figure 7.3 shows the general GA scheme.

It is proposed to use genetic operators accelerated by cumulative estimates of the reliability boundaries to optimize the network structure by the reliability criterion. Cumulative estimates allow genetic operators to work faster by eliminating

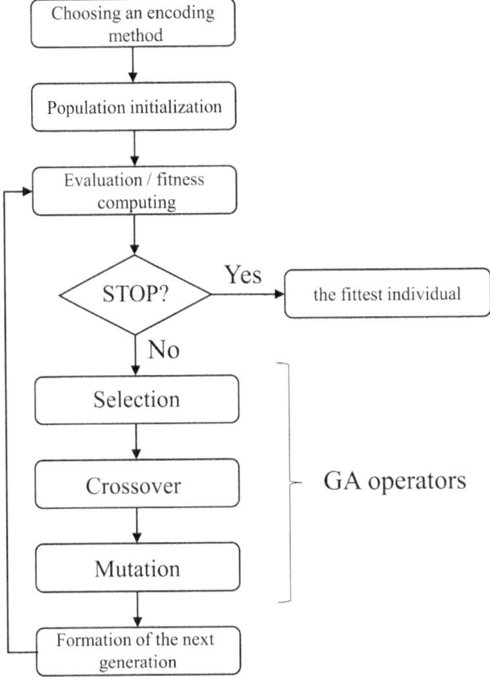

Figure 7.3 Genetic algorithm scheme.

pre-bad chromosomes with inappropriate values of the fitness function (not reliable enough network structures). Operators in this case will be as follows.

Mutation operator. We assume that A_0 is the given chromosome and its reliability equals to R_0. By *Rel* we denoted the reliability measure under consideration: ATR or DCNR. The fitness of the new chromosome A_1 will be calculated according to the following rule:

$$Feas(A_1, P_0) = \begin{cases} 1, \text{ if } Rel(A_1) > R_0; \\ 0, \text{otherwise.} \end{cases} \tag{7.7}$$

Crossover operator. Let A_0 and A_1 be the selected parents for crossing. Let R_{max} be the reliability value for the most reliable of the parents. The fitness of the child A_2 is calculated according to the following rule:

$$Feas(A_2, R_{max}) = \begin{cases} 1, \text{ if } Rel(A_2) > R_{max}; \\ 0, \text{otherwise.} \end{cases} \tag{7.8}$$

In this way, the reliability boundaries of a new individual are refined when the fitness function is calculated. If the upper limit has crossed the threshold value (R_0 or R_{max}), then the calculation stops, and a decision is made about the fitness of this individual. If the lower limit has crossed the threshold value, then the same decision is made about an individual and the reliability value is calculated to the end.

The theory of artificial immune systems (AIS) emerged from theoretical immunology in the mid-1980s. Initially, immune algorithms were employed by Bersini to address various problems [23]. The primary objective of AIS is to apply principles from immunology to develop systems for solving various problems of optimization [24].

One prominent algorithm within the AIS family is the CSA [25]. This algorithm is based on the natural B-cell response mechanism. When an antigen, such as a virus, enters the bloodstream, B-cells initiate the secretion of antibodies. Each cell produces a single type of antibody specifically targeting the antigen. Throughout this process, B-cells undergo cloning and mutation to achieve the best possible match with the antigen. B-cells that exhibit superior matching rapidly disseminate antibodies, transforming into plasma cells. Some of them become memory cells, circulating in the bloodstream until a new invasion occurs.

The algorithmic scheme of the CSA is shown in Figure 7.4.

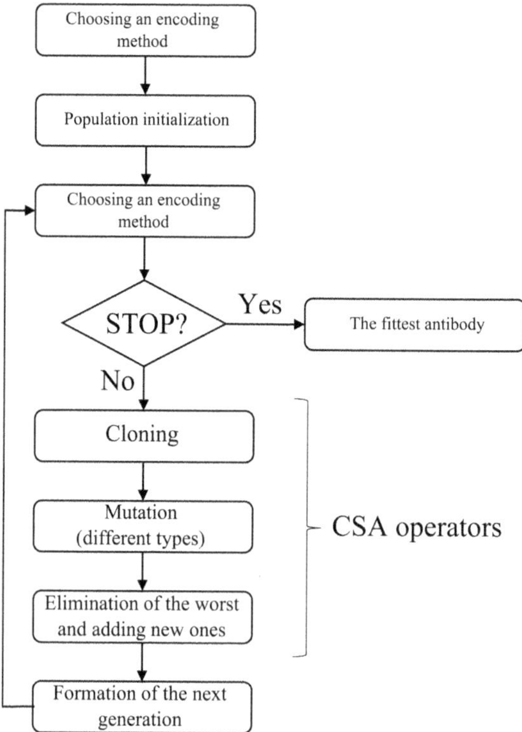

Figure 7.4 Clonal selection algorithm scheme.

Similar to chromosomes in GAs, the antibodies are represented by bit strings or collections of elements. A population is formed by selecting an arbitrary set of these antibodies. Subsequently, we evaluate the affinity of each antibody within the population. The next step involves cloning antibodies based on their affinity values. Each clone undergoes mutation to enhance its solution quality. After eliminating a portion of the least effective clones, a new batch of random antibodies, equal in number, is added to the population. Then, tree algorithm iteratively operates on this updated population until it meets specified termination criteria, such as an expired time or a given number of generations.

Just as in GAs, the operators of the clonal selection can be improved by removing the weakest antibodies through cumulative updating.

7.6 NUMERICAL EXPERIMENTS

This section presents a series of experiments aimed at demonstrating relation between given data and running time. First, we investigate the effectiveness of the proposed GA in the case of the ATR optimization criterion in comparison with a similar GA without using cumulative updating. Optimal structures were searched for three combinations of the number of nodes, the number of edges, and the diameter value. The corresponding values are shown in the top line of the table; edge reliability was 0.9. Ten runs of both algorithms were performed for each of these three cases. The same parameters were chosen for both algorithms in each run: population size, mutation probability, number of generations, tournament selection. Table 7.1 shows the average calculation time for each combination of the number of nodes, edges, and diameter value for each algorithm.

Results show that in all cases, the modified GA is faster. The difference in calculation time was approximately 2%, 90%, and 98% for the input data shown in the second, third, and fourth columns, accordingly. The best results of GA using cumulative estimates for the fitness function were shown on sparse graphs (98%). In this case, many unfitted offspring are produced by crossover and mutation operators, elimination of such solutions is much faster than a full reliability calculation. Calculation computer settings: quad-core Intel(R) Core(TM) i3-4130 CPU 3.40 GHz, RAM = 6 Gb.

Table 7.1 Computational results for GA in the case of ATR optimization criterion

Size	$\|V\| = 10, \|E\| = 20$	$\|V\| = 20, \|E\| = 30$	$\|V\| = 50, \|E\| = 70$
Diameter	d = 4	d = 5	d = 8
Genetic Algorithm			
Time	8s 411ms	4m 46s 662ms	1h 52m 58s 643ms
Genetic Algorithm + Cumulative Updating			
Time	8s 240ms	25s 374ms	1m 40s 607ms

Table 7.2 Computational results for GA in the case of DCNR optimization criterion

Graph $\|V\| = 15$, $\|E\| = 22$		$d = 5$	$d = 8$	$d = 10$
r = 0.1	GA	1m 55s 789ms	2m 43s 671ms	3m 50s 336ms
	GA with CU	1m 56s 70ms	2m 43s 563ms	3m 50s 374ms
	Reliability	0.053377031	0.035480363	0.042657732
r = 0.5	GA	39s 247ms	2m 13s 754ms	12m 18s 777ms
	GA with CU	39s 321ms	2m 11s 322ms	12m 21s 311ms
	Reliability	0.801047325	0.796380579	0.830094337
r = 0.9	GA	1m 59s 10ms	3m 5s 699ms	4m 1s 559ms
	GA with CU	1m 54s 263ms	2m 41s 897ms	3m 27s 677ms
	Reliability	0.999966901	0.999951435	0.999883982

Let us evaluate the performance of the optimization techniques proposed with respect to the DCNR criterion. For the experiment, non-dense 15-vertex graphs were chosen for their expedited computational processing. The results of the GA and CSA performance are presented in Tables 7.2 and 7.3, respectively.

We test various diameter values (5, 8, 10) and edge reliabilities (0.1, 0.5, 0.9), which remain uniform across all edges. Several other parameters are held constant: the probability of the mutation is set at 0.1, the population size is 50, and there are 10 populations in total. Consequently, the reliability values of 500 graphs should be calculated, considering both mutation and crossover operations. The number of terminal nodes is set at 3.

Table 7.4 shows experimental results when the number of terminals equals to 3 and 7.

Our observations reveal that the combination of the GA or the CSA with the cumulative updating yields noteworthy results, particularly when dealing

Table 7.3 Computational results for CSA in case of DCNR optimization criterion

Graph $\|V\| = 15$, $\|E\| = 22$		$d = 5$	$d = 8$	$d = 10$
r = 0.1	CSA	42s 939ms	1m 18s 686ms	1m 20s 586ms
	CAS with CU	43s 484ms	1m 19s 501ms	1m 20s 227ms
	Reliability	0.033715333	0.035687813	0.035632029
r = 0.5	CSA	46s 870ms	1m 46s 51ms	2m 14s 788ms
	CSA with CU	46s 965ms	1m 44s 935ms	2m 5s 659ms
	Reliability	0.771484375	0.756820679	0.652519226
r = 0.9	CSA	44s 178ms	1m 52s 241ms	2m 50s 238ms
	CSA with CU	43s 616ms	1m 45s 308ms	2m 31s 871ms
	Reliability	0.999604889	0.999946036	0.999672344

Table 7.4 Calculation time depending on the number of terminals

	3 terminals	*7 terminals*
GA	3m 5s 699ms	39m 20s 844ms
GA with CU	2m 41s 897ms	33m 29s 853ms
Reliability	0.999951434	0.998523183
CSA	1m 52s 241ms	13m 34s 306ms
CSA with CU	1m 45s 308ms	9m 53s 614ms
Reliability	0.999946036	0.992177548

with non-dense graphs with high reliability. The most effective use of cumulative updating was for the larger number of terminals.

7.7 CONCLUSION

We have described bio-inspired algorithms for network topology optimization in the case of unreliable network edges. As an optimization criterion, the network reliability is considered. The main problem of GA development is the selection of genetic operators (selection, crossover, mutation). The right ones leave solutions in the desired class and represent individuals (in this case, network structures) in a suitable form. Another important aspect that affects the effectiveness of a GA is the selection of calculating or evaluating method for the fitness function (objective function). This becomes especially relevant in the case of an NP-hard problem, including the problem of calculating the network reliability.

In order to speed up the elimination of unfit individuals, it is proposed to use the method of cumulative updating of network reliability boundaries, which allows us to decide about the network reliability (or unreliability) with respect to a given threshold without performing the exhaustive reliability calculation. New operators allow to stop calculating the fitness function for new individuals at an earlier stage of the algorithm with use of cumulative estimates of the network reliability and eliminate unfit individuals. The new bio-inspired algorithms for the network optimization problem with modified operators significantly improve runtime, just as the numerical experiment has shown.

REFERENCES

1. Colbourn C.J. The Combinatorics of Network Reliability. New York: Oxford University Press, p. 160, 1987.
2. Koide T., Shinmori S., Ishii H. Topological Optimization With a Network Reliability Constraint. Discrete Applied Mathematics.. Vol. 115, n. 1–3, November 2001, p. 135–149.

3. Liu B., Iwamura K. Topological Optimization Models for Communication Network With Multiple Reliability Goals. Computers & Mathematics with Applications. Vol. 39, n. 7–8, April 2000, p. 59–69.
4. Won J.-M., Karray F. Cumulative Update of All-Terminal Reliability for Faster Feasibility Decision. IEEE Transactions on Reliability. Vol. 59, n. 3, 2010, p. 551–562.
5. Rodionov A., Migov D., Rodionova O. Improvements in the Efficiency of Cumulative Updating of All Terminal Network Reliability. IEEE Transactions on Reliability. Vol. 61, n. 2, 2012, p. 460–465.
6. Migov D., Nechunaeva K., Nesterov S., Rodionov A. Cumulative Updating of Network Reliability with Diameter Constraint and Network Topology Optimization. Springer Lecture Notes in Computer Science (in ICCSA 2016, Part 1). Vol. 9786, 2016. P. 141–152.
7. Rodionov A., Migov D. Chapter 5: Obtaining and Using Cumulative Bounds of Network Reliability. In: System Reliability. Edited by Constantin Volosencu. Rijeka, Croatia: InTech, p. 93–112, 2017. ISBN 978-953-51-3705-4.
8. Migov D.A., Volzhankina K.A., Rodionov A.S. Genetic Algorithms for Drain Placement in Wireless Sensor Networks Optimal by the Reliability Criterion. Optoelectronics Instrumentation and Data Processing. Vol. 57, n. 3, 2021, p. 240–249.
9. Valiant L.G. The Complexity of Enumeration and Reliability Problems. SIAM Journal on Computing. Vol. 8, n. 3, 1979, p. 410–421.
10. Cancela H., Petingi L. Diameter Constrained Network Reliability: Exact Evaluation by Factorization and Bounds. In: Int. Conf. on Industrial Logistics, Okinawa, Japan. p. 359–356, 2001.
11. Page L.B., Perry J.E. A Practical Implementation of the Factoring Theorem for Network Reliability. IEEE Transactions on Reliability. Vol. 37, n. 3, 1988, p. 259–267.
12. Youssef M., Khorramzadeh Y., Eubank S. Network Reliability: The Effect of Local Network Structure on Diffusive Processes. Physical Review, E. Vol. 88, n. 5), 2013, p. 052810.
13. Imai H., Sekine K., Imai K. Computational Investigations of All-Terminal Network Reliability via BDDs. IEICE Transactions on Fundamentals of Electronics, Communications and Computer Sciences. Vol. E82-A, n. 5, 1999, p. 714–721.
14. Chaturvedi S.K. Network Reliability Measures and Evaluation. Scrivener Publishing LLC, p. 237, 2016.
15. Yeh W.C. Novel Binary-Addition Tree Algorithm (BAT) for Binary-State Network Reliability Problem. Reliability Engineering and System Safety. Vol. 208, 2020, p. 107448.
16. Wood R.K. Triconnected Decomposition for Computing K–terminal Network Reliability. Networks. Vol. 19, 1989, p. 203–220.
17. Migov D.A., Rodionova O.K., Rodionov A.S., Choo H. Network Probabilistic Connectivity: Using Node Cuts. EUC Workshops, Springer Lecture Notes in Computer Sciences. Vol. 4097, 2006, p. 702–709.
18. Migov D.A. Computing Diameter Constrained Reliability of a Network with Junction Points. Automation and Remote Control. Vol. 72, n. 7, 2011, p. 1415–1419.
19. Satyanarayana A., Wood R.K. A Linear-Time Algorithm for Computing K-Terminal Reliability in Series-Parallel Networks. SIAM Journal on Computing. Vol. 14, 1985, p. 818–883.
20. Rodionova O.K., Rodionov A.S., Choo H. Network Probabilistic Connectivity: Exact Calculation with Use of Chains. In: ICCSA 2004. LNCS, vol. 3045. P. 315–324.

21. Nesterov S., Migov D. Series-Parallel Transformation for Diameter Constrained Network Reliability Computation. Proc. of the IEEE 2017 Int. Multi-Conf. on Engineering, Computer and Information Sciences (SIBIRCON). Novosibirsk, Russia. p. 121–125, 2017.
22. Darwin C.R. On the Origin of Species by Means of Natural Selection, or the Preservation of Favoured Races in the Struggle for Life. London: John Murray, 1859.
23. Bersini H., Varela F.J. Hints for Adaptive Problem Solving Gleaned from Immune Networks. In: Parallel Problem Solving from Nature, First Workshop PPSW 1. 1991. P. 343–354.
24. Ishida Y. The Immune System as a Self-Identification Process: A Survey and a Proposal. In: Proc. of the ICMAS Int. Workshop on Immunity-Based Systems. p. 2–12, 1996.
25. De Castro L.N., Von Zuben F.J. The Clonal Selection Algorithm with Engineering Applications. In: Workshop Proceedings of GECCO. Workshop on Artificial Immune Systems and Their Applications, Las Vegas, USA. p. 36–37, 2000.

Chapter 8

Quantitative analysis of reliability and cost factors in pulse radar systems

Shivani, S. Rana, P. Singh, A. Kumar, R. Pandey, A. Kumar, N. Goyal, M. Ram, A. S. Bhandari, and S. Bisht

8.1 INTRODUCTION

In today's world, setting up an aviation radio service is a challenging endeavour that must take into account organisational and human aspects so as to reduce the possibility of undesirable events like air accidents and flight mishaps. For a variety of uses where accurate object detection and ranging are crucial, radar systems are essential. A pulse radar system is a type of radar system that uses short pulses of electromagnetic radiation to detect and locate objects in its vicinity. This technology is widely used in a variety of applications, including military radar systems, weather monitoring, and air traffic control. Depending on their design and intended use, they can operate at frequencies ranging from a few megahertz to several gigahertz and have a variety of characteristics that help to improve their accuracy and resolution. For example, some systems use multiple pulses with different frequencies to create a more detailed image of the target object, while others use advanced signal processing algorithms to filter out noise and interference from the environment.

Pulse radar technology was first developed in the early 20th century as a result of the diligent efforts of several researchers and engineers who were seeking insight into the characteristics of electromagnetic waves. The basic principle of a pulse radar system is that it transmits a brief burst of radio frequency (RF) energy, or a pulse, towards a target object. When the pulse encounters the object, it reflects some of the energy back towards the radar receiver. By measuring the time it takes for the reflected signal to return to the receiver, the system can determine the range, or distance, to the object. In addition to range, pulse radar systems can also determine the bearing, or direction, of an object by using an antenna to scan the surrounding area. By combining range and bearing information, a pulse radar system can create a map of its environment and detect the presence of any objects within it. Pulse radar technology is now employed in a variety of fields, such as air traffic management, weather tracking, security applications, aviation and navigation, military and civilian radar systems, and others. The continued development of pulse radar systems is helping to improve our ability to detect and track objects in our environment and is playing an important role in many areas of science and technology.

DOI: 10.1201/9781003546214-8

In order to make sure that a pulse radar system operates precisely as intended, is cost-effective, and complies with safety and legal requirements, reliability analysis is crucial. The design and modelling of the system play a key role in determining the reliability of pulse radar systems. With a long-life expectancy and low risk of failure, pulse radar systems are generally regarded as being very reliable. This is due to the robustness and resistance of these systems to environmental factors like vibration, temperature changes, and electromagnetic interference. To ensure the reliability of these systems, it is important to follow best practices in system design, component selection, and maintenance. This may include regular inspections and calibration of the system, as well as replacing components that are known to have a limited lifespan. Additionally, it is important to test the system regularly to ensure that it is functioning correctly and that any issues are identified and addressed promptly. This may involve running diagnostic tests on the system as well as conducting field tests to evaluate its performance under real-world conditions. The accuracy and reliability of pulse radar systems make them ideal for a variety of applications where these qualities are crucial. It is possible to make sure that these systems continue to provide accurate and reliable performance over a long service life by adhering to best practises in system design, maintenance, and testing.

8.2 LITERATURE REVIEW

Reliability plays a crucial role in numerous fields, including engineering, manufacturing, healthcare, and software development. Understanding the factors that contribute to reliability and exploring the existing research in this area are essential for enhancing performance, reducing failures, and improving overall system efficiency. From the past investigation, it has been observed that several studies on reliability analysis of complex systems have been conducted. Bartko and Carpenter (1976) examined the most commonly employed and misapplied reliability measures in the mental health field. They provided instances that demonstrated the advantages and disadvantages of each type of data format. The reliability of man-machine interaction was discussed in a study by Hollnagel (1992), which was dominated by an engineering view, as human reliability is seen as analogous to the reliability of technical or engineering systems. An approach to investigate timed Bayesian networks (BNs) and to find a suitable reliability framework for dynamic systems was demonstrated by Boudali and Dugan (2005) in their research. The features of the modelling framework which make BNs well-suited for reliability applications were discussed in a study by Langseth and Portinale (2007). Also, a probabilistic reliability evaluation of a power system, including solar/photovoltaic cells, was carried out by Park et al. (2009), which states that the reliability evaluation of solar cell generators is different from conventional generators. Maintenance scheduling of generating units was very important for reliable operation of units, and in this area, Reihani et al. (2010) presented a hybrid evolutionary algorithm to tackle the GMS (generator maintenance scheduling) problem, assuming reliability as an objective function. A rainwater harvesting system reliability model based on nonparametric stochastic rainfall generator was developed by Basinger et al. (2010).

There have been numerous prior studies done to determine the reliability of radar system. Ram (2013), in his study, provided a brief survey on the system reliability. It surveys the reliability methods used in various engineering and physical science fields. It also provides the major areas, i.e. past, present, and future trends of reliability methods and applications for the readers. Li and Wang (2014), in their research, proposed a mixed testing method (which combines the operational profile with directed testing) to test the reliability of a certain radar system simulation software. The first phase is to carry out testing based on operational profile, and the second phase is to execute the direct testing. Cai et al. (2015) developed a real-time reliability evaluation methodology by combining a root cause diagnosis phase based on BNs and a reliability evaluation phase based on dynamic BNs (DBNs). An investigation by Kozhokhina et al. (2016) provided the information regarding reliability of radar system, and the main attention is on human-operator and informational reliability, and the factors that decrease it. Research carried out by Ridder and Narayanan (2018) offers the operational reliability of the radar system, which is defined as the measure of its ability to act reliably in various operational situations and uses formulations based on probabilities of false alarm and detection to allow the optimisation of radar performance under various target and clutter scenarios. Zhang et al. (2019) proposed a new reliability analysis method for the load-sharing k-out-of-n: F system, which was based on the load-strength model. As reliability was already discussed in previous research, there was also a need to make improvements in this concept. In their study, El-Faheem et al. (2022) offered two key techniques (reduction and redundancy methods) for enhancing the performance of a radar system based on the Rayleigh distribution. Gamma fractiles and the mean time to failure (MTTF) have been calculated for contrasting the upgraded methods.

Talking about the pulse-radar system, Marcum (1960) in his study presented data from which the probability of the system to detect a given target in any range can be obtained, using three different variables. In another study related to the system, Gill (1979) used the UI Octoson and a frequency-offset pulsed Doppler system to obtain fully quantitative blood flow measurements in deep-lying vessels. Accuracy levels for in vitro flow measurement experiments show 14% rms error and 32% maximum error. The performance of a weather Doppler radar having a staggered pulse repetition time is compared with a radar having a random (but known) phase, as studied by Zrinc and Mahapatra (1985). In the study, a theoretical expression for the signal-to-noise ratio caused by recohering-filtering-recohering for the random phase radar was established, and a statistical analysis of the spectral moment estimations for the staggered scheme was developed. O'Neill and Arcone (1991) presented and overviewed recent activities and results of the use of commercially available short-pulse UHF radar, which is used to survey ice conditions in freshwater bodies. A matched filter system was used to resolve thin ice layers down to a few centimetres and differentiate them from an ice-free condition. A general framework for calculating and comparing the dynamic range of radar systems was developed by Hamran et al. (1995), in which the dynamic range of an impulse radar was contrasted with that of a synthetic pulse radar. Sakamoto and Sato (2004) proposed a phase compensation algorithm for high-resolution phase radar systems, which works well with the seabed algorithm, which estimates shapes of target and is basically based on a reversible transform. Zhang et al. (2007) in their work described a software-defined measurement system that combined pseudorandom (PN) code pulse and frequency modulation continuous wave (FMCW) techniques in novel ways for automotive applications. Yang and Fathy (2009) developed a real-time ultra-wideband (UWB) see-through-wall (STW) imaging radar system. It shows that a workable stand-alone STW system could be created using readily available technologies and commercially available parts with a manageable amount of design work and expenses, which will help in better understanding of the difficult issues related to STW technology. Kim et al. (2011) in their study design a full-digital pulse-Doppler radar-based FPGA and DSP system and implement it for automotive applications, also verifying the 24-GHz pulse-Doppler radar system.

Research by Zheng et al. (2017) focused on how a communication system and a pulsed radar can coexist while using the same bandwidth. The region of the achievable communication rates with and without interference was also discussed in this study. As the technology increases, the risks involved also increase, and they need to be analysed accurately. Cohen et al. (2019) estimated the risks associated with radar systems, and these were discussed along with any potential consequences. Potential future research directions were also pointed out in order to safeguard sophisticated radar systems from cyberattacks. Ridder and Narayanan (2019), in their paper, examined the degradation of typical RF parts which are used in radar system. Further, an overall reliability model of a radar system was developed by using the degradation of the components along with the operational reliability of the radar. Cho and Park (2022) provided a radar-based motion recognition method in which a detailed study of human body movement was recognised by using UWB radar pulses, which were reflected from the body and expressed in micro-range components. These ranges are further converted into images, which are provided with specific motions using the pretrained convolutional neural networks (CNNs). Xiahou et al. (2023) in their study propose a new system structure called modular k-out-of-n system with functional dependency (FDEP), in which the failure of some specific components will disable some components in the k-out-of-n structure. Also, a DBN is developed to update the reliability of a modular k-out-of-n system dynamically, as well as real-world case is studied to check the effectiveness of the proposed method.

8.3 DESCRIPTION OF THE MODEL

The pulse radar system is composed of up to three major components that are linked in series: the sender, transmitter, and receiver. A detailed explanation of how a pulse radar system operates is discussed below, and the configuration of the system is displayed in Figure 8.1.

The pulse modulator initially takes in a low-power radar signal that is later transmitted into space. These modulators generate high-power pulses with high voltage for a transmitter tube when transmitting. Further, modulating the radio wave that the radio transmitter emits results in the formation of the pulse train. In order to provide a sufficient range, the transmitter also amplifies the signal to a high-power level. It is then connected to a duplexer, a microwave switch that sporadically connects the antenna to the transmitter and receiving components. After being connected to the transmitter by the duplexer, the antenna transmits the pulse-modulated signal. The signal received by the antenna is then sent to the amplifier, known as the low-noise RF amplifier. It enhances the poor RF signal that is received by the antenna, and the output is connected to the mixer. It produces an IF (intermediate frequency) pulse out of the RF pulse it receives from the

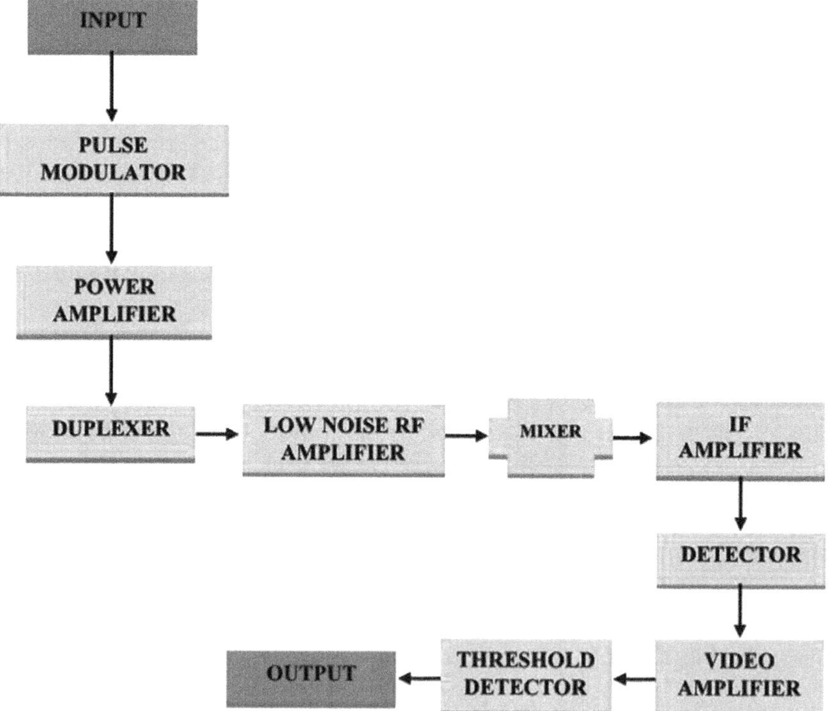

Figure 8.1 Configuration of the system.

low-noise RF amplifier. In the receiver section, the RF amplifier typically acts at the input stage, but occasionally the mixer takes the place of the RF amplifier. The IF amplifier amplifies the IF pulse produced by the mixer circuit. Also, it improves the capacity of the receiver section for echo detection by reducing the impact of unwanted signals. After that, the IF amplifier is linked to the detector. It is a crystal diode that demodulates the signal by separating the carrier from the transmitted signal. The received signal is then amplified by the video amplifier to a level where it can be observed on the screen. Further, the threshold detector determines whether the target has been detected in space. In essence, it has a threshold limit that is established and compared with the magnitude of the signal received. The presence of the target is indicated if the output signal exceeds the threshold value. If not, it is assumed that the space only contains the noise component. The final outcome is then put into action. The amplified video is displayed on the cathode ray tube (CRT) screen.

8.4 ASSUMPTIONS

i. The system is initially thought to be in good operational condition.
ii. The system functions just as new after being repaired.
iii. All the components are assumed to have different failure rates.

8.5 DESCRIPTION OF STATES

The different states depicted in Figure 8.2 are presented in Table 8.1.

8.6 NOTATIONS

The notations that pertain to the work are provided in Table 8.2.

8.7 FORMULATION AND SOLUTION OF THE MATHEMATICAL MODEL

In the following section, the system's transition state probabilities are discussed. The performance indices of the proposed system are assessed using these probabilities. With the help of a schematic of state transitions and the developed Markov model, the following differential equations are generated, and necessary initial and boundary conditions are also obtained:

$$\left(\frac{\partial}{\partial t}+\lambda_A+\lambda_B+\lambda_C+\lambda_D+\lambda_E+\lambda_F+\lambda_G+\lambda_H+\lambda_I\right)P_0(t)=\int_0^\infty\sum_{i=1}^{9}\mu P_i(x,t)dx. \quad (8.1)$$

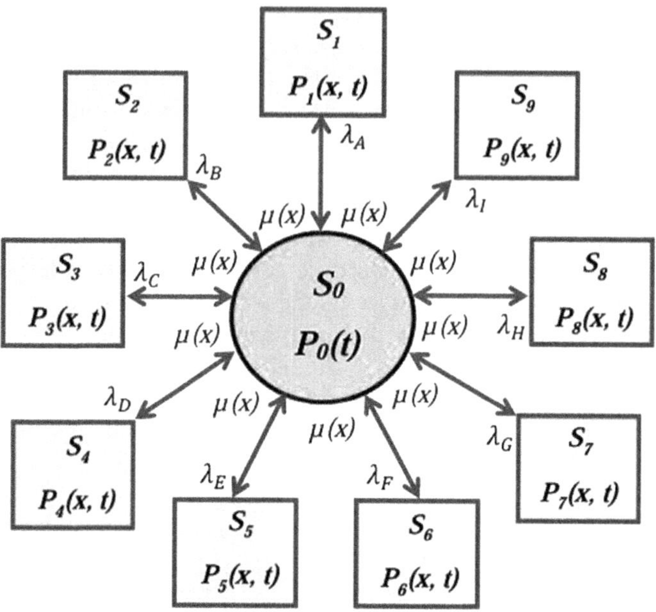

Figure 8.2 Schematic of state transitions.

Table 8.1 Description of States

State	*Description*
S_0	Good state due to proper functioning of all the components.
S_1	Failed state due to breakdown of component 1, i.e. pulse modulator.
S_2	Failed state due to breakdown of component 2, i.e. power amplifier.
S_3	Failed state due to breakdown of component 3, i.e. duplexer.
S_4	Failed state due to breakdown of component 4, i.e. low-noise RF amplifier.
S_5	Failed state due to breakdown of component 5, i.e. mixer.
S_6	Failed state due to breakdown of component 6, i.e. if amplifier.
S_7	Failed state due to breakdown of component 7, i.e. detector.
S_8	Failed state due to breakdown of component 8, i.e. video amplifier.
S_9	Failed state due to breakdown of component 9, i.e. threshold detector.

Boundary conditions,

$$P_1(0,t) = \lambda_A P_0(t). \tag{8.2}$$

$$P_2(0,t) = \lambda_B P_0(t). \tag{8.3}$$

$$P_3(0,t) = \lambda_C P_0(t). \tag{8.4}$$

Table 8.2 Notations

Notation	*Description*
$P_0(t)$	Probability of the system at S_0 state.
$P_i(x,t)$	Probability of the system at S_i state, where i = 1,2,3,4,5,6,7,8,9.
S	Variable of Laplace transform.
$\bar{P}(t)$	Laplace transformation of $P(t)$.
$\mu(x)$	Repair rate of failed states.
λ_A	Failure rate of component 1, i.e. pulse modulator.
λ_B	Failure rate of component 2, i.e. power amplifier.
λ_C	Failure rate of component 3, i.e. duplexer.
λ_D	Failure rate of component 4, i.e. low noise RF amplifier.
λ_E	Failure rate of component 5, i.e. mixer.
λ_F	Failure rate of component 6, i.e. if amplifier.
λ_G	Failure rate of component 7, i.e. detector.
λ_H	Failure rate of component 8, i.e. video amplifier.
λ_I	Failure rate of component 9, i.e. threshold detector.
$P_{up}(t)$	Upstate probability.
$P_{down}(t)$	Downstate probability.

$$P_4(0,t) = \lambda_D P_0(t). \tag{8.5}$$

$$P_5(0,t) = \lambda_E\ P_0(t). \tag{8.6}$$

$$P_6(0,t) = \lambda_E\ P_0(t). \tag{8.7}$$

$$P_7(0,t) = \lambda_G P_0(t). \tag{8.8}$$

$$P_8(0,t) = \lambda_H\ P_0(t). \tag{8.9}$$

$$P_9(0,t) = \lambda_I\ P_0(t). \tag{8.10}$$

Initial condition,

$$P_i(0) = \begin{cases} 1, & i = 0 \\ 0, & i \geq 1 \end{cases} \tag{8.11}$$

Eqs. (8.1)–(8.10) are altered with the help of the Laplace transformation, and using the initial condition, we get,

$$\left(s + \lambda_A + \lambda_B + \lambda_C + \lambda_D + \lambda_E + \lambda_F + \lambda_G + \lambda_H + \lambda_I\right)\overline{P_0}(s) = 1 + \int_0^\infty \sum_{i=1}^{9} \mu \bar{P}_i(x,s)\,dx. \tag{8.12}$$

Boundary conditions,

$$\overline{P_1}(0,s) = \lambda_A \overline{P_0}(s). \tag{8.13}$$

$$\overline{P_2}(0,s) = \lambda_B \overline{P_0}(s). \tag{8.14}$$

$$\overline{P_3}(0,s) = \lambda_C \overline{P_0}(s). \tag{8.15}$$

$$\overline{P_4}(0,s) = \lambda_D \overline{P_0}(s). \tag{8.16}$$

$$\overline{P_5}(0,s) = \lambda_E \overline{P_0}(s). \tag{8.17}$$

$$\overline{P_6}(0,s) = \lambda_F \overline{P_0}(s). \tag{8.18}$$

$$\overline{P_7}(0,s) = \lambda_G \overline{P_0}(s). \tag{8.19}$$

$$\overline{P_8}(0,s) = \lambda_H \overline{P_0}(s). \tag{8.20}$$

$$\overline{P_9}(0,s) = \lambda_I \overline{P_0}(s). \tag{8.21}$$

Now, transition state probabilities are determined by resolving Eq. (8.12) with the aid of Eqs. (8.13)–(8.21). Thus,

$$\overline{P_0}(s) = \frac{1}{\left[(s+C) + C \times \overline{S_\mu}(s)\right]}. \tag{8.22}$$

$$\overline{P_1}(s) = \lambda_A \left(\frac{1-\overline{S_\mu}(s)}{s}\right) \overline{P_0}(s). \tag{8.23}$$

$$\overline{P_2}(s) = \lambda_B \left(\frac{1-\overline{S_\mu}(s)}{s}\right) \overline{P_0}(s). \tag{8.24}$$

$$\overline{P_3}(s) = \lambda_C \left(\frac{1-\overline{S_\mu}(s)}{s}\right) \overline{P_0}(s). \tag{8.25}$$

$$\overline{P_4}(s) = \lambda_D \left(\frac{1-\overline{S_\mu}(s)}{s}\right) \overline{P_0}(s). \tag{8.26}$$

$$\overline{P_5}(s) = \lambda_E \left(\frac{1-\overline{S_\mu}(s)}{s}\right) \overline{P_0}(s). \tag{8.27}$$

$$\overline{P_6}(s) = \lambda_F\left(\frac{1-\overline{S_\mu}(s)}{s}\right)\overline{P_0}(s). \tag{8.28}$$

$$\overline{P_7}(s) = \lambda_G\left(\frac{1-\overline{S_\mu}(s)}{s}\right)\overline{P_0}(s). \tag{8.29}$$

$$\overline{P_8}(s) = \lambda_H\left(\frac{1-\overline{S_\mu}(s)}{s}\right)\overline{P_0}(s). \tag{8.30}$$

$$\overline{P_9}(s) = \lambda_I\left(\frac{1-\overline{S_\mu}(s)}{s}\right)\overline{P_0}(s). \tag{8.31}$$

Now, the Laplace transformations of upstate and downstate system probabilities are computed as

$$\overline{P_{up}}(s) = \overline{P_0}(s). \tag{8.32}$$

$$\begin{aligned}\overline{P_{down}}(s) &= \overline{P_1}(s)+\overline{P_2}(s)+\overline{P_3}(s)+\overline{P_4}(s)+\overline{P_5}(s)+\overline{P_6}(s)+\overline{P_7}(s)+\overline{P_8}(s)+\overline{P_9}(s)\\ \overline{P_{down}}(s) &= \left[\left(\frac{1-\overline{S_\mu}(s)}{s}\right)(\lambda_A+\lambda_B+\lambda_C+\lambda_D+\lambda_E+\lambda_F+\lambda_G+\lambda_H+\lambda_I)\overline{P_0}(s)\right]\end{aligned}. \tag{8.33}$$

8.8 COMPUTATION OF IMPORTANT RELIABILITY MEASURES

8.8.1 Computation of availability

Availability of a system is a vital reliability characteristic that measures the amount of time over which a system is operational and able to carry out its intended function. It is typically expressed as a percentage of the total time that the system ought to be readily available. For the availability of this system, putting the failure rates as $\lambda_A = 0.014$, $\lambda_B = 0.022$, $\lambda_C = 0.013$, $\lambda_D = 0.017$, $\lambda_E = 0.021$, $\lambda_F = 0.030$, $\lambda_G = 0.028$, $\lambda_H = 0.015$, $\lambda_I = 0.016$, and repair rate, $\mu = 1$ in Eq. (8.32) and taking inverse Laplace transformation, we obtain

$$A(t) = 0.85034 + 0.14965\ e^{(-1.17600t)}. \tag{8.34}$$

For obtaining numerical values of availability, alter time interval t from 0 to 30 in Eq. (8.34), which is provided in Table 8.3 and displayed graphically in Figure 8.3.

Table 8.3 Availability of radar system

Time (t)	*Availability*
0	1.00000
1	0.89651
2	0.86458
3	0.85473
4	0.85169
5	0.85075
6	0.85046
7	0.85037
8	0.85035
9	0.85034
10	0.85034
11	0.85034
12	0.85034
13	0.85034
14	0.85034
15	0.85034
16	0.85034
17	0.85034
18	0.85034
19	0.85034
20	0.85034
21	0.85034
22	0.85034
23	0.85034
24	0.85034
25	0.85034
26	0.85034
27	0.85034
28	0.85034
29	0.85034
30	0.85034

8.8.2 Computation of reliability

Reliability is the measure of the consistency and stability of a system or component's performance over time and the ability to maintain that performance under changing conditions or demands. Here, the reliability of the pulse radar system is given as

$$R(t) = e^{\left(-(\lambda_A+\lambda_B+\lambda_C+\lambda_D+\lambda_E+\lambda_F+\lambda_G+\lambda_H+\lambda_I)t\right)}. \tag{8.35}$$

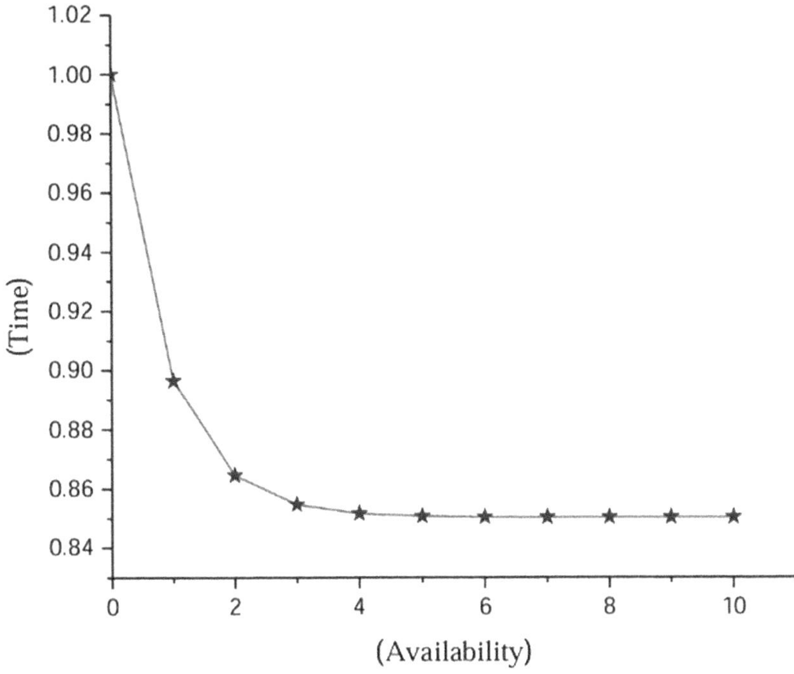

Figure 8.3 Availability as a function of time.

Setting repair rate as $\mu = 0$ and failure rates as $\lambda_A = 0.014$, $\lambda_B = 0.022$, $\lambda_C = 0.013$, $\lambda_D = 0.017$, $\lambda_E = 0.021$, $\lambda_F = 0.030$, $\lambda_G = 0.028$, $\lambda_H = 0.015$, and $\lambda_I = 0.016$ in Eq. (8.32), then taking inverse Laplace of it, we get

$$R(t) = e^{(-0.17600\ t)}. \tag{8.36}$$

Now, for acquiring reliability metrics, alter time interval t in Eq. (8.36) as $t = 0, 1, 2, \ldots, 30$. The observed values are shown in Table 8.4, and time vs. reliability relationship is demonstrated in Figure 8.4.

8.8.3 Computation of mean time to failure (MTTF)

MTTF is a measure of the reliability of a system or component, calculated by dividing the total operating time by the number of failures that occur during that time. For calculating the MTTF of pulse radar system, putting repair rate as $\mu = 0$ and taking limit tends to zero in Eq. (8.32), we get

$$MTTF = \lim_{s \to 0} \overline{P_{up}}(s).$$

Table 8.4 Reliability of radar system

Time (t)	*Reliability*
0	1.00000
1	0.83861
2	0.70328
3	0.58978
4	0.49460
5	0.41478
6	0.34784
7	0.29170
8	0.24463
9	0.20515
10	0.17204
11	0.14427
12	0.12099
13	0.10146
14	0.08509
15	0.07136
16	0.05984
17	0.05018
18	0.04208
19	0.03529
20	0.02959
21	0.02482
22	0.02081
23	0.01745
24	0.01463
25	0.01227
26	0.01029
27	0.00863
28	0.00724
29	0.00607
30	0.00509

Therefore,

$$MTTF = \frac{1}{\lambda_A + \lambda_B + \lambda_C + \lambda_D + \lambda_E + \lambda_F + \lambda_G + \lambda_H + \lambda_I}. \tag{8.37}$$

For determining MTTF with regard to all the failure rates, setting them as $\lambda_A = 0.014$, $\lambda_B = 0.022$, $\lambda_C = 0.013$, $\lambda_D = 0.017$, $\lambda_E = 0.021$, $\lambda_F = 0.030$, $\lambda_G = 0.028$, $\lambda_H = 0.015$, and $\lambda_I = 0.016$ and varying all of them one by one as 0.10, 0.15, 0.20, 0.25, 0.30, 0.35, 0.40, 0.45, 0.50, 0.55, 0.60, 0.65, 0.70, 0.75,

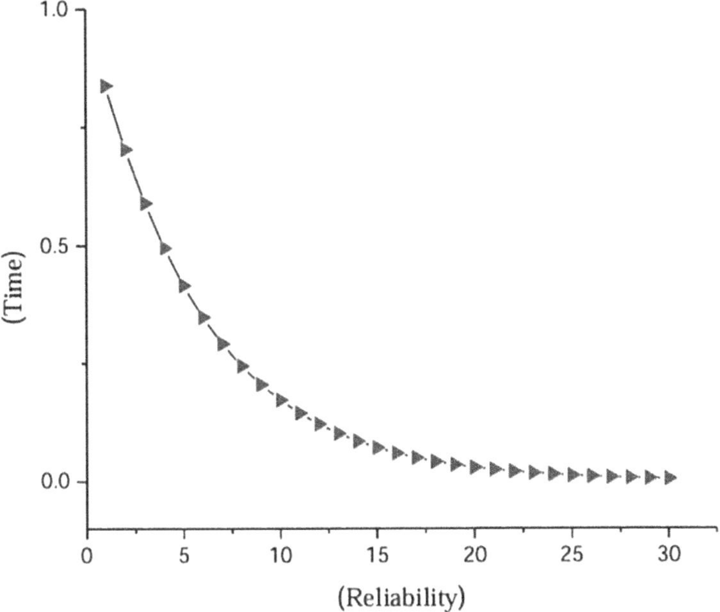

Figure 8.4 Reliability as a function of time.

0.80, 0.85, 0.90, and 0.95, respectively, in Eq. (8.32). The values obtained are provided in Table 8.5 and graphically shown in Figure 8.5.

8.9 COST ANALYSIS

Cost serves a significant role in determining how reliable a system is. Expected profit is of the utmost importance for maintaining the reliability of the system. Assuming that the service facility remains available, the expected profit over the interval [0, t) is given as follows:

$$E_p(\mathrm{t}) = k_1 \int_0^t P_{up}(t)dt - tk_2. \tag{8.38}$$

where k_1 and k_2 represent, respectively, the revenue cost and the service cost per unit of time.

For computing the expected profit of the system, using Eq. (8.32) in Eq. (8.38), we obtain

$$E_p(\mathrm{t}) = k_1\left(0.85034\ t\ -\ 0.12726\ e^{(-1.17600)} + 0.12726\right) - tk_2. \tag{8.39}$$

Table 8.5 MTTF of radar system with failure rates

Variation in failure rates	*MTTF*								
	λ_A	λ_B	λ_C	λ_D	λ_E	λ_F	λ_G	λ_H	λ_I
0.10	12.69841	3.93700	3.80228	3.86100	3.92156	4.06504	4.03225	3.83141	3.84615
0.15	9.10620	3.28947	3.19488	3.23624	3.27868	3.37837	3.35570	3.21543	3.22580
0.20	7.06168	2.82485	2.75482	2.78551	2.81690	2.89017	2.87356	2.77008	2.77777
0.25	5.75293	2.47524	2.42130	2.44498	2.46913	2.52525	2.51262	2.43309	2.43902
0.30	4.84713	2.20264	2.15982	2.17864	2.19780	2.24215	2.23214	2.16919	2.17391
0.35	4.18459	1.98412	1.94931	1.96463	1.98019	2.01612	2.00803	1.95694	1.96078
0.40	3.67965	1.80505	1.77619	1.78890	1.80180	1.83150	1.82481	1.78253	1.78571
0.45	3.28242	1.65562	1.63132	1.64203	1.65289	1.67785	1.67224	1.63666	1.63934
0.50	1.51057	1.52905	1.50829	1.51745	1.52671	1.54798	1.54320	1.51285	1.51515
0.55	1.40449	1.42045	1.40252	1.41043	1.41843	1.43678	1.43266	1.40646	1.40845
0.60	1.31233	1.32625	1.31061	1.31752	1.32450	1.34048	1.33689	1.31406	1.31578
0.65	1.23152	1.24378	1.23001	1.23609	1.24223	1.25628	1.25313	1.23304	1.23456
0.70	1.16009	1.17096	1.15874	1.16414	1.16959	1.18203	1.17924	1.16144	1.16279
0.75	1.09649	1.10619	1.09529	1.10011	1.10497	1.11607	1.11358	1.09769	1.09890
0.80	1.03950	1.04821	1.03842	1.04275	1.04712	1.05708	1.05485	1.04058	1.04166
0.85	0.98814	0.99601	0.98716	0.99108	0.99502	1.00401	1.00200	0.98911	0.99009
0.90	0.94161	0.94876	0.94073	0.94428	0.94786	0.95602	0.95419	0.94250	0.94339
0.95	0.89928	0.90579	0.89847	0.90171	0.90497	0.91240	0.91074	0.90009	0.90090

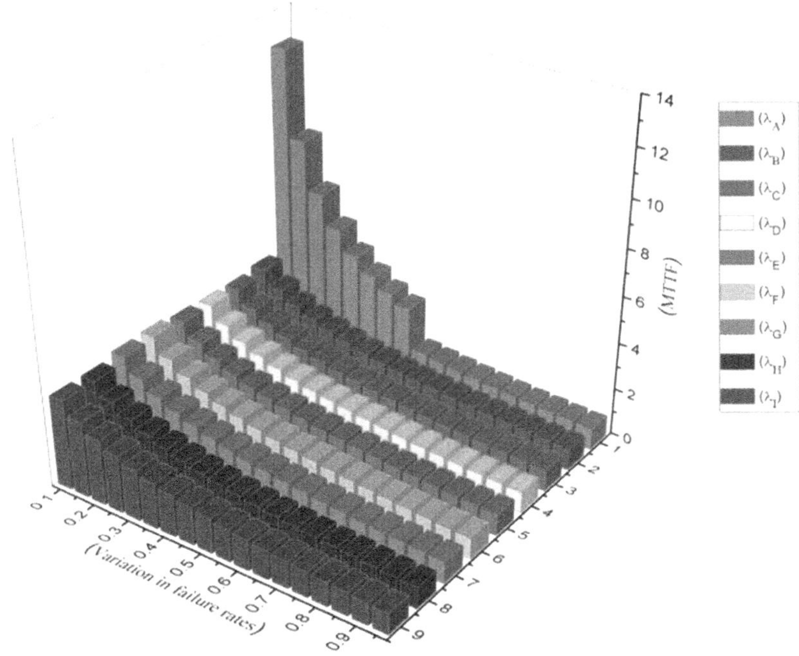

Figure 8.5 MTTF of radar system with respect to failure rates.

Now, putting $k_1 = 1$, $k_2 = 0.1, 0.2, 0.3, 0.4, 0.5$ and taking $t = 0, 1, 2, 3, 4, 5, 6, 7, 8, 9$ units, respectively, in Eq. (8.39). The computed values of $E_p(t)$ are provided in Table 8.6, and the corresponding graph is shown in Figure 8.6.

Table 8.6 Cost as a function of time

	Expected profit				
Time(t)	$k_2 = 0.1$	$k_2 = 0.2$	$k_2 = 0.3$	$k_2 = 0.4$	$k_2 = 0.5$
0	0	0	0	0	0
1	0.83834	0.73834	0.63834	0.53834	0.43834
2	1.61582	1.41582	1.21582	1.01582	0.81582
3	2.37454	2.07454	1.77454	1.47454	1.17454
4	3.12746	2.72746	2.32746	1.92746	1.52746
5	3.87860	3.37860	2.87860	2.37860	1.87860
6	4.62919	4.02919	3.42919	2.82919	2.22919
7	5.37960	4.67960	3.97960	3.27960	2.57960
8	6.12997	5.32997	4.52997	3.72997	2.92997
9	6.88031	5.98031	5.08031	4.18031	3.28031
10	7.63066	6.63066	5.63066	4.63066	3.63066

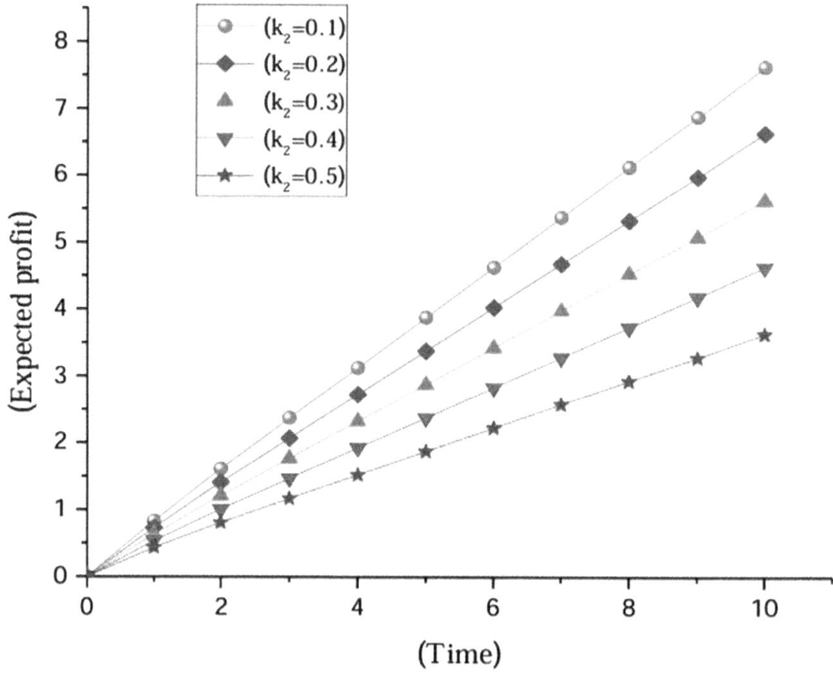

Figure 8.6 Expected profit as a function of time.

8.10 RESULT DISCUSSION

In this present investigation, authors have examined the important reliability measures of a pulse radar system. Following implementation of our thorough approach to the considered system, authors have obtained the following significant outcomes, which are discussed as follows:

- The probability of failure rises for all failure rate values taken into consideration, causing a decline in the pulse radar system's availability. From the values obtained in Table 8.3, it is observed that initially, the availability decreases very strictly, and after $t = 9$ units, it becomes constant.
- Table 8.4 demonstrates that a change in time results in a change in the system's reliability. Figure 8.4, i.e. the time-reliability graph, depicts how the system's reliability declines exponentially consistently over time and eventually tends to zero.
- By considering the variations in all the failure rates as shown in Table 8.5, the MTTF of the system is computed. The graphs in Figure 8.5 show that the MTTF of the pulse radar system with respect to λ_A initially increases and then falls until 0.45 units, and after that it almost becomes constant. In contrast, it gradually declines in the context of other failure rates.

- The analysis of expected profit in relation to service cost as a function of time is depicted in Table 8.6. A critical analysis of the graph in Figure 8.6 reveals that the system yields profits that increase over time, but they decrease as service costs rise.

8.11 CONCLUSION

The present research investigates the reliability indices of a pulse radar system. To model the entire system, the authors employed the Laplace transform and the supplementary variable approach. From the results discussed in Section 8.10, it can be concluded that the reliability and availability of a pulse radar system are affected by the failure rates of its components. The probability of failure increases with time, causing a decline in the system's availability. The reliability of the system also decreases exponentially over time, eventually tending to zero. The MTTF of the system varies with different failure rates, increasing initially and then falling until a certain point, after which it becomes almost constant. The analysis of expected profit shows that the system generates increasing profits over time, but this trend is affected by the rising service costs. Overall, these observations suggest that the pulse radar system's performance is impacted by its component failure rates and maintenance costs over time, and these factors need to be considered when making decisions about its maintenance and operation.

REFERENCES

Bartko, J. J., & Carpenter, W. T. (1976). On the methods and theory of reliability. *Journal of Nervous and Mental Disease*, *163*(5), 307–317

Basinger, M., Montalto, F., & Lall, U. (2010). A rainwater harvesting system reliability model based on nonparametric stochastic rainfall generator. *Journal of Hydrology*, *392*(3–4), 105–118.

Boudali, H., & Dugan, J. B. (2005). A discrete-time Bayesian network reliability modeling and analysis framework. *Reliability Engineering & System Safety*, *87*(3), 337–349.

Cai, B., Liu, Y., Ma, Y., Liu, Z., Zhou, Y., & Sun, J. (2015). Real-time reliability evaluation methodology based on dynamic Bayesian networks: A case study of a subsea pipe ram BOP system. *ISA Transactions*, *58*, 595–604.

Cho, H. S., & Park, Y. J. (2022). Classification of human body motions using an ultra-wideband pulse radar. *Technology and Health Care*, *30*(1), 93–104.

Cohen, S., Gluck, T., Elovici, Y., & Shabtai, A. (2019, November). Security analysis of radar systems. In *Proceedings of the ACM Workshop on Cyber-Physical Systems Security & Privacy*, 3–14.

El-Faheem, A. A., Mustafa, A., & Abd El-Hafeez, T. (2022). Improving the reliability performance for radar system based on Rayleigh distribution. *Scientific African*, *17*, 1–12.

Gill, R. W. (1979). Pulsed Doppler with B-mode imaging for quantitative blood flow measurement. *Ultrasound in Medicine & Biology*, *5*(3), 223–235.

Hamran, S. E., Gjessing, D. T., Hjelmstad, J., & Aarholt, E. (1995). Ground penetrating synthetic pulse radar: Dynamic range and modes of operation. *Journal of Applied Geophysics*, *33*(1–3), 7–14.

Hollnagel, E. (1992). The reliability of man-machine interaction. *Reliability Engineering & System Safety*, *38*(1–2), 81–89.
Kim, S. D., Ju, Y., & Lee, J. H. (2011, January). Design and implementation of a full-digital pulse-Doppler radar system for automotive applications. In *2011 IEEE International Conference on Consumer Electronics (ICCE)*, 563–564. IEEE.
Kozhokhina, O. V., Blahaia, L. V., Rudas, S. I., & Alexeiev, O. M. (2016, May). Informational reliability of radar system operator. In *2016 17th International Radar Symposium (IRS)*, 1–4.
Langseth, H., & Portinale, L. (2007). Bayesian networks in reliability. *Reliability Engineering & System Safety*, *92*(1), 92–108.
Li, X., & Wang, L. (2014, October). Research on the reliability testing method of radar system simulation software. In *2014 12th International Conference on Signal Processing (ICSP)*, 2411–2415. IEEE.
Marcum, J. (1960). A statistical theory of target detection by pulsed radar. *IRE Transactions on Information Theory*, *6*(2), 59–267.
O'Neill, K., & Arcone, S. A. (1991). Investigations of Freshwater and Ice Surveying Using Short-Pulse Radar. Hanover, NH, Cold Regions Research and Eng. Lab. (CRREL) Rep. 91–15.
Park, J., Liang, W., Choi, J., El-Keib, A. A., Shahidehpour, M., & Billinton, R. (2009, July). A probabilistic reliability evaluation of a power system including solar/photovoltaic cell generator. In *2009 IEEE Power & Energy Society General Meeting*, 1–6.
Ram, M. (2013). On system reliability approaches: A brief survey. *International Journal of System Assurance Engineering and Management*, *4*, 101–117.
Reihani, E., Najjar, M., Davodi, M., & Norouzizadeh, R. (2010, December). Reliability based generator maintenance scheduling using hybrid evolutionary approach. In *2010 IEEE International Energy Conference*, 847–852.
Ridder, T. D., & Narayanan, R. M. (2019, May). Total reliability of radar systems: incorporating component degradation effects in operational reliability. In *Radar Sensor Technology XXIII, 11003*, 270–279.
Ridder, T. D., & Narayanan, R. M. (2018, July). Operational reliability of radar systems. In *NAECON 2018-IEEE National Aerospace and Electronics Conference*, 561–567.
Sakamoto, T., & Sato, T. (2004). A phase compensation algorithm for high-resolution pulse radar systems. *IEICE Transactions on Communications*, *87*(11), 3314–3321.
Xiahou, T., Zheng, Y. X., Liu, Y., & Chen, H. (2023). *Reliability Engineering & System Safety*, *233*, 109120.
Yang, Y., & Fathy, A. E. (2009). Development and implementation of a real-time see-through-wall radar system based on FPGA. *IEEE Transactions on Geoscience and Remote Sensing*, *47*(5), 1270–1280.
Zhang, H., Li, L., & Wu, K. (2007, October). 24GHz software-defined radar system for automotive applications. In *2007 European Conference on Wireless Technologies*, 138–141.
Zhang, J., Zhao, Y., & Ma, X. (2019). A new reliability analysis method for load-sharing *k*-out-of-*n*: F system based on load-strength model. *Reliability Engineering & System Safety*, *182*, 152–165.
Zheng, L., Lops, M., Wang, X., & Grossi, E. (2017). Joint design of overlaid communication systems and pulsed radars. *IEEE Transactions on Signal Processing*, *66*(1), 139–154.
Zrinc, D., & Mahapatra, P. (1985). Two methods of ambiguity resolution in pulse Doppler weather radars. *IEEE Transactions on Aerospace and Electronic Systems*, *21*(4), 470–483.

Chapter 9

Reliability analysis of multiple cylinder system with MADM techniques in diesel engine, employed in 3 out of 5 subsystems

Yamuna B, Radha Gupta, Geetha N K, and Suresh H N

9.1 INTRODUCTION: BACKGROUND

Diesel Engines are responding to the legislation on emission norms and duly fulfilling the needs of society; diesel engines keep themselves side by side in the modern world. A high level of performance, reliability and optimized consumption are an added advantage to diesel engines.

In this study, we have used 12 different diesel engine cylinders to analyse the cylinder efficiency to get the better performance of the Diesel Engine. It is quite difficult to increase a complicated system's reliability just by using high-quality components.

It is often desirable to process redundancy and backup modules and subsystems in order to improve or enhance the overall reliability of the systems. A redundant system having "n" components connected in parallel of which minimum of "k" should be operational for the success of the system is referred as *k out of n* systems, where "$(n–k)$" components are the redundant components. Parallel redundancy is a specific case of *k out of n* configuration. For the system to function with this kind of configuration, at least *k of the n* parallel components must be successful. The case study considered here is of a diesel engine with 12 cylinders. In a 12 Diesel Engine cylinders, it is highly impossible to drive the vehicle if less than 3 cylinders are firing, but if more than 3 cylinders fire it will be possible to drive the vehicle.

While incorporating redundancy into the system, the main goal is to find the values of n & k under the restrictions of Cost & Weight of the components.

This primary goal of this study is to examine *3 out of 5* configurations to maximize the number of cylinders in diesel engine system with optimal reliability while keeping costs and weight as constraints. These constraints have been considered as polynomial function as well as exponential functions. The use of the GRG nonlinear, Evolutionary and Graph theory matrix techniques are offered for the investigation. The comparison between these techniques is helpful in assessment of the best possible set of operating parameters for a diesel engine. Graph Theory Matrix Approach is taken into consideration to gauge optimal combination of operating parameters i.e., Cylinder combination from Cost and Weight constraints. A computer program was generated to simplify the

DOI: 10.1201/9781003546214-9

outcome Graph Theory Matrix Approach helps to find the performance index of the cylinders of a diesel engine. This performance index is used to determine the system reliability of diesel engine comprising of *3 out of 5* cylinders. The GRG nonlinear, Evolutionary techniques work with population of solutions and provide the best amongst the many. The diesel engine reliability obtained by all the three techniques is comparable however GRG nonlinear and Evolutionary techniques give more promising results whereas GTMA gives the optimal combination of operating parameters. On the other hand, after identifying the operating parameters using GTMA, the reliability model is solved with the help of Excel Solver using GRG Nonlinear and Evolutionary techniques.

The diesel engine that is considered here has identical and statistically independent components. Hence the Binomial distribution is used to determine the reliability. If p is the probability of success of each component, then the probability of x success out of n components is

$$P(n,x) = C_x^n p^x (1-p)^{n-x}$$

If k is the minimum number of components needed for system to be successful then the system will be successful if k, $k + 1$, $k + 2$,... components are successful. Hence the system reliability is given by $R = \Sigma_{i=k}^{n} C_i^n p^i (1-p)^{n-i}$.

Reliability characteristics of a system and individual components of the subsystems form the basis for the stability of a system. The reliability of a component can be increased by adding one or more similar components in parallel.

9.2 LITERATURE SURVEY

The optimization of redundancy allocation and graph theory approaches is continued with various research activity. Chiang and Niu [1] introduces a formula of recursive for computing the exact reliability of a consecutive system, along with sharp upper and lower bounds. Barlow and Heidtmann [2] the reliability calculations for a system with k out of n subsystem were initially proposed by them. Derman, Lieberman and Ross [3] which entails a sequence of n ordered components where failure occurs if consecutive k components fail. This system exhibits series and parallel characteristics when k equals 1 and n, respectively, and can be represented in linear or circular topologies, necessitating the distinction between the two variants. Hwang [4] introduces a recursive formula to compute the exact reliability of consecutive- *k out of n*:F systems, providing precise upper and lower bounds. Shantikumar [5] the reliability of consecutive systems, offering explicit solutions for k = 2 in linear or cyclic configurations under the i.i.d. assumption and providing precise lower and upper bounds for k ≥ 3, demonstrated across various parameters, with exact solutions given for specific values of r.

Jain and Gopal [6] In a *k*-to-1-out of *n*:G system, success relies on having precisely between k and l units out of n operational, with failure occurring if fewer than k or more than l units operate simultaneously. Risse [7] various methods for determining the reliability of systems, particularly focusing costs compared to existing algorithms. Sarje and Prasad [8] computationally efficient algorithm was used reliability of the system. Ge and Wang [9] to compute the reliability of system with homogeneous Markov dependence, emphasizing the reliability of systems based solely on the state of each component. Suich and Patterson [10] optimize redundancy levels to minimize total costs and average losses due to failure, thereby aiding in decision-making for subsystem design.

Pham and Malon [36] determining optimal configurations to minimize mean system cost and studying the behaviour of optimal k in relation to failure probabilities and maximizing majority system reliability. Yamamoto and Miyakawa [12] it establishes a crucial result regarding the system's reliability in large-scale scenarios, shedding light on its behaviour as the number of components increases infinitely. Coit and Smith [13] genetic algorithm is used to determining optimal design configurations while considering various constraints such as cost and weight. Pham [11] mathematically identify the cost efficient competent to minimize the mean cost

Koutras [14]. Incorporate dual failure modes and offers recursive formulas to evaluate its reliability. Venkatasamy and Agrawa [15], Huang, Zuo and Wu [16] defining conditions for system states based on the minimum number of components in each state and providing presenting algorithms for reliability evaluation. Rao [17] using Graph Theory Matrix Approach, which is used to find ranking of the technical institutions.

Gandhi and Agrawal, FMEA [18]. Grover, Agrawal and Khan [19] considered human factors, behavioural factors.

Chen and Yang [20] worked with calculation and minimal cuts & paths generation, along with investigating reliability bounds for systems with s-dependent component failures.

Zuo and Tian [29] introduces FMCI approach to evaluate system state distributions in generalized multi-state *k out of n*:F systems.

Tian, Zuo, and Yam [22] Accommodating multiple performance levels of components and systems, with proposed models allowing for varying requirements and practical engineering applications, demonstrated through interpretation.

Anand and Wani [23] outlines a step-by-step procedure for evaluating product life-cycle design at the conceptual stage, utilizing digraph and matrix methods to assess life-cycle design attributes and calculate a corresponding index. Sridhar et al. [24] addresses the optimization of a redundant Integrate Reliability Model (IRM) for *k out of n* configuration systems, incorporating multiple constraints beyond conventional cost considerations, such as weight, volume, size and space, to comprehensively optimize system reliability in real-life scenarios. Lakshminarayana and Vijay Kumar [25] their work lies in extending the optimization of integrated reliability models beyond cost constraints to include volume and weight constraints for redundant systems, employing both Lagrangean multiplier method and dynamic programming to achieve more accurate results, while also integrating failure modes effects and criticality analysis (FMECA) for comprehensive assessment [26].

Watada and Melo [27] introduces an exact branch-and-price algorithm for the redundancy allocation problem (RAP) with the combination of redundancy strategy.

Singh and Kumar [28] introduces a parallel big bang big crunch approach for reliability optimization, validated in MATLAB with two examples, showcasing faster convergence and higher accuracy compared to seven other algorithms, including CPLEX, Genetic Algorithm, Ant Colony Optimization, Immune Algorithm, Tabu Search, Hybrid Parallel Genetic Algorithm and the traditional Big Bang Big Crunch approach.

Geetha and Sekhar [21] demonstrates how Graph theory and matrix approach serve as decision-making tools, applied to finding the optimal operating parameters for a diesel engine, with further analysis using Fuzzy set theory.

Ramesh and Gupta [30] demonstrate solving Redundancy allocation problems with a more generalized approach, accommodating *k out of n* subsystems and maximizing system reliability while considering cost and weight constraints, utilizing an integrated reliability model to compute system reliability based on stage and component reliabilities, employing Lagrange's method of undetermined multipliers, and comparing results with the Random Search Technique for global optimization (RST). Rykov, Kochueva and Farkhadov [31] used a mathematical model of a *k out of n* system for decision-making in preventive maintenance of underwater pipeline monitoring systems, considering failure location dependence and proposing a novel solution method based on distribution of failing components, with numerical examples demonstrating theoretical findings applied to specific equipment. Hu, Wang and Xing [32] introduces a new method for analyzing the reliability of non-repairable phased-mission systems (PMS) with varying k-out-of-n structure and phase-AND requirements, offering a multi-valued decision diagram (MDD)-based approach that considers all mission phases. Rykov, Ivanova, and Kochetkova [33] concentrated on the reliability of load-sharing, where component failures increase load on others, altering the function of hazard rate and calculates functional reliability and other characteristics. Yamuna, Gupta, Ramesh and Geetha [35] employed both Evolutionary algorithms and the Generalized Reduced Gradient method (GRG nonlinear Solver) in Excel to optimize the redundancy allocation problem for mechanical systems with *k out of n* subsystems, aiming to maximize system reliability while considering constraints on cost and weight [34].

9.3 METHODOLOGY

The Generalized Reduced Gradient (GRG) nonlinear approach is primarily used for resolving nonlinear optimization issues. For nonlinear functions, it is assumed that both the objective function and the restrictions are to variables. The GRG approach is quicker than the evolutionary approach [35]. While solving a particular problem, the solver produces three distinct types of reports. These reports provide boundaries, sensitivity and answers. The values of the objective function, choice variables and restrictions limiting the problem's specifics are displayed in the response report. The Lagrange multiplier variables and the restrictions reduced gradient are displayed in the sensitivity report. The variables upper and lower bounds are displayed in the limit report.

Excel's evolutionary approach: Genetic algorithm is modelled by the solver. This approach looks for solutions that are either optimal or nearly optimal. Comparable to the GRG nonlinear approach, the evolutionary method is slow yet successful because it deals with a large population of solutions and a large search space [36]. The solver looks in two distinct report types—the population report and the answer—for the solution to the given problem [37]. The values of the problem's objective function, decision variables, constraints and binding information are provided in the response report. The population report provides the mean value and standard deviation of the variables and constraints.

Graph Theory Matrix Approach (GTMA): Graph Theory Matrix Approach is adopted to find the optimal combination of operating parameters of Diesel Engine's cylinder combination from Cost and Weight Constraints. This is attained by determining the performance index of the system parameters. Further the approach can also be used for finding the system reliability.

9.4 RELIABILITY MODEL I

The reliability model pertaining to diesel engine having 12 cylinders with the objective function of maximizing the reliability of system with the cost and weight constraints is

$$\begin{aligned}
&\text{Maximize } Q_s = \prod_{j=1}^{n} Q_{i}, \text{ where } Q_i = \sum_{k=2}^{x_j} \binom{x_j}{k} q_i^k (1-q_i)^{x_j-k} \\
&\text{Subject to: } \sum_{j=1}^{n} c_j x_j \le C_0 \\
&\sum_{j=1}^{n} w_j x_j \le W_0 \\
&x_j \ge 0 \ \forall \ j
\end{aligned} \tag{9.1}$$

Where Q_s System Reliability; Q_j Reliability of each Phase q_j Reliability of the Component x_j No. of components in phase, j is assumed to be constant c_j Coefficient of cost of each component in j^{th} stage w_j is the Coefficient of weight of each component in j^{th} stage C_0 is the Highest permissible system expenditure W_0 is the Upper limit on system weight.

9.5 PROBLEM FORMULATION

This study delves into the decision factors involved in optimizing reliability, including the overall number of redundancies, the reliability levels of individual components quantified as real numbers or a blend of these elements. Specifically, it focuses on developing integrated reliability models for 12-cylinder diesel engines. The analysis includes examining 3 out of 5 cylinders, where the phases are statistically independent when arranged in series and the components x_j within each phase are also statistically independent.

Case (i): The cost coefficient of each component in j^{th} stage is determined by the relationship between reliability and cost expressed as $q_j = \left(\frac{c_j}{g_j}\right)^{\frac{1}{h_j}}$ *ie* $c_j = g_j q_j^{h_j}$ under the cost constraint. Similarly, the relationship between reliability and weight is given by $w_j = u_j q_j^{v_j}$ as the weight constraint. Given that x_j is linear in cost constraint, substituting these relations into mathematical model I yields following equations for the constraints:

$$\sum_{j=1}^{n}[(g_j q_j^{h_j})x_j] - C_0 \leq 0 \text{ And } \sum_{j=1}^{n}[(u_j q_j^{v_j})x_j] - W_0 \leq 0$$

Hence, Model I is transformed to as **Reliability Model II:**

$$Q_s = \prod_{j=1}^{n}\sum_{k=2}^{x_j}\binom{x_j}{k} q_j^k (1-q_j)^{x_j-k}$$

$$\textit{Subjected to} \quad \sum_{j=1}^{n}[(g_j q_j^{h_j})x_j] - C_0 \leq 0 \tag{9.2}$$

$$\sum_{j=1}^{n}[(u_j q_j^{v_j})x_j] - W_0 \leq 0$$

with non negativity restriction $q_j \geq 0$

Case (ii): the cost coefficient of each component in j^{th} phase is determined by the relationship between reliability and cost expressed as $c_j = g_j e^{\left(\frac{h_j}{1-q_j}\right)}$ under the cost constraint. Similarly, the relationship between reliability and weight is given by $w_j = u_j e^{\left(\frac{v_j}{1-q_j}\right)}$ as the weight constraint.

Substituting these relations into mathematical model I yields following equations for the constraints:

$$\sum_{j=1}^{n}\left[\left(g_j e^{\left(\frac{h_j}{1-q_j}\right)}\right)x_j\right]-C_0 \le 0$$

$$\sum_{j=1}^{n}\left[\left(u_j e^{\left(\frac{v_j}{1-q_j}\right)}\right)x_j\right]-W_0 \le 0$$

Hence, Model I is transformed to **Reliability Model III:**

$$Q_s = \prod_{j=1}^{n}\sum_{k=2}^{x_j}\binom{x_j}{k} q_j^k (1-q_j)^{x_j-k}$$

$$\textit{Subject to} \quad \sum_{j=1}^{n}\left[\left(g_j e^{\left(\frac{h_j}{1-q_j}\right)}\right)x_j\right]-C_0 \le 0 \qquad (9.3)$$

$$\sum_{j=1}^{n}\left[\left(u_j e^{\left(\frac{v_j}{1-q_j}\right)}\right)x_j\right]-W_0 \le 0$$

Non negativity restriction $q_j \ge 0$

Case (i): The polynomial function $c_j = g_j q_j^{h_j}$ used in solving reliability Model II lab data in Table 9.1 has been taken to find the reliability of the system by using excel solver.

Case (ii): The exponential function $c_j = g_j e^{\left(\frac{h_j}{1-q_j}\right)}$ used in solving Reliability Model III lab data in Table 9.2 has been taken to find the reliability of the system by using excel solver.

Table 9.1 Cost and Weight coefficients for polynomial function

	Cost coefficients		*Weight coefficients*	
STAGE(j)	g_j	h_j	u_j	v_j
1	22	2	3126	2
2	29	3	4168	3
3	36	4	5210	4

Table 9.2 Cost and Weight coefficients for exponential function

	Cost coefficients		Weight coefficients	
STAGE(j)	g_j	h_j	u_j	v_j
1	100	0.6	3126	0.6
2	120	0.5	4168	0.5
3	94	0.4	5210	0.4

9.6 NUMERICAL RESULTS

The aim of this work is to maximize the reliability of a diesel engine systems performance by targeting $3\,out\,of\,5$ subsystems, which are the cylinders in this context. To achieve this, mathematical relationships are utilized to link constraints like cost and weight with component reliability, thereby enhancing the overall system reliability. The reliability of the system, individual components and stages for Models II & III is determined through the application of Evolutionary and GRG nonlinear Solver methods within Excel Solver, alongside GTMA.

The polynomial function $c_j = g_j q_j^{h_j}$ used in Reliability Model II, maximum for the system cost of Rs 90 and weight 14,000 kgs is considered, to determine optimum reliability of each component (q_j), Reliability of Phase (Q_i) and the Reliability of System (Q_s).

Table 9.3 & 9.4 gives the System Reliability of Diesel Engines using Excel solver.

Table 9.3 System reliability for reliability model II using Evolutionary method

STAGE(j)	q_j	Q_j	x_j	c_j	w_j	x_jc_j	x_jw_j
1	0.8069	0.9473	2	14.325	2035.414	28.649	4070.83
2	0.7190	0.8612	3	10.781	1549.431	32.342	4648.29
3	0.6699	0.7950	4	7.252	1049.550	29.009	4198.20
Total cost and total weight						90.000	12917.32
System reliability (Q_s)						0.6485	

Table 9.4 System reliability for reliability model II using GRG nonlinear method

STAGE(j)	q_j	Q_j	x_j	c_j	w_j	x_jc_j	x_jw_j
1	0.8074	0.948	2	14.34	2037.92	28.69	4075.83
2	0.7124	0.8529	3	10.48	1506.78	31.45	4520.33
3	0.6686	0.7930	4	7.195	1041.26	28.78	4165.02
Total cost and total weight						88.92	12,761.18
System reliability (Q_s)						0.6409	

Table 9.5 System reliability for reliability model III using Evolutionary method in (0.5, 0.95)

STAGE(j)	q_j	Q_j	x_j	c_j	w_j	x_jc_j	x_jw_j
1	0.7379	0.8833	3	986.76	30846.17	2960.29	92538.51
2	0.7422	0.8881	4	834.8	28995.51	3339.21	115982.03
3	0.7712	0.9176	5	540.10	29935.51	2700.52	149677.57
Total cost and total weight						9000.02	358198.11
System reliability (Q_s)						0.719848	

Table 9.6 System reliability for reliability model III using GRG non linear method in (0.5, 0.95)

STAGE(j)	q_j	Q_j	x_j	c_j	w_j	x_jc_j	x_jw_j
1	0.7378	0.8832	3	986.15	30,827.12	2958.46	92,481.37
2	0.7423	0.8881	4	834.95	29,000.59	3339.80	116,002.34
3	0.7713	0.9177	5	540.35	29,949.13	2701.75	149,745.66
Total cost and total weight						9000.00	358,229.37
System reliability (Q_s)						0.719847	

The exponential function $c_j = g_j e^{\left(\frac{h_j}{1-q_j}\right)}$ used in Reliability Model III, maximum value of the system cost of Rs 9000 and weight 40,000 kgs, is considered to determine optimum reliability of each component (q_j), Reliability of Phase (Q_i) and the Reliability of system (Q_s).

Tables 9.5 & 9.6 give the System Reliability of diesel engine using excel solver.

9.6.1 Graph Theory Matrix Approach (GTMA)

GTMA is a legitimate and precise methodology. The high level hypothesis of charts and its applications are very irrefutable [21]. Diagram hypothesis grid approach comprises the accompanying parts:

1. Directed graph
2. Representation of Matrix
3. Permanent function

9.6.2 Digraph

Directed Graph model portrayal has ended up being valuable in demonstrating and dissecting different sorts of frameworks in areas of science and innovation. The exhibition credits digraph utilized in the current review comprises Chambers, Cost requirements, Weight imperatives as hubs and

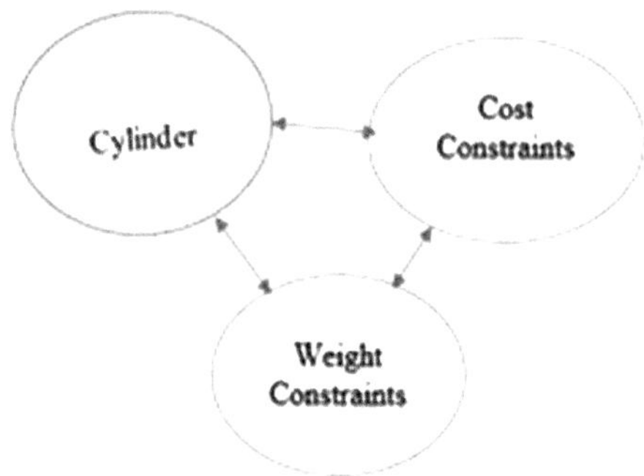

Figure 9.1 Digraph.

the interrelations between them are addressed as edges. The exhibition ascribes digraph is displayed in figure 9.1 for guaranteed visual evaluation. As the quantity of hubs and their overall significance increment, the digraph becomes complicated. To conquer this trouble, the digraph is addressed in lattice structure.

9.6.3 Permanent Index

It is an extent of the effortlessness which gives the best blend of working limits of an engine Chambers. It contains the extents of factors and their relative importance. As the very tough ability contains only the positive terms, the higher upsides of q_j and ($aij = c_j$ and 〖w〗$_j$) will bring about expanded worth of execution file. The upsides of q_j are gotten from the trial results which are displayed in Table 9.7.

Table 9.7 Experimental results

S. No	*Cylinders*	*Cost*	*Weight*
1	C1-SS316	60	4.36
2	C2-SS316	65	1.76
3	C3-SS316	70	4.88
4	C4-SS316	75	2.4
5	C5-SS316	80	5.48
6	C6-SS316	85	1.4
7	C7-SS316	60	1.8
8	C8-SS316	65	1.76
9	C9-SS316	70	2.64
10	C10-SS316	75	3.4
11	C11-SS316	80	2.04
12	C12-SS316	85	3.6

$$A = \begin{bmatrix} a_{11} & a_{12} & a_{13} & a_{14} & a_{15} \\ a_{21} & a_{22} & a_{23} & a_{24} & a_{25} \\ a_{31} & a_{32} & a_{33} & a_{34} & a_{35} \\ a_{41} & a_{42} & a_{43} & a_{44} & a_{45} \\ a_{51} & a_{52} & a_{53} & a_{54} & a_{55} \end{bmatrix}$$

The diagonal components of the matrix A, q_j are quantitative; they are normalized in the range of 0 and 1. Normalization is to set the attribute data on same scale so that comparisons can be made easier. Let x_{ij}—Cylinder is the normalized value of y_{ij} cost and weight for attribute i then,

$$\begin{aligned} x_{ij} &= \frac{y_{ij}}{\max_j(y_{ij})}; \text{ if } j^{th} \text{ attribute is beneficial} \\ x_{ij} &= \frac{\min_j(y_{ij})}{y_{ij}}; \text{ if } j^{th} \text{ attribute is non-beneficial} \end{aligned} \tag{9.4}$$

The normalized values of q_j are shown in Table 9.8.

The relative importance of the matrix A, a_{ij} are assigned between 0 and 10 on a qualitative scale, as shown in Table 9.9. If a_{ij} represent the relative importance of the ith attribute over the jth attribute, then the relative importance of the jth attribute over the ith attribute is evaluated using the Eq. (9.5). The relative importance between i, j and i is given as

$$a_{ji} = 1/a_{ji} \tag{9.5}$$

Table 9.8 Normalized values

S. No	*Cylinders*	*Cost*	*Weight*
1	C1-SS316	50	1.36
2	C2-SS316	55	1.76
3	C3-SS316	50	1.08
4	C4-SS316	55	2.4
5	C5-SS316	85	5.48
6	C6-SS316	80	5.30
7	C7-SS316	70	5.00
8	C8-SS316	65	4.76
9	C9-SS316	50	2.64
10	C10-SS316	55	3.4
11	C11-SS316	60	4.04
12	C12-SS316	55	3.6

Table 9.9 Relative importance of performance attributes

Class description	*Relative importance of attributes* a_{ji}	$a_{ji} = 1/a_{ji}$
Two attributes are equally important	1	1.00
One attribute is strongly more important over the other	2	0.50
One attribute is very strongly more important over the other	3	0.33
One attribute is extremely important over the other	4	0.75
One attribute is exceptionally more important over the other	5	0.60

Table 9.10 Consistency evaluation

	Consistency Measure	
q_1	7.010	
q_2	6.004	
q_3	5.002	
Component Reliability q_1		**0.781**
	q_2	**0.680**
	q_3	**0.619**

The mathematical mean methodology of AHP (Logical Pecking order Cycle) is utilized to decide the overall standardized loads of the qualities, and the consistency look at is conveyed. It is expected that the consistency proportion worth of the overall significance of qualities ought to be under 0.10 [21]. The consistency proportion in the current review is viewed as 0.019. The consistency assessment is shown in Table 9.10.

The boundary list for each analysis displayed in Table 9.10 is assessed utilizing the Eq. (9.4) subbing the upsides of q_j and a_{ji}. The presentation list values for all tests are organized in dropping requests to rank them. The analysis for which the worth of boundary record is most elevated is the ideal mix of working boundaries (Table 9.11).

9.7 RESULTS AND DISCUSSION

From Table 9.12, it is observed that the component reliabilities obtained for cylinder C5, C6, C7 by GRG nonlinear & Evolutionary are 0.81, 0.72 & 0.67, respectively. Whereas, the performance index obtained for cylinders C5,

Table 9.11 Parameter index values and Rank

S. No	Material (Cylinder)	Parameter Index Cost	Parameter Index Weight (X100)	Rank
1	C5-SS316	72.7509	7.84	1
2	C6-SS316	64.9739	6.48	2
3	C7-SS316	62.1073	5.48	3
4	C8-SS316	53.1012	4.36	4
5	C11-SS316	51.2007	3.96	5
Component Reliability			q_1	0.781
			q_2	0.680
			q_3	0.619
System Reliability (Q_s)			**0.6**	

Table 9.12 System Reliability

		Graph Theory Matrix Approach	Evolutionary	GRG nonlinear	Evolutionary	GRG nonlinear
			Polynomial function		Exponential function	
Component Reliability	q_1	0.781	0.8069	0.8074	0.7379	0.7378
	q_2	0.680	0.7190	0.7124	0.7422	0.7423
	q_3	0.619	0.6699	0.6686	0.7712	0.7713
System Reliability	Q_s	0.600	0.6485	0.6409	0.71984	0.71984

C6, C7 using GTMA is 0.78, 0.68 & 0.61, respectively. These component reliabilities result in System Reliability of 0.6 by all the three methods, when the cost and the weight constraints are polynomial function of q_i, whereas when the cost and the weight constraints are exponential function of q_i, the system reliability is found to be 0.72 using GRG nonlinear & Evolutionary method and 0.6 using GTMA.

9.8 CONCLUSION

Observations from Tables 9.3 and 9.4 reveal that the reliability of the system achieved through the Excel solver is 0.5676 for reliability model II. This approach yields the component reliability comparable to the Graph Theory Matrix Approach method. The total cost and weight are either similar to or lower than the specified values.

Furthermore, analysis from Table 9.5 & 9.6, indicates that the reliability of system derived from the Excel solver is 0.7198 for reliability model III. This methodology results in component that surpasses the reliability obtained from the Graph Theory Matrix Approach method.

DECLARATIONS

Conflict of Interest: On behalf of all authors, the corresponding author states that there is no Conflict of interest.

ACKNOWLEDGEMENT

The authors are very much thankful to the reviewer for the valuable comments. The authors are also thankful to the Management, Principal and Head of the Department of Mathematics, Dayananda Sagar College of Engineering, and Bangalore for their constant encouragement.

REFERENCES

1. D. T. Chiang and S. C. Niu.: Reliability of a consecutive-k-out-of-n: F system, IEEE Trans Reliab, 30, 87–89, (1981).
2. R. E. Barlow and K. D. Heidtmann: Computing k-out-of-n system Reliability, IEEE Trans Reliab, 33, 322–323, (1984).
3. D. Derman, G. Lieberman and S. Ross.: On the consecutive-k-out-of-n: F system, IEEE Trans Reliab, 31, 57–63, (1982).
4. F. K. Hwang.: Fast Solutions for Consecutive-k-out-of-n:F System, IEEE Trans Reliab, R-31(5), 447–448, Reliability of a Mechanical System with k-out-of-n Subsystems 7187 www.tjprc.org SCOPUS Indexed Journal editor@tjprc.org. (1982).
5. J. G. Shantikumar.: Recursive algorithm to evaluate the reliability of consecutive k-out-of-n:F system networks, IEEE Trans Reliab, 31, 84–87, (1982).
6. S. P. Jain and K. Gopal.: Reliability of k-to-l-out-of-n systems, Reliab Eng, 12, 175–179, (1985).
7. T. Risse.: On the evaluation of the reliability of k-out-of-n systems, IEEE Trans Reliab, 224–234, (1987).
8. A. K. Sarje and E. V. Prasad.: An efficient non-recursive algorithm for computing the reliability of k-out-of-n systems, IEEE Trans Reliab, R-38(2), 234–235, (1989).
9. G. Ge and L. Wang.: Exact Reliability Formula for Consecutive-k-out-of-n:F Systems with Homogeneous Markov Dependence, IEEE Trans Reliab, 39(5), (1990).
10. R. C. Suich and R. L. Patterson.: *k*-out-of-*n*:G systems; Some Cost Considerations, IEEE Trans Reliab, 40, 259–264, (1991).
11. H. H. Pham.: On the Optimal Design of k-out-of-n:G Subsystems, IEEE Trans Reliab, 45(2), 254–260, (1996).
12. H. Yamamoto and M. Miyakawa.: Reliability of a circular connected -(r,s)-out-of-(m,n):F lattice system, J Oper Res, 39(3), (1995).

13. D. W. Coit and A. E. Smith.: Reliability optimization of series-parallel systems using a genetic algorithm, IEEE Trans Reliab, 45(2), 254–260, (1996).
14. M. V. Koutras.: Consecutive k,r-out-of-n: DFM systems, Microelectron Reliab, 37 597–603, (1997).
15. R. Venkatasamy and V. P Agrawal.: System and structural analysis of an automobile vehicle – a graph theoretic approach, J Veh Des, 16, 477–505, (1997).
16. J. Huang, M. J. Zuo and Y. H. Wu.: Generalized multi-state k-out-of-n:G system, IEEE Trans Reliab, (1), 105–111, (2000).
17. R. V. Rao.: Graph theory and matrix approach for the performance evaluation of technical Institutions, Ind J Tech Educ, 23(2), 27–33, (2000).
18. O. P. Gandhi and V. P. Agrawal.: FMEA – A digraph and matrix approach, Reliab Eng Syst Saf, 35, 147–158, (2002).
19. S. Grover, V. P. Agrawal and I. A. Khan.: A digraph approach to TQM evaluation of an industry, Int J Prod Res, 42(19), pp. 4031–4053, (2004).
20. Y. Chen and Q. Yang.: Reliability of two-stage weighted k-out-of-n systems with components in common, IEEE Trans Reliab, 54(3), 431–440, (2005).
21. N. K. Geetha and P. Sekar.: Graph theory matrix approach with fuzzy set for optimization of operating parameters on a diesel engine, Mater Today Proc, 4(8), 7750–7759, (2017).
22. Z. Tian, M. J. Zuo and R. C. M. Yam.: Multi-state *k*-out-of-*n* systems and their performance evaluation, IIE Trans, 41, 32–44, (2009).
23. A. Anand and M. F. Wani.: Product life cycle modeling and evaluation at the conceptual design stage a digraph and matrix approach, J Mec Des, 132(9), (2010).
24. A. Sridhar, S. Pavan Kumar, Y. Raghunatha Reddy, G. Sankaraiah and C. Umashankar.: The k-out-of-n Redundant IRM Optimization with multiple constraints, IJRSAT, 2(1), (2013).
25. K. S. Lakshminarayana and Y. Vijay Kumar.: Reliability optimization of integrated reliability model using dynamic programming and failure modes effects and criticality analysis, J Acad Indus Res, 1(10), 622–625, (2013).
26. R. V. Rao.: Decision making in the manufacturing environment: using graph theory and fuzzy multiple attribute decision making methods, Springer, London, 2007.
27. J. Watada and H. Melo.: An Estimation of distribution Algorithm approach to redundancy allocation problem for a high security system, IEEJ J Ind Appl, 3(4), 358–367, (2014).
28. A. Singh and S. Kumar.: Reliability Optimization Using Hardware Redundancy: a Parallel BB-BC Based Approach, Int J Electron Commun Eng, 12, 2278–9901, (2015).
29. M. Zuo and Z. Tian.: Performance Evaluation of Generalized Multi-State k-out-of-n Systems, IEEE Trans Reliab, 55(2), (2006).
30. K. Ramesh and R. Gupta.: Reliability of a Mechanical System with k-out of n subsystems, Int J Mech Prod Eng Res Develop, 10, 7179–7188, (2020).
31. V. Rykov, O. Kochueva and M. Farkhadov.: Preventive Maintenance of a k-out-of-n System with Applications in Subsea Pipeline Monitoring, J Mare Sci Eng, 9, 85, (2021). https://doi.org/10.3390/jmse9010085
32. Y. Hu, C. Wang and L. Xing.: Reliability analysis of k-out-of-n Phased-Mission systems with Phase-AND requirement, Conf Ser: Mater Sci Eng, 1043, 032038, (2021).
33. V. Rykov, N. Ivanova and I. Kochetkova.: Reliability Analysis of a Load-Sharing k-out-of n System Due to Its Components Failure, Mathematics, 2457, (2022). https://doi.org/10.3390/math10142457
34. M. Mohan, O. P. Gandhi and V. P. Agrawal.: Systems modeling of a coal based steam power plant: a systems approach, Proc Inst Mech Eng, Part A, 217, 259–277, (2003).

35. B. Yamuna, R. Gupta, K. Ramesh and N. K. Geetha.: Reliability Analysis of a Mechanical System with 3 Out of 5 Subsystems, In book: Third Congress on Intelligent Systems DOI: 10.1007/978-981-19-9225-4_39, (2023).
36. H. Pham and D. M. Malon.: Optimal design of systems with competing failure modes, IEEE Trans Reliab, 43(2), pp. 251–254, (1994).
37. S. V. Suresh Babu, D. Maheswar, G. Ranganath, Y. Vijaya Kumar and G. Sankaraiah.: Redundancy Allocation for Series Parallel Systems with Multiple Constraints and Sensitivity Analysis, IOSR J Eng, 2(3), 424–428, (2012).

Chapter 10

Reliability and sensitivity analysis of a network structure

Shristi Kharola, Mangey Ram, and Nupur Goyal

10.1 INTRODUCTION

Today, technological innovations have an influence across every evolving industry and marketplace throughout the globe. The expansion of network and communication technology has been widely used for several years and has proven to be a great benefit for local and long-distance networking. Conventional wired network systems are still anticipated to be completely replaced by retrofittable wireless network systems but are expected to be integrated with wireless systems in the near future for hybrid systems (Underberg et al., 2018). For instance, the fourth industrial revolution supports existing industrial applications based on cyber-physical systems (CPSs) and will continue to drive industrial growth employing CPS. Therefore, in order to support the emerging systems of the future, it is crucial to work on and develop wired networks in such a way that they keep the system running smoothly and efficiently.

The advancement of science and innovation has consistently wandered about the efficiency of networking systems. To keep the systems working, it is important to keep the state-of-the-art with the goal that the systems become reliable. Reliability is a part of designing systems that manages the improvement in failures of a system or a structure under explicit conditions and a given time-frame (Zio, 2009).

10.2 NETWORK SYSTEMS

Network theory in system design studies the science of connectivity and interaction between multiple components in a system. It consists of all kinds of hardware and software components that work together as a whole system. Various types of configurations of a networking system can be defined on the basis of the physical or logical structure of the network system, known as the topology. Where physical topology focuses on the placement of devices that are connected/inter-linked within the network, logical topology focuses on the flow of data inside the network. For any two separate network topologies, the distance, transmission rates, signal types and physical infrastructure

DOI: 10.1201/9781003546214-10

might vary or even be similar depending on the logical topological structure. One such topology that is commonly used in industries and organizations is the ring network topology (Liu and Liu, 2014).

10.3 RING TOPOLOGY NETWORK SYSTEM

Unlike other network topologies, a ring topology does not require a central system or a central node. In this topological network, at least three computers identified as nodes are connected to one another through cables to form a ring like structure. The data transferred is unidirectional, i.e., in one direction, either clockwise or counterclockwise. Unidirectional progression keeps the crash of information trivial (Yoon et al., 2006). To prevent information loss, repeaters are introduced to each node to support signal strength. It is easy to reconfigure a ring topology network because of the immediate connectivity of nodes. So, if any node fails, it can immediately be replaced or removed to keep the system successfully working. But this means that a node failure can lead the whole ring to shut down and lead to its reconfiguration. This can cause communication delays and can affect the entire network system. Also, increasing the number of nodes means slower network systems, which can create sudden, unexpected problems due to failure in any industrial system. Also, in such systems, there may emerge various sorts of failures in the general availability of the system regardless of whether there is a little break in the system. Failure of any node, cable or repeater can prompt the structure to fail partially or completely.

10.4 RELIABILITY ANALYSIS

The term 'reliability analysis of engineering and industrial systems' refers to the process of preventing certain failure modes in systems and reducing the frequency of failures. The analysis is based on probability estimation, reliability testing and modeling techniques. Since no failure in a system can be completely guaranteed, reliability is always expressed in terms of probability. The primary goal of reliability analysis in systems includes identifying failures and their causes and applying methods, along with redesigning systems to correct or prevent these failures. Reliability can be calculated based on functions depending upon failure rates and time-based quantities, such as mean-time-to-failure (MTTF), mean-time-to-repair (MTTR) and mean-time-between-failures (MTBF). Aside from this, the availability and expected cost can also be determined for any system (Birolini, 2013).

10.5 LITERATURE REVIEW

Several research papers in the past have explained reliability analysis of various network topologies (Shooman, 1995; Lin and Yeh, 2011). The reliability of computer networks is critical to modern civilization. One approach to achieving this aim is to identify the best redundancy allocation in order to optimize system reliability. Yeh and Fiondella (2017) employed a correlated binomial distribution to define the state distribution of all network edges and a redundancy optimization strategy incorporating simulated annealing (SA), minimum routes and the correlated binomial distribution. Kumar et al. (1995) described the creation and use of a novel approach for extending current computer networks. A genetic-algorithm-based computer network expansion methodology was created in order to optimize a certain goal function. Yunus et al. (2018) employed an enhanced augmented SEN by repeating the network to build a redundant route and improve reliability. The primary source of concern in system performance was the reliable operation of interconnection networks. The reliability of multistage interconnection networks was determined by their architecture, network design and number of system stages. The terminal reliability study revealed that the method of employing a replicated network resulted in the best reliability performance. Singh et al. (2021) presented the results of a study of reliability metrics for a complicated system composed of three computer laboratories connected to a server in parallel under a 2-out-of-3: G operation strategy. Watcharasitthiwat and Wardkein (2009) studied a novel efficiency method based on traditional ant colony optimization (ACO) to solve the communication network design while taking both economics and reliability into account. presented a method for creating an optimum network topology that provides the highest s-t reliability. The algorithm is demonstrated using an example. A modern distributed network for data transmission was studied by Huang et al. (2020), which included PCs, the IoT (Internet of Things), cloud servers and edge servers. A scattered network can be represented graphically as a graph with nodes and arcs, with each arc representing a transmission line and each node representing an IoT device, edge server or cloud server. Ahmad et al. (2017) investigated several modeling and analytic approaches that might assist in the research of communication network reliability. Kounev et al. (2016) discussed the smart power grid's communications reliability needs, with an emphasis on communications in support of broad area situational awareness, to address the challenge of making the substation network and wide area network fulfill reliability needs while reducing network costs. Zuev et al. (2015) proposed a stochastic framework for quantitatively assessing network service reliability, established a generic network reliability issue and then demonstrated how to compute service reliability using subset simulation.

10.6 AIMS AND STRUCTURE OF THE WORK

In this work, the authors highlight the problems faced during the failure of components and systems in a ring network topology. To deal with this, authors use reliability as an estimation of the probability to understand the science of the network system and preserve its effectiveness for as long as possible. The vital function of reliability in a ring topology network is the execution of the whole network system within the given time-frame without total failure of the system (Lin et al., 2001). For the designed ring topology network, system reliability factors, such as MTTF, availability, expected profit and reliability, have been analyzed, and the outcomes obtained are discussed with reference to time to establish the overall reliability of the system. As a distinction, this work's effort primarily adds to the network system's sensitivity analysis. The measures are calculated and analyzed using the Markov mathematical modeling.

10.7 MATHEMATICAL METHODS

To assess overall reliability metrics, this study employs Markov modeling, the supplementary variable technique and Laplace transformation. The Markov process is a mathematical representation of state transition rates or probabilities as transition rates or probability matrices, and state probabilities are finally calculated by solving a set of equations. When a system can be characterized by a collection of discrete states and the probability of transitioning to a new state is only dependent on the current state, a Markov model may be used to represent it. Because distribution systems contain a large number of possible states, various simplifying assumptions must be made in order for the Markov model to be understood. Markov modeling has been used effectively in transmission and distribution network systems (Gonen, 1986). The Laplace transform is a mathematical integral transform that converts a function of a real variable t (usually time) to a function of a complex variable s (complex frequency). The transform has various applications in science and engineering since it is a tool for solving differential equations (Schiff, 1999). Non-Markovian states can be investigated using either embedded Markov chains or the supplementary variable approach. By adding one or more variables to the state space, the supplementary variable approach aids the Markovian process. This supplementary variable is a continuous variable. This results in a continuous state space and continuous time Markov process (Alfa and Rao, 2000).

10.8 SYSTEM DESCRIPTION

Computers or nodes are connected to one another through cables to form a ring-like structure in ring network topology. Data is transmitted unidirectionally, that is, in only one direction, either clockwise or counterclockwise. Repeaters are installed at each node to support signal strength and avoid information loss. Any node, cable or repeater failure can cause the structure

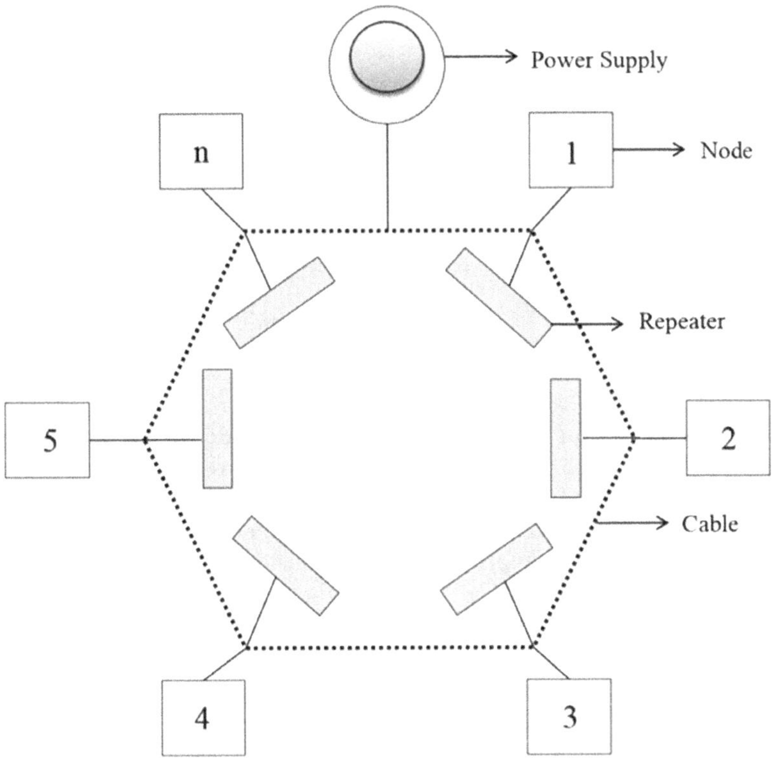

Figure 10.1 Ring network topology (configuration diagram).

to collapse partially or entirely. Along with this, a primary power supply provides electricity to the whole ring structure. As seen in Figure 10.1, the complete configuration diagram.

10.9 ASSUMPTIONS

Before evaluating the network model, the author considers the following assumptions:

a. In the initial stage, the authors assume that the framework is in a good working state.
b. The system has three states: completely functional, partially functional and failed.
c. In this model, constant failure and repair rates are considered.
d. Fixing facility is accessible when a failure happens.
e. The system configuration consists of n units connected in series.

10.10 STATE TRANSITION DIAGRAM

State transition diagram of the ring topology system is shown in Figure 10.2. The states of the system are described further below.

System states	*Description*	*Symbol*
S_0	Perfectly working state.	○
S_1, S_2, S_3, S_4	Partially working states.	▱
S_5, S_6, S_7, S_8	Completely failed states.	□

The developer of this model took into account four different types of failure rates: α, β, λ and γ. If any node fails by λ rate of failure, the system enters partial operation mode. If a repeater fails at failure rate β, there are no chances of data being received or retransmitted, which means the node has utterly collapsed. If the power supply fails with failure rate α, the system fails completely. If the cables fail with failure rate γ, the complete system fails. If all the nodes fail, the entire system fails. ϕ represents the repair rate of the systems from completely failed to perfectly working.

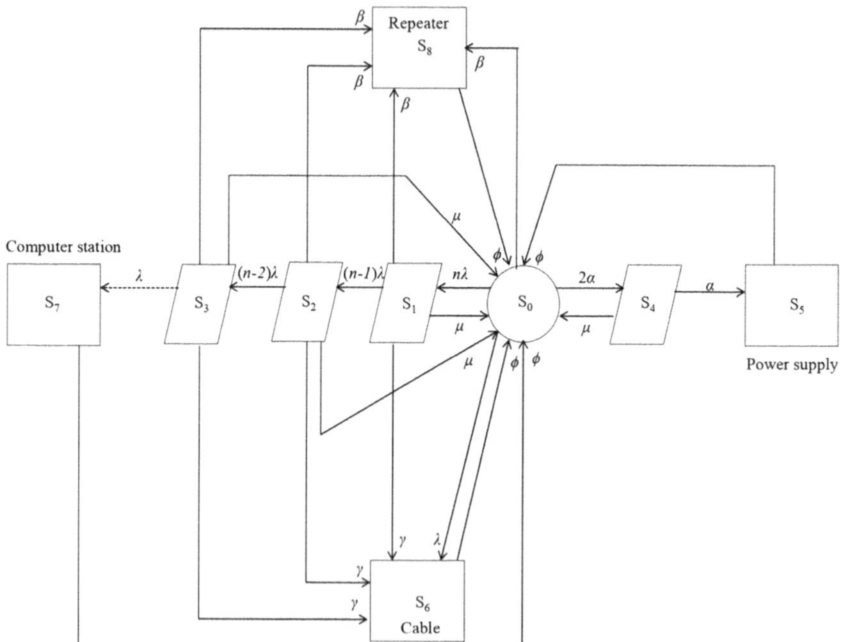

Figure 10.2 State transition diagram.

10.11 NOTATIONS

$P_0(t)$	Probability, at time (t) and state 0.
$P_i(t)$	Probability of being in the 'i^{th}' state at time t; $i = 1, 2, 3, 4$
$P_j(x, t)$	Probability-density-function (PDF), that the system at epoch-time t is in state i^{th} and has an elapsed repair time of x. where $j = 5, 6, 7, \ldots, n$.
t	Time scale
s	Laplace transform variable
λ	It reflects any node's failure rate.
β	It reflects the repeater's failure rate.
γ	It represents the failure rate of cable.
ϕ	Repair rate from completely failed to perfectly working state.
α	It reflects the power-supply rate of failure.
n	It denotes the number of nodes.
A(t)	Availability
R(t)	Reliability
E_p	Expected profit

10.12 THE MODEL'S MATHEMATICAL FORMULATION

With the assistance of the state transition diagram, a set of differential equations by the Markov process has been developed. For a viable solution, the Laplace transform and supplementary variable technique are also used.

The probability of different states is given by

$$\left\{\frac{\partial}{\partial t}+\beta+\gamma+2\alpha+n\lambda\right\}P_0(t)-\mu\left\{P_1(t)+P_2(t)+P_3(t)+P_4(t)\right\}=\sum_{j}\int_{0}^{\infty}P_j(x,t)\phi dx$$

where $j = 5$ to 8, (10.1)

$$\left\{\frac{\partial}{\partial t}+\mu+\beta+\gamma+(n-1)\lambda\right\}P_1(t)=n\lambda P_0(t), \tag{10.2}$$

$$\left\{\frac{\partial}{\partial t}+\mu+\beta+\gamma+(n-2)\lambda\right\}P_2(t)=(n-1)\lambda P_1(t), \tag{10.3}$$

$$\left\{\frac{\partial}{\partial t}+\mu+\beta+\gamma+(n-3)\lambda\right\}P_3(t)=(n-2)\lambda P_2(t), \tag{10.4}$$

$$\left\{\frac{\partial}{\partial t}+\mu+\alpha\right\}P_4(t)=2\alpha P_0(t), \tag{10.5}$$

$$\left\{\frac{\partial}{\partial t}+\frac{\partial}{\partial x}+\phi(x)\right\}P_j(x,t)=0;\ \mathrm{j}=5\ to\ 8. \tag{10.6}$$

The boundary conditions are as follows:

$$P_5(0, t) = \alpha P_4(t), \tag{10.7}$$

$$P_6(0, t) = \gamma\{P_0(t) + P_1(t) + P_2(t) + P_3(t)\}, \tag{10.8}$$

$$P_7(0, t) = \lambda P_6(t), \tag{10.9}$$

$$P_8(0, t) = \beta\{P_0(t) + P_1(t) + P_2(t) + P_3(t)\}. \tag{10.10}$$

The initial conditions are

$$\{P_0(0) = 1\} \text{ and } (\text{all other probabilities} = 0).$$

Taking Laplace transformation from Eqs. (10.1) to (10.10), we get,

$$\{s + \beta + \gamma + 2\alpha + n\lambda\}\bar{P}_0(s) - \mu\{\bar{P}_1(s) + \bar{P}_2(s) + \bar{P}_3(s) + \bar{P}_4(s)\}$$
$$= 1 + \sum_j \int_0^\infty \bar{P}_j(x, s)\phi dx; \; j = 5 \; to \; 8 \tag{10.11}$$

$$\{s + \mu + \beta + \gamma + (n-1)\lambda\}\bar{P}_1(s) = n\lambda\bar{P}_0(s), \tag{10.12}$$

$$\{s + \mu + \beta + \gamma + (n-2)\lambda\}\bar{P}_2(s) = (n-1)\lambda\bar{P}_1(s), \tag{10.13}$$

$$\{s + \mu + \beta + \gamma + (n-3)\lambda\}\bar{P}_3(s) = (n-2)\lambda\bar{P}_2(s), \tag{10.14}$$

$$\{s + \mu + \alpha\}\bar{P}_4(s) = 2\alpha\bar{P}_0(s), \tag{10.15}$$

$$\left\{s + \frac{\partial}{\partial x} + \phi(x)\right\}\bar{P}_j(x, s) = 0; \; j = 5 \; to \; 8. \tag{10.16}$$

From boundary conditions,

$$\bar{P}_5(0, s) = \alpha\bar{P}_4(s), \tag{10.17}$$

$$\bar{P}_6(0, s) = \gamma\{\bar{P}_0(s) + \bar{P}_1(s) + \bar{P}_2(s) + \bar{P}_3(s)\}, \tag{10.18}$$

$$\bar{P}_7(0, s) = \lambda\bar{P}_6(s), \tag{10.19}$$

$$\bar{P}_8(0, s) = \beta\{\bar{P}_0(s) + \bar{P}_1(s) + \bar{P}_2(s) + \bar{P}_3(s)\}. \tag{10.20}$$

From Eq. (10.16), we get,

$$\bar{P}_j(x, s) = \bar{P}_j(0, s)exp\left\{-sx - \int_0^x \phi dx\right\};\ j = 5\ to\ 8. \tag{10.21}$$

After solving these differential equations, we find the following transition state probabilities:

$$\bar{P}_0(s) = \frac{1}{\left[X_1(s) - X_2(s)H_2 - \left\{\bar{S}_\phi(s)\alpha + \mu\right\}\left\{\frac{2\alpha}{s+\mu+\gamma}\right\} - \left\{\bar{S}_\phi(s)\lambda\left(\frac{1}{s}\right)\left(1-\bar{S}_\phi(s)\right)\gamma H_1\right\}\right]}, \tag{10.22}$$

$$\bar{P}_1(s) = \frac{n\lambda\bar{P}_0(s)}{s+c1}, \tag{10.23}$$

$$\bar{P}_2(s) = \frac{(n-1)n\lambda^2\bar{P}_0(s)}{(s+c1)(s+c2)}, \tag{10.24}$$

$$\bar{P}_3(s) = \frac{(n-2)(n-1)n\lambda^3\bar{P}_0(s)}{(s+c1)(s+c2)(s+c3)}, \tag{10.25}$$

$$\bar{P}_4(s) = c7\bar{P}_0(s). \tag{10.26}$$

where

$$c1 = ((n-1)\lambda + \gamma + \beta + \mu),$$

$$c2 = ((n-2)\lambda + \beta + \mu),$$

$$c3 = ((n-3)\lambda + \beta + \mu),$$

$$c4 = \frac{n\lambda}{s+c1},$$

$$c5 = \frac{(n-1)n\lambda^2}{(s+c1)(s+c2)},$$

$$c6 = \frac{(n-2)(n-1)n\lambda^3}{(s+c1)(s+c2)(s+c3)},$$

$$c7 = \frac{2\alpha}{s+\mu+\alpha},$$

$$c8 = \gamma + \lambda\gamma + \beta,$$

$$X_1(s) = s + \beta + \gamma + 2\alpha + n\lambda - \bar{S}_\varnothing(s)\gamma - \bar{S}_\varnothing(s)\beta,$$

$$X_2(s) = \bar{S}_\phi(s)\gamma + \bar{S}_\phi(s)\beta + \mu,$$

$$H_1 = 1 + c4 + c5 + c6,$$

$$H_2 = c4 + c5 + c6.$$

Now, calculate the Laplace transformation of the probabilities that the system is in the up-state or down-state at any given time.

$$\begin{aligned} \bar{P}_{up}(s) &= \bar{P}_0(s) + \bar{P}_1(s) + \bar{P}_2(s) + \bar{P}_3(s) + \bar{P}_4(s), \\ \bar{P}_{up}(s) &= (1 + c4 + c5 + c6 + c7)\bar{P}_0(s), \end{aligned} \tag{10.27}$$

$$\begin{aligned} \bar{P}_{down}(s) &= \sum_j \bar{P}_j(s); j = 5 \text{ to } 8, \\ \bar{P}_{down}(s) &= (1 + c4 + c5 + c6 + \alpha c7)(c8)\bar{P}_0(s). \end{aligned} \tag{10.28}$$

10.13 CALCULATION OF MEASURES

The reliability measures are taken into consideration in the following sections to observe the validity of the network system at various time and failure rates using equations, graphs and tables. The measurements were computed using the MAPLE software.

10.13.1 Availability analysis

The probability or the chance of the system operating satisfactorily at any time is referred to as availability, and it is determined by the system's reliability and maintainability. It is another measure of a system's performance (Li, 2018). In their study, Heimann et al. (1990) have described in detail availability and its different forms. In this system, the Markov state transition diagram is used to analyze the availability or up-time of the system. On using the values of various failure limits, $\lambda = 0.02$, $\gamma = 0.03$, $\beta = 0.01$, $\alpha = 0.05$ and $\mu = 1$ (Nagiya and Ram, 2014) and n = 50 in Eq. (10.27), as well as using the inverse LT, one can determine the availability by

$$\begin{aligned} A(t) = {} & 0.08348e^{(-1.05\,t)} - 0.11511e^{(-2.77438\,t)} + 0.02206e^{(-2.08553\,t)}\cos(0.782208\,t) \\ & - 0.23312e^{(-2.08553\,t)}\sin(0.782208\,t) - 0.006586e^{(-1.071021\,t)} - 0.003635e^{(-1.01048\,t)} \\ & + 1.01979e^{(-0.05304\,t)} \end{aligned} \tag{10.29}$$

Putting the value of t from 0 to 15, the authors acquire Table 10.1 and Figure 10.3 as a representation of the availability in a time function.

Table 10.1 Availability

Time (t)	*Availability A(t)*
0	0.99999
1	0.96709
2	0.92207
3	0.87253
4	0.82593
5	0.78262
6	0.74195
7	0.70354
8	0.66717
9	0.63270
10	0.60001
11	0.56901
12	0.53962
13	0.51174
14	0.48531
15	0.46024

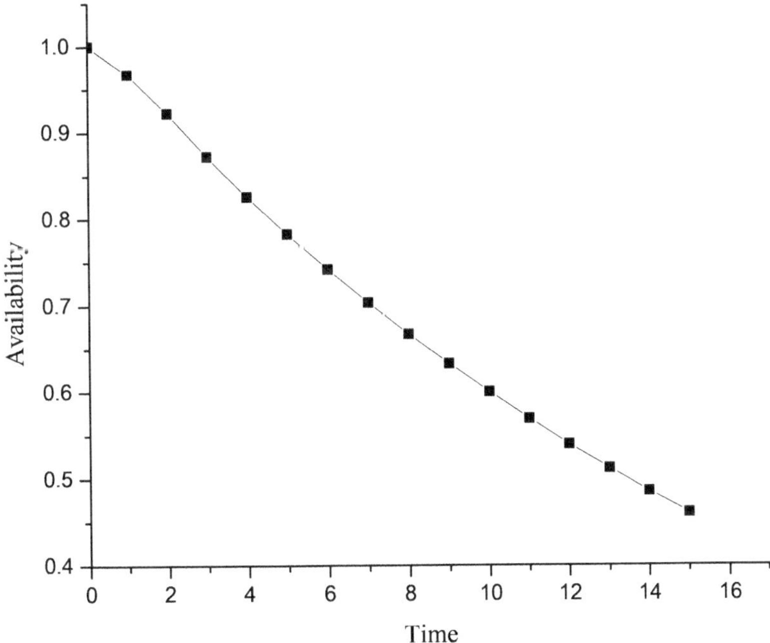

Figure 10.3 Availability as a function of time.

10.13.2 Reliability analysis

Reliability calculates the capability of a system to perform its intended function for a specific interval without failure in given time (Kuo and Prasad, 2000; Chopra and Ram, 2019). On using the value of rate of repair = 0 in Eq. (10.27) and taking inverse-LT, the network system's reliability is stated as

$$R(t) = 2085e^{(-1.02\ t)} - 5418.82e^{(-0.97\ t)} + 3536.84e^{(-0.95\ t)} + 0.091743e^{(-0.05\ t)} - 202.1103e^{(-1.14\ t)} \tag{10.30}$$

The system's numerical values and reliability graph are shown in Table 10.2 and Figure 10.4, respectively.

10.13.3 MTTF analysis

MTTF is the average length of time that a non-repairable asset operates before failing, and it is determined by dividing the entire operating duration of the tested units by the total number of failures. Several research papers have measured the reliability growth using MTTF (Ryu and Chang, 2005; Kumar and Ram, 2019). In Equation (10.27), setting repair rate = 0

Table 10.2 Reliability with respect to time

Time (t)	*Reliability R(t)*
0	1.00000
1	0.94684
2	0.81442
3	0.62015
4	0.43139
5	0.28599
6	0.18868
7	0.12911
8	0.09452
9	0.07486
10	0.06353
11	0.05665
12	0.05207
13	0.04867
14	0.04590
15	0.04349

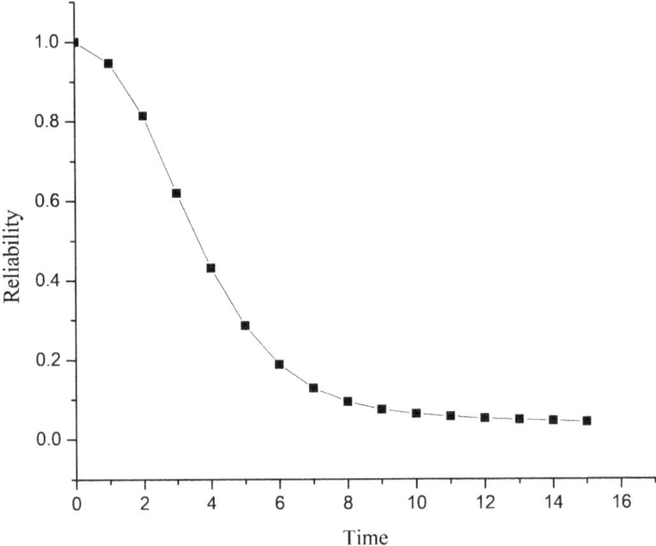

Figure 10.4 Reliability as a function of time.

and allowing the 'Laplace variable s' to approach zero yields the mean time to failure:

$$MTTF = \frac{1}{(\beta+\gamma+2\alpha+\mathrm{n}\lambda)}\left[3+\frac{n\lambda}{(n-1)\lambda+\gamma+\beta}+\frac{(n-1)n\lambda^2}{((n-1)\lambda+\gamma+b)((n-2)\lambda)+\beta}\right.$$
$$\left.+\frac{(n-2)(n-1)n\lambda^3}{((n-1)\lambda+\gamma+\beta)((n-2)\lambda+\beta)(\beta+(n-3)\lambda)}\right] \tag{10.31}$$

Again, on putting n = 50, β = 0.01, γ = 0.03 and λ = 0.02 in Equation (10.31), as shown in Table 10.3 and Figure 10.5, the MTTF is obtained in relation to the rate of failures.

Table 10.3 MTTF as a function of failure-rate

Variating values of λ, β, γ, a	*MTTF with respect to failure rates*			
	Λ	*β*	*γ*	*a*
0.01	9.10966	5.23843	5.38504	5.63378
0.02	5.23843	5.14140	5.31083	5.52946
0.03	3.67178	5.04773	5.23843	5.42892
0.04	2.82579	4.95726	5.16779	5.33197
0.05	2.29646	4.86983	5.09885	5.23843
0.06	1.93409	4.78531	5.03154	5.14811
0.07	1.67047	4.70355	4.96582	5.06086
0.08	1.47007	4.62444	4.90163	4.97651
0.09	1.31260	4.54786	4.83893	4.89493

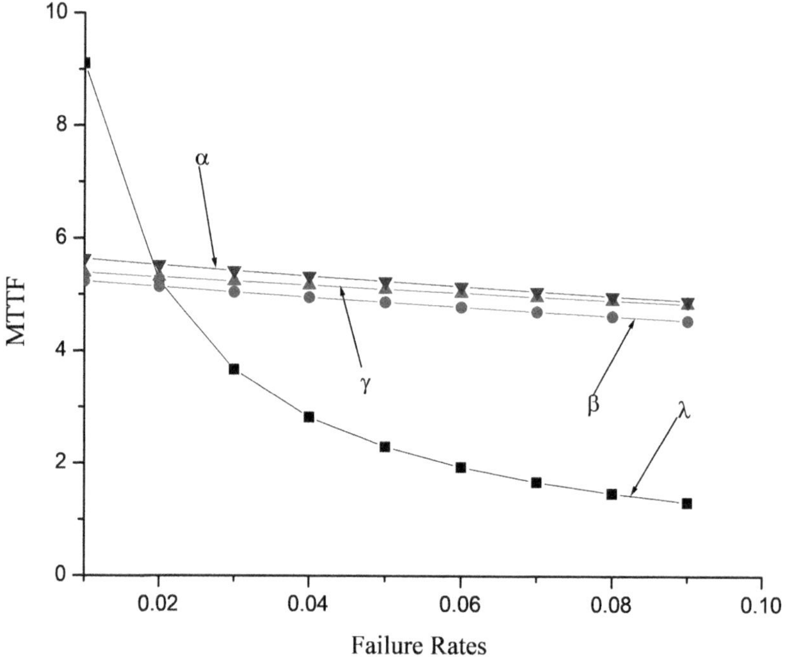

Figure 10.5 MTTF w.r.t. to failure rates.

10.13.4 Expected profit (Ep)

If the service facility is always available, then expected profit during the interval [0, t) is given by the formula with revenue per unit time K1 and K2 service cost per unit time:

The Ep for the interval [0, t) is calculated as follows:

$$E_p(t) = K_1 \int_0^t P_{up}(t)dt - tK_2$$

$$E_p(t) = K_1\left(-0.07951e^{(-1.05000\ t)} + 0.041493\right)e^{(-2.77438\ t)} + 0.02748e^{(-2.08553\ t)}\cos(0.78220\ t)$$
$$+ 0.10147e^{(-2.08553\ t)}\sin(0.78220\ t) + 0.00614e^{(-1.07102\ t)} + 0.00359e^{(-1.01048\ t)}$$
$$- 19.2267e^{(-0.05304\ t)} + 19.2275) - K_2 \tag{10.32}$$

On using K1 = 1 and K2 = 0.01, 0.03, 0.05, 0.07, 0.1, 0.3, 0.6, 0.9, respectively, in Equation (10.32), we get Table 10.4 and Figure 10.6.

Table 10.4 Ep (time varying)

	Expected profit Ep(t)							
Time (t)	*K2=0.01*	*K2=0.03*	*K2=0.05*	*K2=0.07*	*K2=0.1*	*K2=0.3*	*K2 = 0.6*	*K2 = 0.9*
0	0	0	0	0	0	0	0	0
1	0.97349	0.95349	0.93349	0.91349	0.88349	0.68349	0.38349	0.08349
2	1.90911	1.86911	1.82911	1.78911	1.72911	1.32911	0.72911	0.12911
3	2.79629	2.73629	2.67629	2.61629	2.52629	1.92629	1.02629	0.12629
4	3.63523	3.55523	3.47523	3.39523	3.27523	2.47523	1.27523	0.07523
5	4.42926	4.32926	4.22926	4.12926	3.97926	2.97926	1.47926	–0.02073
6	5.18135	5.06135	4.94135	4.82135	4.64135	3.44135	1.64135	–0.15864
7	5.89392	5.75392	5.61392	5.47392	5.26392	3.86392	1.76392	–0.33607
8	6.56912	6.40912	6.24912	6.08912	5.84912	4.24912	1.84912	–0.55087
9	7.20891	7.02891	6.84891	6.66891	6.39891	4.59891	1.89891	–0.80108
10	7.81512	7.61512	7.41512	7.21512	6.91512	4.91512	1.91512	–1.08487
11	8.38950	8.16950	7.94950	7.72950	7.39950	5.19950	1.89950	–1.40049
12	8.93369	8.69369	8.45369	8.21369	7.85369	5.45369	1.85369	–1.74630
13	9.44925	9.18925	8.92925	8.66925	8.27925	5.67925	1.77925	–2.12074
14	9.93767	9.65767	9.37767	9.09767	8.67767	5.87767	1.67767	–2.52232
15	10.4003	10.1003	9.80033	9.50033	9.05033	6.05033	1.55033	–2.94966

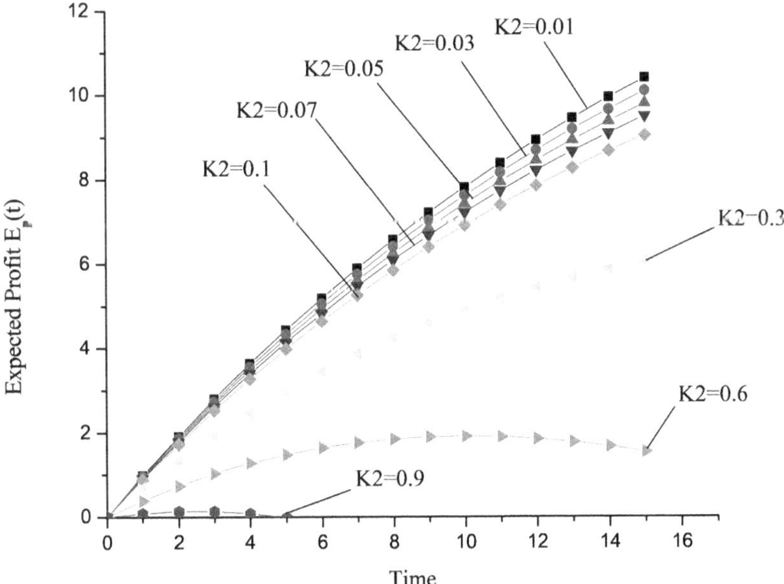

Figure 10.6 Expected profit.

10.13.5 Sensitivity analysis

The partial-derivative of the reliability function to a factor is known as sensitivity to that factor (Du and Nicholson, 1997). In the following analysis, these variables are failure rates.

1. Analysis of Sensitivity of R(t)
   The sensitivity of R(t) is measured as the percentage change as a result of a change in one of the system variables/parameters. On differentiating reliability function in relation to rate of failures, the reliability sensitivity can be computed by substituting values of λ, β, γ and α as 0.02, 0.01, 0.03 and 0.05, respectively, also '$n = 50$'. Varying $t = 0–15$, one can acquire Table 10.5 and Figure 10.7.
2. Sensitivity Analysis of MTTF
   The percentage change produced by a change in one of the system parameters is known as MTTF sensitivity analysis. Differentiating the equation of MTTF yields the sensitivity analysis of MTTF (31) in relation to failure rates α, β, λ and γ. On using values of $\alpha = 0.05$, $\beta = 0.01$, $\lambda = 0.02$, $\gamma = 0.03$ and $n = 50$, the authors obtain Table 10.6 and Figure 10.8.

Table 10.5 Sensitivity (reliability)

Time (t)	$\frac{\partial R(t)}{\partial \alpha}$	$\frac{\partial R(t)}{\partial \beta}$	$\frac{\partial R(t)}{\partial \lambda}$	$\frac{\partial R(t)}{\partial \gamma}$
0	–0.00001	0.100681^{-6}	0.0001	0
1	–0.03584	–0.91272	–2.45342	–0.82547
2	–0.04204	–1.53037	–14.8125	–1.19450
3	0.06235	–1.69325	–27.7296	–1.17888
4	0.23488	–1.49317	–32.4112	–0.96030
5	0.39900	–1.13799	–29.4641	–0.69477
6	0.50960	–0.78661	–23.0527	–0.46657
7	0.55849	–0.51052	–16.4782	–0.30119
8	0.55750	–0.32061	–11.2434	–0.19336
9	0.52356	–0.20095	–7.61035	–0.12783
10	0.47124	–0.12997	–5.29499	–0.08978
11	0.41071	–0.08948	–3.89481	–0.06818
12	0.34825	–0.06683	–3.06882	–0.05588
13	0.28732	–0.05411	–2.57933	–0.04862
14	0.22967	–0.04671	–2.27798	–0.04401
15	0.1605	–0.04209	–2.07868	–0.04079

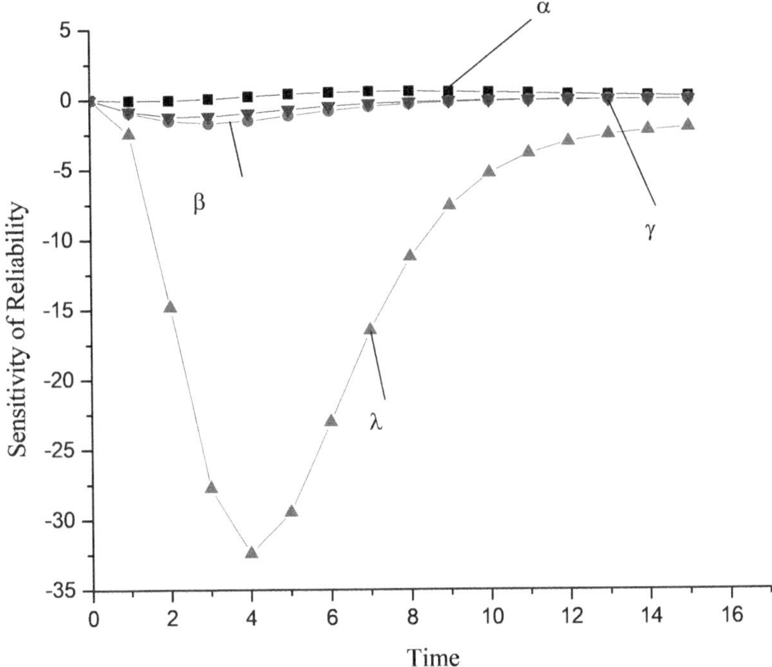

Figure 10.7 Sensitivity of reliability.

10.14 RESULTS

Figures 10.3–10.8 explain the different measures in their graphical formation. The authors used this information to interpret the tables to study the measures further.

Table 10.6 Sensitivity (MTTF)

Failure rate $(\alpha, \beta, \lambda, \gamma)$	$\frac{\partial MTTF}{\partial \alpha}$	$\frac{\partial MTTF}{\partial \beta}$	$\frac{\partial MTTF}{\partial \lambda}$	$\frac{\partial MTTF}{\partial \gamma}$
0.01	−10.6297	−9.87596	−669.230	−7.51455
0.02	−10.2397	−9.53208	−223.281	−7.32952
0.03	−9.87077	−9.20466	−109.887	−7.15086
0.04	−9.52139	−8.89273	−65.1236	−6.97829
0.05	−9.19023	−8.59537	−43.0229	−6.81156
0.06	−8.87606	−8.31176	−30.5212	−6.65041
0.07	−8.57773	−8.04110	−22.7701	−6.49460
0.08	−8.29419	−7.78266	−17.6357	−6.34390
0.09	−8.02447	−7.53575	−14.0605	−6.19811

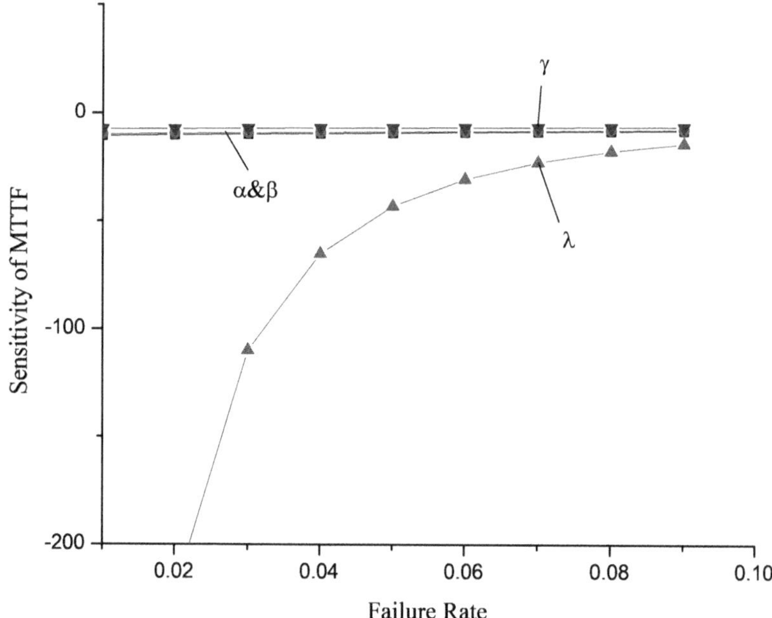

Figure 10.8 Sensitivity (MTTF).

From the graph of availability (Figure 10.3), one can fundamentally inspect that the availability decreases as time increases. This proves that it is inversely proportional to time. Hence, for any system, the availability is maximum at time $t = 0$. The reliability (time function) is depicted in Figure 10.4, in an interval of 0–15. Efficiency decreases or the risk of failure rises as time passes. As seen in the graph, reliability is a non-increasing function.

Figure 10.5 shows the system's MTTF in relation to failure rates in the gap of a numeric value of 0.01. For α, β and γ, the MTTF is nearly equal. In the case of nodes, the MTTF is a sudden u-curve, as seen from the figure. MTTF increases with an increase in the rate of failures for this network system. Figure 10.6 depicts the expected profit as time progresses, in an interval of 0–15 at the rate of service cost. It can be seen from the graph that the system's expected profit declines as time passes. So, to maximize profit, it is necessary to keep service costs under control.

The sensitivity of reliability as a time-function is depicted in Figure 10.7. The sensitivity of parameter α has positive sign, which means a little increase in parameter's value will improve $R(t)$. The sensitivity of parameters β, λ and γ has a negative sign, which implies that a minor increment in parameter's value will cause a decline in $R(t)$. The sensitivity of MTTF is shown in Figure 10.8 as a function of failure rates. The curves so obtained show that the sensitivity of MTTF is nearly similar in the case of the repeater and cable, as the failure rates differ by 0.01 in Table 10.6. The curve of sensitivity of MTTF differs from other failure rates in the case of nodes. The sensitivity of parameters λ has a negative sign, which means a somewhat increment in parameter's value will abate MTTF. The parameters α, β and γ rarely affect the MTTF.

10.15 CONCLUSION

This research is valuable for deciding the reliability measures of a ring topology system, including computers, repeaters and cables. In this chapter, apart from availability, MTTF, reliability and cost, the authors have investigated the network system's sensitivity of reliability and sensitivity of MTTF as a distinction from other models. The availability, reliability, MTTF, cost and sensitivity examination of networks in ring topology by employing Markov process has been investigated. These reliability measures offer necessary conditions by which substitute design plans and strategies can be judged and compared and benefit the system designer to choose the one that fulfills the objectives well under specific technical/economic constraints. The graphs of reliability and availability indicate that as the time increases, the availability and reliability diminish. Also, the reliability and availability of a system are maximum at time $t = 0$. Henceforth, availability and reliability are inversely proportional to time, as can be seen by the curves obtained in Figures 10.3 and 10.4. By improving the design of any system, its reliability can be improved, as shown in this study. The graph of MTTF displays the change of MTTF in relation to different rates of failures. A variation of 0.01 has been taken among the failure rates as shown in Table 10.3. As the rate of failure increases, the MTTF of the system decreases. Also, from Figure 10.6, it is clear that with an increase in time, the rate of service cost increases as expected profit decreases. So, for maximum profit, regulating service costs is essential. Finally, one can see that as service costs fall, profit grows with the time. In general, the projected profit for low service costs is higher than for high service costs. The sensitivity of reliability illustrates the deviation as time increases in intervals of 0, 1, 2, ..., 15 as shown in Table 10.5. Finally, a sensitivity analysis for system reliability and the MTTF for specific values of λ, γ, α, and β have been performed. The graph of sensitivity of MTTF depicts the change in MTTF w.r.t the failure rates. Figure 10.8 conveys the sensitivity of MTTF and how it rises through each failure, ranging 0.01–0.09. The graph clearly shows the structure's MTTF is extra sensitive in case of the node in the network since it follows a steep curve. This study's thorough investigation of reliability analysis proves that failure of a node is extra sensitive to the system's failure and acts as a catalyst for improving reliability techniques for future work.

CONFLICTS OF INTEREST

The author(s) declare no conflict of interest.

ACKNOWLEDGEMENTS

The authors are grateful to 'Graphic Era Deemed to be University, Dehradun, Uttarakhand, India', for support. This article was written in good faith using information available at the time of publication. The views, findings and conclusion articulated are of the author.

REFERENCES

Ahmad, W., Hasan, O., Pervez, U., & Qadir, J. (2017). Reliability Modeling and Analysis of Communication Networks. *Journal of Network and Computer Applications*, *78*, 191–215.

Alfa, A. S., & Rao, T. S. (2000). Supplementary Variable Technique in Stochastic Models. *Probability in the Engineering and Informational Sciences*, *14*(2), 203–218.

Birolini, A. (2013). Reliability Engineering: Theory and Practice. *Springer Science & Business Media*.

Chopra, G., & Ram, M. (2019). Reliability Measures of Two Dissimilar Units Parallel System Using Gumbel-Hougaard Family Copula. *International Journal of Mathematical, Engineering and Management Sciences*, *4*(1), 116–130.

Du, Z. P., & Nicholson, A. (1997). Degradable Transportation Systems: Sensitivity and Reliability Analysis. *Transportation Research Part B: Methodological*, *31*(3), 225–237.

Gonen, T. (1986). Electric Power Distribution System Engineering, *McGraw Hill*.

Heimann, D. I., Mittal, N., & Trivedi, K. S. (1990). Availability and Reliability Modeling for Computer Systems. In Advances in Computers, 31, 175–233. *Elsevier*. doi:10.1016/s0065-2458(08)60154-0.

Huang, C. F., Huang, D. H., & Lin, Y. K. (2020). Network Reliability Evaluation for a Distributed Network with Edge Computing. *Computers & Industrial Engineering*, *147*, 106492.

Kounev, V., Lévesque, M., Tipper, D., & Gomes, T. (2016). Reliable Communication Networks for Smart Grid Transmission Systems. *Journal of Network and Systems Management*, *24*(3), 629–652.

Kumar, A., Pathak, R. M., & Gupta, Y. P. (1995). Genetic-Algorithm-Based Reliability Optimization for Computer Network Expansion. *IEEE Transactions on Reliability*, *44*(1), 63–72.

Kumar, A., & Ram, M. (2019). Computation Interval-Valued Reliability of Sliding Window System. *International Journal of Mathematical, Engineering and Management Sciences*, *4*(1), 108–115.

Kuo, W., & Prasad, V. R. (2000). An Annotated Overview of System-Reliability Optimization. *IEEE Transactions on Reliability*, *49*(2), 176–187.

Li, J. (2018). SIL Implementation on Safety Functions in Mass Transit System. *International Journal of Mathematical, Engineering and Management Sciences*, *3*(3).

Lin, M. S., Chang, M. S., Chen, D. J., & Ku, K. L. (2001). The Distributed Program Reliability Analysis on Ring-Type Topologies. *Computers & Operations Research*, *28*(7), 625–635.

Lin, Y. K., & Yeh, C. T. (2011). Computer Network Reliability Optimization under Double-Resource Assignments Subject to a Transmission Budget. *Information Sciences*, *181*(3), 582–599.

Liu, Q., & Liu, Q. (2014, October). A Study on Topology in Computer Network. In *2014 7th International Conference on Intelligent Computation Technology and Automation* (pp. 45–48). *IEEE*.

Nagiya, K., & Ram, M. (2014). Performance Evaluation of a Computer Workstation under Ring Topology. *Journal of Engineering Science and Technology*, *9*(1), 91–103.

Ryu, D., & Chang, S. (2005). Novel Concepts for Reliability Technology. *Microelectronics Reliability*, *45*(3–4), 611–622.

Schiff, J. L. (1999). The Laplace Transform: Theory and Applications. *Springer Science & Business Media*.

Shooman, A. M. (1995, June). Algorithms for Network Reliability and Connection Availability Analysis. In *Proceedings of Electro/International*, 309–333. IEEE.

Singh, V. V., Poonia, P. K., & Rawal, D. K. (2021). Reliability Analysis of Repairable Network System of Three Computer Labs Connected with a Server under 2-Out-of-3 G Configuration. *Life Cycle Reliability and Safety Engineering*, *10*(1), 19–29.

Underberg, L., Kays, R., Dietrich, S., & Fohler, G. (2018, May). Towards Hybrid Wired-Wireless Networks in Industrial Applications. In *2018 IEEE Industrial Cyber-Physical Systems (ICPS)*, 768–773. IEEE

Watcharasitthiwat, K., & Wardkein, P. (2009). Reliability Optimization of Topology Communication Network Design Using an Improved Ant Colony Optimization. *Computers & Electrical Engineering*, *35*(5), 730–747.

Yeh, C. T., & Fiondella, L. (2017). Optimal Redundancy Allocation to Maximize Multi-State Computer Network Reliability Subject to Correlated Failures. *Reliability Engineering & System Safety*, *166*, 13–150.

Yoon, G., Kwon, D. H., Kwon, S. C., Park, Y. O., & Lee, Y. J. (2006, October). Ring Topology-Based Redundancy Ethernet for Industrial Network. In 2006 *SICE-ICASE International Joint Conference*, 1404–1407. IEEE. doi:10.1109/sice.2006.315661.

Yunus, N. A. M., Othman, M., Hanapi, Z. M., & Lun, K. Y. (2018). Enhancement Replicated Network: a Reliable Multistage Interconnection Network Topology. *IEEE Systems Journal*, *13*(3), 2653–2663.

Zio, E. (2009). Reliability Engineering: Old Problems and New Challenges. *Reliability Engineering & System Safety*, 94(2), 125–141. doi:10.1016/j.ress.2008.06.002.

Zuev, K. M., Wu, S., & Beck, J. L. (2015). General Network Reliability Problem and Its Efficient Solution by Subset Simulation. *Probabilistic Engineering Mechanics*, *40*, 25–35.

Chapter 11

Generic model and reliability measures of multi-state system analysis with a case study from industry

Sarfraz Ali Quadri, Shrikrishna B Pawar, Varsha Jadhav, and Dhananjay R Dolas

11.1 INTRODUCTION

The study of multi-state system (MSS) reliability surpasses limitations of conventional binary reliability analysis, which assumes that components can only exist in two states: fully functional or completely failed. In reality, numerous systems are composed of components that possess multiple levels of performance and various failure modes, which affect the overall system performance in diverse ways. This is where MSS reliability becomes relevant. MSS models, extensively used in showcasing the system reliability with fixed performance levels/rates, have garnered significant attention in previous research endeavours. These investigations predominantly focus on evaluating the reliability of MSSs, assuming that the transition between different performance states occurs instantaneously and without any time delay. However, it is important to note that certain engineering systems, such as gas supply systems and heating distribution systems, exhibit gradual transitions between different states.

Every system is meticulously designed to effectively carry out their designated work within its specific environment. It is worth mentioning that some systems possess the ability to perform their work with varying levels of effectiveness, i.e. performance rates. If system/systems are able of carrying a number of performance rates, they are designated as an MSS. It is very significant to observe that an MSS is usually formed of components that can also possess multiple states. In fact, the simplest form of an MSS is a binary system, which encompasses two distinct states, i.e. functioning perfectly or in the state of complete failure. The fundamental principles and reliability theories of MSS were mainly presented in the 1970s by Levitin in 2005 [1] and El-Neveihi et al. in 1978 [18]. Later, Hudson and Kapur in 1982 [10] contributed upon the findings of these works. Since then the field of MSS reliability has experienced a considerable rush in development and exploration. For researchers who have an interest in delving into the historical aspects of MSS, a comprehensive literature survey can be found in the works of Lisnianski and Levitin [2], [5], [9].

DOI: 10.1201/9781003546214-11

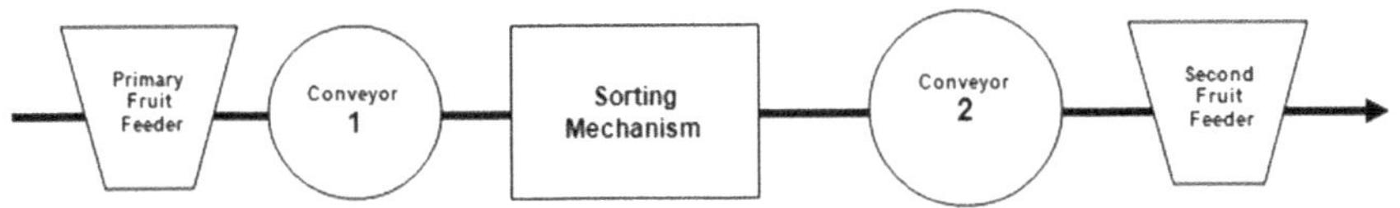

Figure 11.1 Example of multi-state fruit transmission line.

Figure 11.1 presents a fruit packaging line at a food processing industry that continuously supplies the fruits to the jam preparation process and consists of five core elements:

- A primary fruit feeder, which loads the raw fruits to Conveyor 1
- Next is Conveyor 1 which transfer the fruits to the sorting mechanism
- Sorting mechanism which remove any damaged fruits and proceeds the useful fruits to Conveyor 2
- Conveyor 2 loads the fruits in the second fruit feeder
- The second fruit feeder supplies the selected fruits to the boilers for further processing.

The quantity of fruits transferred to the boilers at each iteration proceeds uninterruptedly through every stage of the line [3]. The feeder and the sorter mechanism consist of two stages, i.e. nominal quantity performance or a complete breakdown. The nominal quantity output may change relying mainly on the availability of individual feeders and sorter mechanisms. In real-world scenarios there are numerous diverse considerations that can make the system multi-state [4]

- When various components having binary states jointly impact the overall performance of the system, the system is designated as multiple state system [6]. These systems are dependant on the availability of their elements for perfect performance, since various levels of availability may result in changing performance levels [7]. A simple example for this system is a typical *k-out-of-n system*, which can demonstrate $n + 1$ state and consist of identical n binary components based on availability of number of components.
- The varying performing rates of components in an MSS may vary because of changing environmental conditions or due to deterioration [8]. The failure in performance of such components can deteriorate the performance of the overall system. In many cases, the levels of performance of an individual component may change, i.e. faultless function or a full breakdown.

11.2 THE GENERIC MODEL OF MULTI-STATE SYSTEM (MSS)

A broad outline that comprehends and scrutinizes systems that are able to perform in various states, each categorized by their rate of performance is typically the Generic model of MSS [11]. In distinction to the conventional binary models, which only recognize the state of perfect functionality and the state of complete breakdown. Contrarily, the generic model of MSS allows a more distinguished understanding of system dynamics.

For elaboration one *k-out-of-n* system is discussed where it is typically a 2-out-of-3 system with three binary components [12]. Their rates of performance for i = 1, 2, 3 can be showcased as

$$G_i(t) \in \{g_{i1}, g_{i2}\} = \{1,0\},$$

i.e.

g_{i1} = {1 if ***i*** performs perfect; 0 if ***i*** fails completely} and
G(t) = {2 if all performs perfect, **1** if only one element fails, 0 failed elements are more than one}

The various criteria of system structure functions and their states with possible values are mentioned in Table 11.1.

To analyse the behaviour of multiple state system, it is vital to recognize the individualities of its various elements. Based on their rates of performance, any element ***j*** can have ***kj*** states where $i \in \{1, 2, 3, \ldots, kj\}$ i.e.

$$g_j = \{g_{j1}, g_{j2}, g_{j3}, \ldots\ldots, g_{jk_j}\}$$

where g_{ji} is the rate of performance of ***j*** component in the performance state of *i*.

Therefore, the rate of performance of element ***j*** becomes a stochastic process. Simultaneously, the performance of elements can be dependant of more values mainly vectors and cannot be dependant on a sole value. In such scenarios, the element ***j*** performance is stated as a *Gj*(*t*), i. e. vector stochastic process. The performance rate probability of system element at any given time ***t*** can be showcased as

$$p_j(t) = \left[p_{j1}\{t\}, p_{j2}\{t\}, p_{j3}\{t\}, \ldots\ldots\ldots, p_{jk_j}\{t\}\right] \tag{11.1}$$

Table 11.1 All the possible states of the stated MSS

G_1(t)	0	0	0	1	1	1	1	0
G_2(t)	1	0	0	1	1	0	0	1
G_3(t)	0	1	0	1	0	0	1	1
f **{G_1(t), G_2(t), G_3(t)}**	0	0	0	1	1	0	1	1

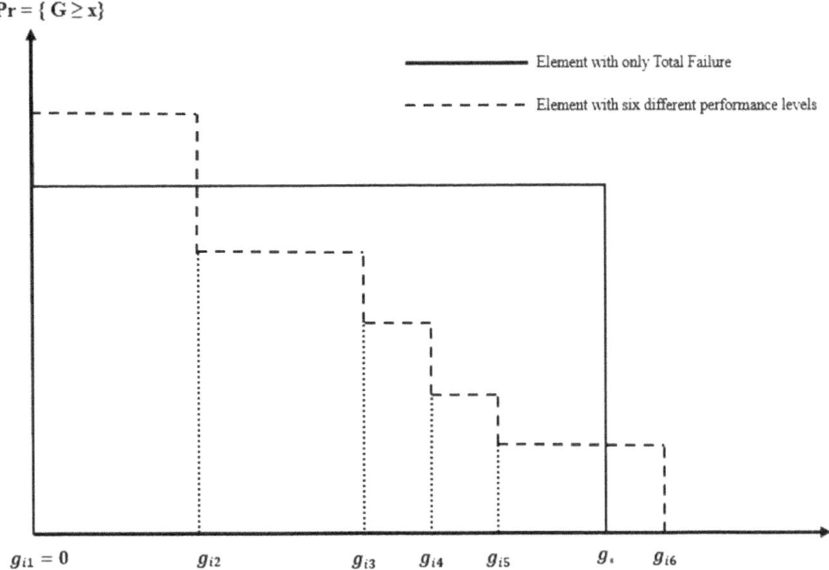

Figure 11.2 Performance curves for binary element and MSS element.

where

$$p_{ji}(t) = \Pr\{G_j(t) = g_{ji}\}$$

The steady-state performance distribution can be graphically represented by cumulative curves [13]. In which, every performance value of x relates to a rate of performance of the component that is more than or equal to their particular level and represented as Pr {Gj ≥ x}.

In order to facilitate comparison, Figure 11.2 displays graphs illustrating the performance distribution of two elements, i.e. multi-state element j consisting of six states of performance and the element b with only binary states of performance. It is worth noting that the cumulative discrete probability distribution consistently exhibits a decremental stepwise pattern.

11.3 THE RELIABILITY MEASURES MULTI-STATE SYSTEM

The reliability measures are quantitative tools used to assess the performance of an MSS with its intended function [14]. Unlike the consideration of conventional binary systems, MSS can exist in multiple states with changing performance rates. These measures help engineers and analysts understand the system's behaviour, identify potential weaknesses, and optimize its performance. Some key MSS reliability measures are as follows:

- **State Probability:** The possibility of the system being in an exact state at a given time that can be calculated by analysing the system's transition behaviour between different states.
- **System Availability:** The possibility of the system being in a state that meets the required performance level at a given time, which is often expressed as a percentage.
- **System Unavailability:** The possibility of the system being not available to execute the intended functions at a given instant, often calculated as 1 – availability.
- **System Mean Time between Failures:** The MTBF is the average time between two consecutive failure within the operating life of system.

  Apart from these some other aspects include System Mean Downtime, i.e. time of failure state of system, state lifetime distribution, system uptime, i.e. perfectly functioning time of system, system efficiency, i.e. ratio of input to output, and transition rates, i.e. the dynamic nature of system to move from perfectly function to failure state due the operating life [15].

The specific measures used depend on the system's characteristics, desired information, and operational context. A permutation of numerous measures might be required to achieve the comprehensive understanding of the reliability of system. By analysing these measures, one can identify critical components that significantly impact system performance, develop maintenance strategies to improve system availability and efficiency, predict system behaviour and performance under different operating conditions, and optimize resource allocation and decision-making for better system management [16].

To understand the behaviour of MSS and its characteristics, it is of utmost necessity to numerically determine its reliability indices which are usually connected to the consequent indices of reliability for a two state conventional system [17]. To represent the stochastic behaviour of MSS performance, an example is shown in Figure 11.3. The system behaviour is characterized by its performance if the performance distribution is not dependant on time.

In the evolution of MSS some arbitrary variables are of prime importance like T_f (time to failure), i.e. time from start of operation to the first start of failure state, T_b (time between failures), i.e. time amongst two successive transitions between working and failure state. The number of failures (N_T) shows failure iterations during the time T.

Figure 11.3 showcases $W(t)$ and $G(t)$ as two processes (stochastic). If the system performance value increases, then the demand value i.e. $F(G(t), W(t)) = G(t) - W(t)$.

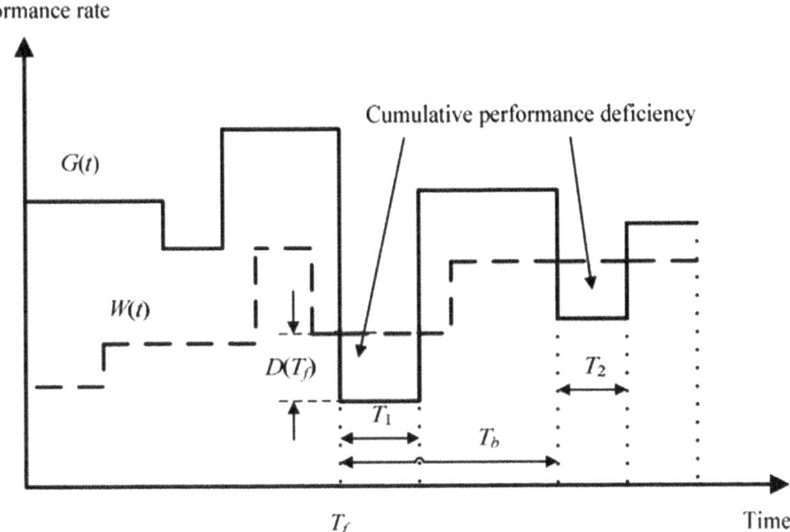

Figure 11.3 MSS behaviour as stochastic process (Lisnianski, Frenkel, Ding 2010).

In that instant, time to MSS failure is determined by the first instant that the performance crosses the demand level which is represented as T_f and can be characterized by the indices as follows:

- The reliability function $\boldsymbol{R(t)}$, i.e. probability of $T_f \geq t$, $(t > 0)$ at the state of t = 0;

$$R(t) = \Pr\left\{T_f \geq t \mid F\left(G(0), W(0)\right)\right\} \geq 0 \tag{11.2}$$

- The *Mean Time To Failure* is when for the first time the system experiences the failure state:

$$MTTF = E\{T_f\}$$

- when the $W(t) = w$, i.e. constant demand level, then the availability of MSS, i.e. A_∞ (w) can be represented as

$$A_\infty(w) = \sum_{k=1}^{k} p_k 1(F(g_k, w) \geq 0 \tag{11.3}$$

where $p_k = \lim_{t \to \infty} p_k(t)$, i.e. MSS $\boldsymbol{k}$ with rate of performance as g_k. It can be represented as

$$A_\infty(w) = \sum_{k=1}^{k} p_k 1(g_k \geq w) = \sum_{g_k \geq w} p_k \tag{11.4}$$

- For representing the mean value of MSS when the rate of performance at time t

  $G_{mean}(t) = E\{G(t)\}$

  As $p_k = \lim_{t \to \infty} p_k(t)$, performance expectation is represented as

$$G_{\infty} = \sum_{k=1}^{k} p_k g_k \tag{11.5}$$

- For representing the mean value of MSS when the rate of deficiency in performance at time t

  $D_m(t,w) = E[D(t,w)]$

  The rate of deficiency in performance can also be obtained from the system performance if the demand is constant:

$$D_{\infty} = \sum_{k=1}^{k} p_k \max\{w - g_k, 0\} \tag{11.6}$$

11.4 CASE STUDY

For the study purpose two hydraulic pumps with a nominal capacity of 100 litre per minute (LPM) from an industry located in Marathwada region of Indian state of Maharashtra are considered. Both the pumps are considered as separate MSSs. The typical failure reduces the delivery volume of first pump to 70 LPM and other issues put pump to a whole failure. Similarly, the typical failure reduces the delivery volume of second pump to 70 LPM and other types of failures reduce pump capacity to 30 LPM; furthermore some types of failures lead the pump to a complete failure.

For the first pump we have assumed the probabilities as $P_{11} = 0.1$ and $P_{12} = 0.5$ and $P_{13} = 0.2$; consecutively, for the second pump the assumed probabilities are $P_{21} = 0.05$ and $P_{22} = 0.25$ and $P_{23} = 0.4$ and $P_{24} = 0.3$.

The nominal capacity can be represented as the ratio of capacity and demand. For the first pump here are three possible relative capacities, i.e. $g_{11} = 0.0$, $g_{12} = 70 \div 100 = 0.7$, $g_{13} = 100 \div 100 = 1.0$.

Similarly, for the second pump there are four relative capacity levels, i.e. $g_{21} = 0.0$, $g_{22} = 30 \div 100 = 0.3$, $g_{23} = 70 \div 100 = 0.7$, $g_{24} = 100 \div 100 = 1.0$.

The delivery volume necessary is 60 LPM, i.e. $w = 60/100 = 0.6$.

The stationary availability of the MSS from Eq. (11.4) is

$$A_{\infty 1}(w) = A_1(0.6) = \sum_{g_{1k} \geq 0.5} p_{1k} = 0.5 + 0.2 = 0.7$$

$$A_{\infty 2}(w) = A_2(0.5) = \sum_{g_{2k} \geq 0.5} p_{2k} = 0.4 + 0.2 = 0.6$$

The rate of performance expected is taken from Eq. (11.5), i.e.

$$G_{1\infty} = \sum_{k=1}^{3} p_{1k} g_{1k} = 0.1 \times 0 + 0.5 \times 0.7 + 0.2 \times 1.0 = 0.55$$

This shows that for first pump there is 55% capacity of the nominal generating capacity:

$$G_{2\infty} = \sum_{k=1}^{4} p_{2k} g_{2k} = 0.05 \times 0 + 0.25 \times 0.3 + 0.4 \times 0.7 + 0.3 \times 1.0 = 0.65$$

This shows that for second pump there is 65% capacity of the nominal generating capacity.

The deficiency in rate of performance is taken from Eq. (11.6)

$$D_{1\infty}(0.6) = \sum_{g1k-w<0} p_{1k}\left(W - g_{1k}\right) = 0.1 \times \left(0.6 - 0.0\right) = 0.06$$

$$D_{2\infty}(0.6) = \sum_{g2k-w<0} p_{2k}\left(W - g_{2k}\right)$$

$$= 0.05 \times \left(0.6 - 0.0\right) + 0.25 \times \left(0.6 - 0.5\right) = 0.055$$

For first pump the total volume of unfulfilled demand is 6 LPM, whereas for the second pump the total volume of the unfulfilled supply is 5.5 LPM. In this scenario, D_∞ is the expected amount of liquid unsupplied. It is worth noting here that nature of the reliability indices is different so they are not interchanged. When the availability parameter is considered, the first pump performs better than the second pump, i.e. $A_1(0.6) > A_2(0.6)$. Whereas when the expected capacity is considered, second pump performs better than the first pump, i.e. $G_{1\infty} < G_{2\infty}$ and the second pump has more unsupplied demand than the first pump, i.e. $D_{1\infty} = D_{2\infty}$.

REFERENCES

1. Levitin G (2005) *Universal generating function in reliability analysis and optimization*. Springer, London.
2. Lisnianski A, Levitin G (2003) *Multi-state system reliability: assessment, optimization and applications*. World Scientific, Singapore.
3. Malinowski J, Preuss W (1995) Reliability of circular consecutively connected systems with multistate components. *IEEE Trans Reliab* 44:532–534.
4. Marshall G, Jones R (2007) Multi-state models in diabetic retinopathy. *Stat Med* 14(18):1975–1983.
5. Murchland J (1975) Fundamental concepts and relations for reliability analysis of multistate systems. In: Barlow RE, Fussell JB, Singpurwalla N (eds) *Reliability and fault tree analysis: theoretical and applied aspects of system reliability*. SIAM, Philadelphia: pp 581–618.

6. Billinton R, Allan R (1996) *Reliability evaluation of power systems*. Plenum, New York.
7. Doulliez P, Jamoulle E (1972) Transportation networks with random arc capacities. *RAIRO* 3:45–60.
8. El-Neweihi E, Proschan F (1984) Degradable systems: a survey of multistate system theory. Commun Stat Theory Methods 13:405–432.
9. Giard N, Lichtenstein P, Yashin A (2002) A multi-state model for genetic analysis of the aging process. Stat Med 21:2511–2526.
10. Hudson JC, Kapur KC (1982) Reliability theory for multistate systems with multistate components. Microelectron Reliab 22:1–7.
11. Van den Hout A, Matthews F (2008) Multi-state analysis of cognitive ability data. Stat Med, published on line Wiley Interscience (www.interscience.wiley.com) DOI: 10.1002/3360.
12. Xie M, Dai YS, Poh KL (2004) *Computing system reliability: models and analysis*. Kluwer/Plenum, New York.
13. Bier V, Nagaraj A, Abhichandani V (2005) Protection of simple series and parallel systems with components of different values. Reliab Eng Syst Saf 87:315–323.
14. Levitin G (2003) Optimal multilevel protection in series-parallel systems. Reliab Eng Syst Saf 81:93–102.
15. Levitin G, Dai Y, Xie M, Poh KL (2003) Optimizing survivability of multi-state systems with multi-level protection by multi-processor genetic algorithm. Reliab Eng Syst Saf 82:93–104.
16. Korczak E, Levitin G, Ben Haim H (2005) Survivability of series-parallel systems with multilevel protection. Reliab Eng Syst Saf 90(1):45–54.
17. Levitin G (2007) Optimal defense strategy against intentional attacks. IEEE Trans Reliab 56(1):148–157.
18. El-Neveihi E, Prochan F, Setharaman (1978) Multi-state coherent systems. J Appl Probab 15:675–688.

Chapter 12

Reliability-centered maintenance (RCM) for multistate systems

Prathamesh R. Potdar and Santosh B. Rane

12.1 INTRODUCTION

Reliability-centered maintenance (RCM) is a holistic method that enables the efficiency, reliability, and security of multistate systems and processes in different industries. Rather than working according to the decision-based or schedule-based maintenance routines, practicing RCM involves finding within the system means of possible faults, as well as the consequences of them and as a result choosing the most appropriate maintenance approach to protect the system from failures. RCM is therefore the most important tool in modern maintenance strategies which is the fundamental list of steps that leave room for improvement of maintenance strategies. RCM came into being as a response to the increasing aerospace sector's complexity and the need for more effective maintenance techniques that were supposed to take care of it. Before that, however, the original ideas of the company have been adapted and have managed to become not only a vital part of production, transportation, energy production, and healthcare but something much more meaningful to the mentioned industries. RCM is the idea that every component is vital for the whole system and their criticality and preventive maintenance requirements are not the same. By emphasizing putting resources and efforts into areas where consequences of failure are most likely to affect the system operations, RCM permits organizations to deploy their assets properly and make them operate efficiently. The uptime is therefore increased, costs are cut down, and system performance is improved. We determine what specific components or systems are critical for the efficient functioning of a given asset at the first step of RCM. Featured components that may have a worse performance compared to all others if they fail are the ones directly linked to safety, operations, or environment. Important entities should be determined first, and a detailed analysis of possible failure methods is a must after that. Here all causes of component failures need to be established and analyzed and possible effects for each failure mode investigated. RCM identifies all the consequences of failure whether risky and life-threatening, operationally adverse and environmentally harmful, and also of financial causes. These evaluations determine how the work

DOI: 10.1201/9781003546214-12

should be ended in the order of the degree of all possible negative consequences. RCM identifies suitable preventive measures for each design part or system according to the functions, and parts criticality, and by simulating outcomes. Preventive maintenance, predictive maintenance, condition-based maintenance, or a combination of these tactics are examples of potential techniques.

12.2 RCM PRINCIPLES

The rigorous approach to maintenance planning known as RCM concentrates on ensuring the performance, safety, and dependability of complex systems. The fundamental principles of RCM include the following:

a. **Identification of Critical Components:** RCM starts by identifying those key components that, if not working efficiently, might cost a lot of money, affect efficiency, and threaten safety. To accomplish this, the crucial processes and the scheme clarifying the system's operations are revealed in the overview.
b. **Identifying Critical Functions:** Understanding the big roles of the multistate system and their importance in the operation system can lead to the selection of maintenance priorities.
c. **Determination of Failure Modes:** RCM embraces two steps; the first one is identifying the critical components of the system and the second is listing the probable failure modes of each of these components which could cause failure of the system. The particular assignment involves internal as well as external operational audits, evaluating the errors caused in such situations that cause and the signs of the errors.
d. **Assessing Failure Modes:** Elucidating potential fall lines and the latter's effects is paramount to discern the maintenance tactics that should be implemented to minimize the risks to system performance and general safety.
e. **Assessment of Consequences:** All the RCM failure modes, namely operations, safety, environmental compliance, and costs, are also separately assessed. The specific outcome of an error can vary greatly in accordance with the factors of the system, such as the degree of the mistake, the extent of the affected region, and the operational state.
f. **Determining Maintenance Strategies:** Each affected area concerning operations, safety, environmental compliance, and cost will be evaluated by RCM. Different from variables such as the error's magnitude, the portion's importance in the main system, and the current operation status, there could be different results.
g. **Selection of Maintenance Strategies:** An RCM aims at conflicting with failure emergence. According to the nature or outcome of the failure, it chooses the corresponding maintenance strategy for correct

maintenance of operability. Among other strategies, these would include fail-safe predictive maintenance, which tracks conditions and tends to root out the sources of the problems, proactive maintenance, which sees systems anew, and preventive maintenance which often involves poking and probing or frequent reconditioning and part replacement.

h. **Optimizing Maintenance Intervals:** The maintenance schedules optimally are devised according to the fact some different parameters are considered such as component reliability, wear trend, and resource use. This reduces the probability of unscheduled failures and increases resource availability at the same time.
i. **Continuous Improvement:** The trustworthiness and smoothness of maintenance as well as livelihood efficiency are pursued and achieved by reviewing and improving maintenance schedules based on performance data, feedback from operational experience as well as technological developments.

The goal is that instead of being overwhelmed by Risk-Contingency Management, organizations are to prevent casualties and violations and strive to reduce the issues of the RCM framework itself; this can be achieved by practicing this thought and it will make the entire system more reliable, performant, and stable in multiple states. Here, the RCM implementation process greases the development part of strategic considerations that follow specific norms when adapting suggestions of RCM to solutions of multistate system challenges:

a. **Dynamic Identification of Critical Components:** The operating system components can be different from one state to important in multistate systems which are different in any given state. In this regard, therefore, RCM needs to factor in all of this as only this will allow an effective measurement of impact to be done not only taking into consideration how the system is configured but also on how the system is used and all the components of the system.
b. **State-Based Failure Mode Analysis:** The omission from the RCM analysis of various system fault modes should be taken into account as diverse operations and procedures need to be covered. From these outcomes, custom-defined pinpointing of barriers and trials of the systems' overall efficiency with the changes in each stage of the integrated process has to be done.
c. **Assessment of State-Dependent Consequences:** States operated by multistate systems can lead to different consequences depending on the mode of the operating system. Insufficient quality of the operating mode implies the different adverse effects on every state of system behavior. First, by utilizing fault mode relevance at various stages of operation as well as implementing maintenance tasks in their order of priority, RCM will ensure an effective routine.

d. **Adaptive Maintenance Strategies:** System maintenance ways must take a new path and method of operation because the failure modes may have changed and wind over time. Adjustments can be used differently, either to improve the performance of the system or to remediate an identified issue. They can include changing the time of normal maintenance activities or the schedule, as well as adapting the criteria being monitored.
e. **Integration of Redundancy and Fault Tolerance:** For multistate systems, seeing if the system has redundancy, along with measures of fault tolerance, which are designed to minimize the extent of failures, come in handy. RCM can take the lead and promote this position. The maintenance plan that allows components, rotating parts and other most capable items to be replaced when needed, will be the most effective and adequate if it is formed at the stages of equipment operation.

12.3 MULTISTATE SYSTEMS

The multistate systems like power grids are made of interconnected components that can perform in different states such as normal, in service, and trouble. These states will have different properties, dependencies, and ways of failing. Multistate systems, in particular, serve to move between various operation systems which are a result of factors such as operating conditions, environmental factors, and complex components. The processes may include normal, reduced, and standby functioning and defects. In densely connected systems, subsystems regularly provide prosections and knock-offs to each other, and thus, mutual performances are interdependent. Hereby, the linkages may become powerful which may lead to an uninterrupted chain reaction of the system issues, starting with the failure of one element, and spilling down to other parts of the system. His words spoke volumes for multistate systems that show the most common types of failures, like mechanical failures, electrical failures, software glitches, and environment deterioration. The ways of revealing any particular failure mode, its dependence on the operating conditions of the system, and the interaction between the elements are different within different systems. Fragments of multistate systems may break down over time, but their performance and reliability can change because these are points of stress. Deterioration of such a system can be the thing that may happen without notice affecting its ability to perform its job properly and increasing the chance of malfunction. Dual-state systems have redundancy and fault rejection parameters to mitigate the effects of failure of components. The substitution of overburdened parts or subsystems can appropriately fill the gap of dysfunctional components and lead to non-interrupted functioning and system's ability to attain its de facto significance. The certainty of the mechanism of a molded case circuit breaker (MCCB) is stated through combined stress-strength interferences (SSIs) with degrading analysis discussed by Potdar in 2013. Why air circuit breaker fails and determining what factors cause the failure by using the appropriate tool is vital for the reliability of ACB (Rane & Narvel, 2016).

When managing multistate systems, e.g. intricate equipment or complex structures, multiple barriers develop in the context of the system's complexity and constantly changing operational conditions. The interconnectivity of the multistate system easily generates many possibilities of breakdown and requirements for repair and maintenance. Therefore, regular and quality inspection and assessment are the keys. To cope with the many complexities, it is clear that a proper grasp of the ways the system functions in different states is essential. The system implementation is influenced by external circumstances, internal component relations, and unplanned operational variables. As a result, the predictability of failures and repair maintenance is decreased, making the planning problem even more complex. If we stick with the same conventional procedures, we can be faced with huge expenditures, and while our money is spent on just some ineffective cases, our operational costs again increase. It is a factor of huge importance that is done by maximizing the Life Cycle Cost (LCC) to ensure the maintenance ratio and accurate availability of the system. One of the situations that can't be predicted and leave unexpected failures is the unplanned downtime that interferes with the flow of production, which leads to mass production losses, safety concerns, and dissatisfied customers. Any downtime during maintenance decreases system availability. Therefore, the efficient maintenance plans are a major thing. Multistate systems regularly operate in settings that provide the necessary controls of safety, environment, and internal performance; their performance therefore is measured in terms of compliance to the regulations. Failure to use devices is not only a criminal problem, but it can also result in consequences such as legal prosecution, possible damage to reputation, and misadventure. RCM provides a tactical approach to combating these challenges by consecutively ordering maintenance tasks according to their respective priorities, risk levels, and cost-effectiveness. The reliability of multipartite systems is assessed by assessing the probability of failure in the case of a transition between two working conditions and the ensuing failure modes. The indexes, such as the Mean Time Between Failures (MTBF), will not be able to reveal the "down-to-the-heart" dependencies of multistate systems that stem normally. Programs of complex systems maintenance should be designed to deal with unforeseen operational circumstances and the emergence of possible failure modes of the system. For instance, one approach may be to implement adaptive condition-based maintenance strategies that constantly monitor the state of components and instigate appropriate maintenance based on given thresholds or predictive artificial intelligence. Among other factors, the proper assessment of the situations/operations in which risks are linked to certain failures/malfunctions has to be done to determine the way of carrying out maintenance activities and resource allocation. FMEA (failure mode and effects analysis) can reflect the actual failure trends and which mitigation they should take. It is necessary to mix strength and adaptability to cope with unplanned hicks and conductions. This might involve overhauling the system of duplication and establishing a method to diagnose and isolate faults, to ensure that the cutting power is resumed as early as possible. The Year III redesign of the SICs for ACBs has proven their superior reliability (10.2%) as well as cost savings (4.18 gHLP) compared to the older version (Rane et al., 2017). This work demonstrates the significance of reliability characteristics and the absence of test reports in making MCCBs and their application. It shows a way in which the Boolean logic and the ALT concept are used for the study, thereby improving the overall reliability of the MCCB mechanism from 72.44% to 95.06% as a result of proposed areas of improvements in the manufacturing (Rane et al., 2019). Summing up, reliability engineering and maintenance management are challenged by certain special features of systems with a multistate nature such as transition between operational states and system degrees of freedom.

The need for effective maintenance activities would be based on the ability to constantly adapt due to the changing parameters of system performance. Therefore, performance, reliability, and safety should result in these systems with ongoing performance. State behavior is in a multilevel way; toward the system reliability and maintenance strategies, positive impacts are very needed and they are thought all deeply in engineering design and maintenance strategy creation. A broader reliability analysis is performed if not only a few but also multiple operational states and failure modes are present in a multistate system. An alternative approach called for conventional metrics of reliability that could be insufficient to encompass the volatile ways of such systems. These approaches, for instance Markov models, fault tree analysis (FTA), and reliability block diagrams (RBDs), are numerous, and system reliability is usually evaluated with a total study.

The basic FTA approach for multistate systems is shown in Figure 12.1 by including AND gate and OR gate, which represents top event (T), intermediate event (I), and basic events (B) based on logic of system level, subsystem level, and component level. In FTA, if failure probability of basic events is known, then by using Boolean algebra logics the mathematical equation can be derived to estimate failure probability of top event. The failure probability equations for intermediate events (Eqs. (12.1) and (12.2)) and top event (Eq. (12.3)) are as follows. Let, P (B1), P (B2), P (B3), P (B4) are the failure probabilities of basic events; P (I1), P (I2), and P (T) are the failure probabilities of intermediate events and top event, respectively; the reliability of multistate systems (R_{ms}) (Eq. (12.4)) can be calculated based on failure probability of top event.

$$P(I1) = P(B1) + P(B2) \tag{12.1}$$

$$P(I2) = P(B3) \times P(B4) \tag{12.2}$$

$$P(T) = P(I1) + P(I2) \tag{12.3}$$

$$R_{ms} = 1 - P(T) \tag{12.4}$$

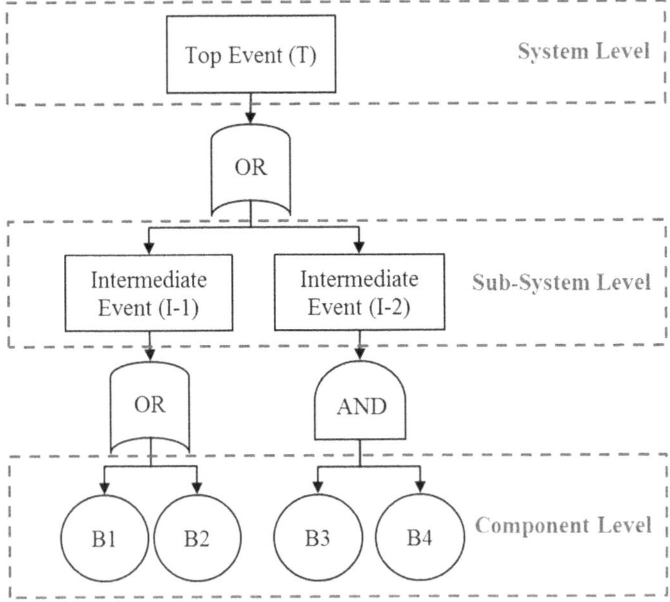

Figure 12.1 Fault tree analysis for multistate systems.

Similarly RBD can be used for reliability calculation of multistate systems by using same logic discussed in the above equation. Machinery maintenance protocols for multistate systems should be flexible enough to adjust to periodic changes in operation and the probable modes of failure. Conventional maintenance plans may have problems because the required maintenance does not conform to the one-time event needed to fix problems that occur based on the system's condition. The overall performance of these maintenance methods largely centers on the continuous or continued observation of the health and operation of a system, and this comes in handy for circumstances where there may be multiple states of the particular system. Through a technique like predictive analytics and the use of machine learning in predictive maintenance, companies can predict falling-outs and run preventative activities on schedule. Coalition detects higher risk in different operational states and failure modes compared with stand-alone systems. This generates the need to decide and concentrate on those resources and exertion based on the criticality and consequences of failure which is the goal of a maintenance strategy. Risk-based maintenance methodologies such as FMEA and risk priority number (RPN) analysis allow for focused mapping of high-risk areas followed by the creation of a budget to be used for maintenance purposes. In multistate systems, there are often built-in elements of duplication and tolerance to errors that result in limited effect of the single component failures. Redundant elements or modules can be a backup in terms of providing a backup operation of the system when it gets out-of-order. The methods of redundant system maintenance should be choosing the redundancy configuration, the reliability of redundant components, and the parallel-to-cost relationship. Forecasting and handling multistate systems by their parts consists of abruptly predicting and preventing unwanted failure and disturbances. Maintenance methods would consider the switch in standby mode in the event of a swift recovery for system restoration. On the other hand, a robust design model centered on certain principles such as module design, standardization, and flexible interchanges enhances resilience by smoothing out repair activities. Last, transit between the different states of the systems can be responsible for a change in reliability and maintenance strategies because it is complex, non-stationary, and unreliable. Maintaining a multistate system requires the active implementation of risk-based techniques, which quite comprehensively analyze the operational state of the system, failure modes of the system, criticality of the system, and many other points as well. The desirability function approach combined with the Taguchi method improves the management of preventive maintenance for machinery and it usually results in equipment utilization and mean time-to-repair to be improved. My work is dedicated to the development of a regression model to correlate the input variables with the overall desirability value (Rane et al., 2021). By drawing up a maintenance strategy supported by relevant computer systems, entities can keep multistate systems efficient and cost-effective with less downtime.

12.4 RELIABILITY-CENTERED MAINTENANCE TO MULTISTATE SYSTEMS

To guarantee the RCM of multistate systems, use a step-by-step system that includes major data gathering, analysis, and decisions (Figure 12.2). Setting about an RCM system, especially in the case of multistate systems, implies a systematic approach of incorporating stage reviews and indicators such as KPIVs and KPOVs to measure performance. First, establish clear milestones and boundaries between system understanding and set objectives, while identifying key milestones for values of information (KPIV) which are system complexity and key position outcomes (KPOV) which are system reliability, goals will get aligned to organizational goals. Subsequent measures (e.g. data collection and critical component identification) are interspersed with Necessary Factors (NFs) like tollgates consisting of several control points to check for completeness and correctness, while KPIVs (e.g. frequency of failures) are used to make the selection of critical components. Failure mode validation and estimating effects are accomplished by evaluating against predetermined thresholds during gate reviews, e.g. failure detection rate as KPIV which demonstrates the accuracy of the process. The maintenance strategies' implementation process is measured against the predefined cost-effectiveness criteria. Key performance indicators, like maintenance costs tracking the effectiveness of the plan, are determined during tollgate assessment to monitor progress.

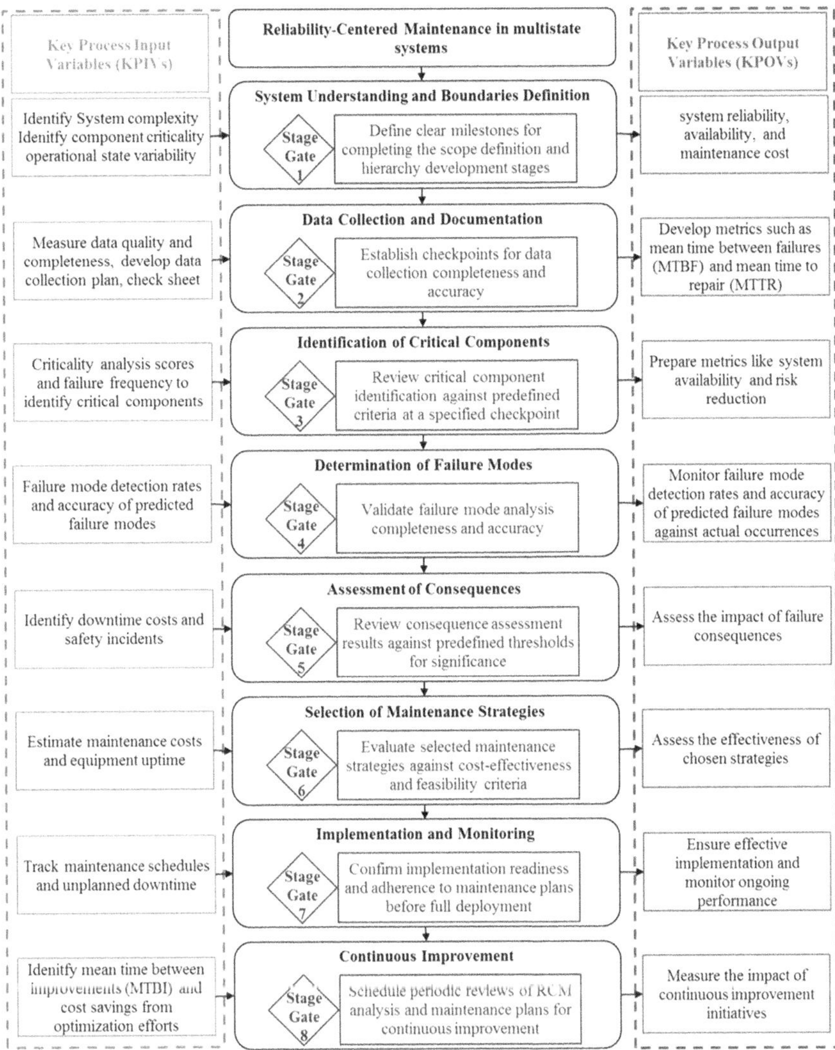

Figure 12.2 Systematic methodology for reliability-centered maintenance (RCM) to multistate systems.

Implementation is broken down into two parts which are tollgate assessment of readiness and plans, which are completed with key indicators allowing to view ongoing performance like schedule adherence. The desire for continuous improvement is motivated through tollgate reviews frequently informed by KPIVs, like mean time between improvements, as such the RCM process is constantly dynamic and aimed at meeting the organizational objectives, thereby increasing system reliability and performance through the application of data-driven decision-making and focused improvement accountability initiatives. Below is an in-depth approach for putting RCM into practice in systems with multiple states:

1. **System Understanding and Boundaries Definition:**
   a. Clearly define the parameters and extent of the multistate system that is being studied.
   b. Identify the crucial roles, working circumstances, and linkages between the various parts.
   c. To classify parts and subsystems based on their relationships and functions, create a system hierarchy or functional breakdown structure.
2. **Data Collection and Documentation:**
   a. Gather a wealth of data on the system's performance, maintenance history, design, and operation.
   b. Document system configurations, state of operation, issues, and results by consulting technical manuals, maintenance records, historical data, and professional opinions.
   c. List the existing maintenance procedures and practices, such as inventory management for spare parts, condition monitoring programs, and scheduled preventive maintenance.
3. **Identification of Critical Components:**
   a. Applying your comprehension of the system and risk assessment methods, specify critical components of the system that have a major effect on costs, security, or efficiency.
   b. Functional relevance, frequency of failure, failure effect, and compliance with the regulations criteria must be considered while evaluating criticality.
4. **Determination of Failure Modes:**
   a. Break down every element that is vital to determine probable failure modes and processes.
   b. Keep in mind that failure scenarios can be triggered randomly due to different operational states and general environmental conditions.
   c. For identification of failure modes that occur more frequently, use FMEA, FTA, and historical data analysis methods.
5. **Assessment of Consequences:**
   a. Carefully study each part separately to develop failure mechanisms as well as operational procedures.
   b. Remember that fault modes appear in various forms depending on the phases of the operation and the operating environment.
   c. Application of FMEA, FTA, and historical data analysis will let you figure out the probable failure modes along with their associated probabilities.
6. **Selection of Maintenance Strategies:**
   a. Concerning failure options consider those significant and, with more effect, plan the corresponding maintenance strategies for the elimination or failure reduction.

   b. Consider incorporating all the proactive, predictive, as well as preventive maintenance techniques, which are suitable to the individual circumstances and features of the multistate system.
   c. For the most efficient use of resources, one must critically examine and match the capital requirement, discreteness, and cost implication of each approach.
7. **Implementation and Monitoring:**
   a. Be particular about the plans and schedules that you set, which should be founded on the objectives and goals set.
   b. Take care of preventive maintenance tasks according to the plan of action and standard protocols of the working industry that meet professional standards and applicable codes.
   c. Follow up on system performance closely and collect data related to routine maintenance to monitor the maintenance schedule's effectiveness and to pinpoint the potential to better them.
8. **Continuous Improvement:**
   a. Review the RCM analysis as frequently as possible for new computing of data, the changed state of the system, and knowledge collected via inspection procedures.
   b. In the long run, pay attention to feedback from the operators, maintenance staff, and reliability engineers to promote the maintenance program, target resource deployment more accurately, and increase systems effectiveness and reliability.

Organizations may apply the RCM system principles for solving such problems in a sequential way for multistate systems so that the systems being intricate, dynamic, and even safe will be ensured. Reduction in downtime and costs of maintenance will therefore result in the end. Progressive tools and methodologies need to be developed to elucidate the multistate behavior, address the key maintenance issues, and identify the best maintenance schedules in complicated systems. The presentation of how modeled systems can be viewed at different operational states that change from one another is given by state-based modeling. Specifically, Markov models are viable to codesign the probability of transitions from distinct states of a process as time goes on. Finite State Machines (FSMs) serve as an alternative form of system representation and show system behavior as a set of statuses, changes, and actions via different states. It is an extremely useful tool as well as offering valuable information on interactions between components and how the functioning of a system as a whole is affected by these elements. RBDs are usually made up of rectangles and arrows, each rectangle representing a piece or subsystem and each arrow representing a relation between two blocks. RBDs represent the reviews of routes as well as advising on the ways to optimize all the ramifications of working perfectly in different operational conditions. FMEA is a structured method of failure data sorting and ranking according to certain criteria – the probability of occurrence and seriousness and being found or discovered, respectively (likelihood, detectability). In the case of multistate FMEA, multiple states of operations are considered and the consequences of each FMEA in these states are being evaluated. The criticality analysis helps in determining the sequence of changing the components by determining the degree of criticality of assets or failure modes concerning system performance, reliability, and cost. Apart from that, working with the customer data platform approach is important as it helps to concentrate on the most urgent maintenance tasks, and the RPN assessment method enables us to rank the priority of repair work. The purpose is to achieve efficiency in the operation of the system development timetable, possibly restricting the work by following priorities and ensuring the system's reliability. The modeling process aims to construct a variety of solution methods making use of mathematical programming, simulation, and heuristic algorithms. Heuristic strategies that rely on simulated annealing, particle swarm optimization, and genetic algorithms can lead to the globally optimal solution of maintenance problems which got multiple solutions. In contrast to periodical-based maintenance, which only engages monitoring for specific time intervals, condition-based monitoring (CBM) monitors performance and health nodes dynamically and activates maintenance because of the condition of the system. This type of maintenance brings about the centered spare parts needs without inactiveness. The benefits are as follows: they enhance uptime, reduce the waste of part life, and ultimately lead to better utilization of maintenance. Analysis of the historical data as well as the sensor readings through the application of forecasting techniques and machine learning algorithms helps in preparation for the future while ensuring that mistakes are correctly remedied. Perfected maintenance procedures by the timely and accurate maintenance programs using these technologies are yet another advantage that these technologies which can produce the patterns and trends in enormous data sizes are capable. They will be envisaging two strategies (reliability performance and low-cost maintenance without data loss) and that will help the organization to fight against these two major problems. Second, they assist with the planning and finish the maintenance and program requirements and they schedule the number of the maintenance works.

12.5 CASE STUDIES AND EXAMPLES

This section incorporates real-time or illustrative instances where RCM has been applied in various multistate systems, including manufacturing facilities, energy systems, and transportation systems, as well as healthcare.

12.5.1 Manufacturing sector

As instability in the car industry changes occur without prior notice which may cause great damage to the entire factory, RCM at big factories can be considered to be a wonder and has been mainly applied in some countries. Actually, with the problem of operation of the hard-to-manage production line equipment that needs to be continuously maintained, the plant management considers the case as a matter of high priority and attempts to elevate the efficiency by introducing the strategic maintenance approach. This led to the creation and implementation of RCM, which is tailored for optimizing the maintenance intention through the understanding of the function of the component and failure types. The process of RCM began with the examination of the equipment in the production line in detail. Each machine was attentively reviewed to recognize the significant parts and dare the mind to think of breaking and interfering with production. This exam was based on a contribution from maintenance personnel, engineers, and operators who had skills and long-term experience in frontline operations with machine behaviors they interacted with. The team was able to accomplish this by taking a unified approach to attaining the information which would enable them to clearly understand the machine's reliability positioning. Thanks to this, informed decisions on the schedule of maintenance work could be made. There was a task that followed the recognition of the critical parts and possible breakdowns: a market-tailored cleaning plan was developed for each asset according to its needs. Rather than following in a footstep of some sort of what-if plans or reactive approach, RCM addresses such risks via appropriate evaluation of the situations and their potential consequences. The plant's main focus was on base activity solving the arising challenges before they became unnecessary, long-term, and accumulating eventualities by establishing the connection between transferred maintenance activities and failure modes of certain components and their criticality. The key thing with the RCM implementation was that it enlightened us on the relevance of including frontline workers in the routine maintenance operations of a plant. Having directly worked with the machines, they are well-versed in what the original operating state was, how it is changing, and what early indicators might indicate impending failures. Utilizing shop designs' incorporated informal knowledge by frontline operators has enabled the plant to harness their experiential knowledge and generated a feeling of agency and responsibility within the employees. Thus, this partnership not only created the improved quality of the data collected but also recognized and approved the new tactics in comparison. The impact of RCM (root cause analysis (RCA)) on the ease of production in the facility is highly notable. The reduction of 30% of such failures will reduce production losses and, thus, lead to better productivity and better net incomes. Moreover, the striking increase of 15% in overall equipment efficiency (OEE) also showed a substantial boost in production resource utilization efficiency. The productivity indicators precisely illustrated what RCM is all about, to achieve maximum effect in the maintenance-based procedures and manufacturing outputs. Indeed, the successful implementation of RCM in a light vehicle factory portrays the outturns of adapting the proactive maintenance approaches which result in improved reliability, productivity, and profitability. By learning via critical faults, hiring onsite operators, and developing novel maintenance routines, this plant not only halted unpredicted downtimes occurring but also created a system of continuous improvement and collective spirit. This case study is a perfect example of the efficacy of RCM since it has proven to be important for the advancement of not only multiple industries but also states in different businesses.

12.5.2 Transportation sector

In order to maintain the required level of safety and to make the operations more efficient, the airline has moved to another kind of maintenance which is named RCM. The airline performed as many checks as possible to cover all the necessary flight systems and parts required for flight safety in the know that preventive maintenance does play a pivotal role in the airline's operations. The experts may perform a thorough analysis and easily sort out the sites that are a high risk of collapse. Consequently, they will work out the maintenance schedule for these sites and include periodic inspections, replacement of faulty parts and other routine actions along with the use of the latest monitoring techniques. Airliner elected to give RCM maintenance winged into its schedules after optimization of its consumption for this component as a care and health tracker. Making use of these sensor data being provided to the aircraft concurrently, they can also keep an eye on the other parameters and do the forecasting so that adverse situations can be avoided as well. Thanks to thorough inspections, they managed to minimize the possibility of in-flight incidents and also to increase the reliability of received planes. This grew the trust among passengers and stakeholders as well. Technology and analytics data integration into their work processes constituted an essential element in the RCM initiative accomplishment. The latest-generation sensor technological indicators provided real-time monitoring of crucial factors; thus, responding quicker to maintenance and repair needs was much more efficient. Furthermore, maintenance crews could be aware of and deal with precursors to observed problems utilizing predictive analytics and data visualization. This leads to the deterrence of breakdowns and makes the operations of airline operations proceed without any break to keep it uninterrupted. In this case, it can be said that the passenger airlines by RCM show that preventive maintenance strategies are slowly laying a foundation for the future development of the industry. Through the usage of real-time data and predictive modeling, the airline experienced reduced operational risks and brainstormed efforts to improve flight safety and fleet availability. This study provides a case in point that the use of technology to conduct maintenance is essential for (a) the protection of assets; (b) higher operational performance; and (c) better experience of the passengers in the aviation activities.

12.5.3 Healthcare sector

Regarding patient safety, the healthcare sector is of the highest level, and the downtime can be life-threatening too. Therefore, RCM plays a crucial strategic role in a large hospital by hearing this. Since the management of various medical devices and facilities was immense i.e. responsible for the performance of many clinics, the hospital administration considered regular preventative maintenance as necessary to eliminate the chances of any glitch that could affect the operation of both clinical activities and health of the patients. Such a decision directed the implantation of the RCM method, which is a systematic way to improve plans of maintenance concerning risk assessment and experience of past maintenance data. The start of the RCM implementation process was to use the risk assessment to identify possible risks related to the medical equipment and facilities at the hospital. It refers to determining the worthiness of the assets measured in giving care to patients, meeting regulations, and ensuring the running of the operations. An analysis of past repairs and failure bulletins helped hospital managers identify assets with especially high additional failure rates or those that can break down with catastrophic consequences. Such a decision helped with the construction of a maintenance plan for each asset considering its specific requirements at the personalized level. The placement of RCM along with the fall in equipment failure during professional surgeries was responsible for keeping patients safe and facilitating the care provision. Overcoming this challenge could be done by emphasizing the maintenance during the criticality of assets and the potential risk of failure that could result in disruption of the clinical process or the sabotage of results. This method had a multi-functional role; from operational resilience improvement, and morale-boosting in the staff to an increase in patient confidence. Moreover, the implementation of RCM had a positive effect on the distribution of resources in the hospital's third department responsible for maintaining infrastructure. The hospital may not only be able to increase the operational efficiency and limit the expenditure but also concentrate on assets that would have the greatest influence on lifesaving operations. This resulted in the more planned and aimed use of maintenance engineers, spare parts as well as diagnostic tools, thus bolstering the continuity and performance of assets. One of the most important lessons I learned regarding RCM implementation is that all clinical staff (physicians, nurses, technicians, and medical records specialists), biomedical experts, and maintenance team workers must closely cooperate and ensure that mechanical systems are functioning correctly. Recognizing that maintenance problems cannot be ignored at a service operation center where saving lives is the highest priority, the hospital had personnel from different departments collaborate to ensure efficient communication channels were in place. The hospital was able to achieve the alignment of maintenance work with patient care needs, and clinical workflow as it did well in ensuring that maintenance tasks were as timely as possible by including clinical staff in the RCM process to help seek their input on the criticality of assets and the prioritization. Taken together, a grand view of RCM at the major hospital emphasizes an overall good contribution of the preventive maintenance approaches in healthcare services. With the risk analysis, records analysis, and the trust of various stakeholders, the hospital achieved a recorded reduction in equipment breakdowns, an improvement in patient safety, and the effective utilization of its resources. The discussed case example illustrates the importance, of which interventions based on RCM can bring about the improvements of reliability, efficiency, and quality of care for healthcare systems.

12.5.4 Energy sector

Operational efficiency and reliability are of paramount importance for the energy sector, that is to say, the ability to service the demand while securing profits. By using RCM models, the power generation company was able to dramatically cut maintenance costs and at the same time raise the efficiency of its systems. With a focus on keeping the operation of turbines dependable, reducing downtime, and maximizing energy output, the company understood a preventive plan with individualized maintenance requirements for this particular asset. Along with this, RCM, an advanced approach navigating toward optimal maintenance measures that rely on parts reliability assessments, was introduced. As the most critical turbine components, namely bearings, blades, and control systems, were thoroughly evaluated before the use of RCM, the reliability ascertainment was made enough. Through the in-depth analysis of the historical data and timing patterns, the organization found certain mechanisms and maintenance requirements for each component. The company that could use monitoring equipment to know how these components behave and why they often fail will consequently manage to adjust its maintenance plans and practices, thereby decreasing downtime and optimizing energy production. The firm utilizes different monitoring techniques, including vibration and oil analysis as the main tools. Thanks to that, the organization was able to detect initial symptoms of decreasing equipment (e.g. increased vibration levels and degraded oil samples, which are representative signs of component condition and function) by regular monitoring. The company's promptness was essential for preventing breakdowns and saving money by acting rapidly to mend problems, like lubricating or replacing components. The power production company went for CBM as well as the RCA approach, which was quite impressive. Through reduced maintenance costs and downtime losses stemming from implementing improved maintenance intervals and techniques based on component reliability analysis, the enterprise becomes inseparably richer. In the meantime, recognition of problems early and the solution to them resulted in wind turbines producing more energy and the income from these turbines uninterruptedly. In conclusion, the effective adoption of RCM shows that the maintenance approaches becoming more advanced and progressive are now affecting the power generating industry. Through fault detection, condition monitoring, and establishing revised maintenance programs on the asset level, expenses were slashed, higher efficiency was achieved, and reliability was improved. The lesson of the case study about RCM implementation for better productivity and profitability in the energy-generating sector is that reliability, cost savings, maintenance, and safety are the key components to success.

12.6 INTEGRATION WITH CONDITION-BASED MAINTENANCE (CBM) AND PREDICTIVE ANALYTICS

The RCM process is a systematic approach to outline key factors of operational conditions and come up with the utmost effective techniques to ensure systems' 100% efficiency and reliability. Despite this, however, applying RCM technology in conjunction with CBM and structured analytics techniques may ensure the best outcomes of maintenance and reliability forecasting for such systems. Respectively, such integration offers positive implications for companies to be able to manage assets in real time and make data-informed decisions. CBM comprises the use of sensors among various methods to sensitively monitor the assets functioning all the time and parent it for early detection of degradation or malfunction. Through this blend of RCM and CBM, organizations can make maintenance more efficient because the merger of on-hand data from monitoring asset health into the management system facilitates informed decisions. The RCM approach is the process of finding key parts and choosing the necessary actions for maintenance based on how crucial each item is and how it may lose functionality. CBM and RCM are the side-by-side functions that create the supply of real-time asset-condition data, where maintenance teams can exponentially detect anomalies timely and think about a system disaster before it occurs. Organizations will be able to do this by pairing it with CBM which will help them move away from the old practice of time-based plans in which companies plan maintenance cases based on time to a sophisticated condition-based method which will not only boost efficiency but will also lower the possibility of unanticipated machine breakdowns. The prediction and forecast make use of mathematical algorithms and artificial intelligence to go through the past results before using the data to prepare for that same future. The maintenance problem is solved if the RCM and CBM are coupled with predictive analytics. A sitting forward look can be used to scrutinize signals from CBM systems and spot patterns or trends that ferret out the fourth time of failure. Through using past maintenance records and fixed asset metrics, predictive logic programs can move forward to estimate future failure circumstances and propose proactive maintenance solutions. With the combination of predictive analysis with RCM and CBM, organizations can shift to predictive maintenance which is predicting the failure probability at which the assets' health will degrade and maintenance tasks are based on scheduled intervals of assets' maintenance preventive task implementation rather than responding to them after they have occurred. Utilization of IoT sensors, cloud analytics, and digital twins delivers extra data and analysis capabilities which will always enable better informed decision-making for maintenance tasks when these factors are working with RCM. IoT sensors facilitate real-time plant asset monitoring that allows for identifying problems and weak signs in advance. By harnessing powerful cloud-based analytics platforms, a more complex set of data analytics features can be used, so that data management becomes easier and users will be able to observe trends that reveal the conditions and performance of the plant. With digital twinning, virtual replicas of real things are being created. The business can use it for modeling different operations and thereby predict the impact on the asset's performance if a maintenance intervention does or does not take place. Through the implementation of the above advanced monitoring technologies, companies can achieve an improvement in the service level of RCM-based maintenance plans which may also result in a change of the timetables for maintenance work, a reduction of downtime, and an increase of asset reliability and performance. Moreover, the development of these systems brings organizations the ability to make smarter maintenance decisions based on real-time asset health data and predictive analytics forecasts that act as the powerful foundations of data-driven decision-making. Ultimately, merging RCM with CBM, predictive analytics, and highly advanced monitoring systems is key since it is only in this way that organizations are able more systematically to oversee asset health and to schedule maintenance with better intelligence. The adaptations in fleet management can be employed using the data of live assets condition, output of sophisticated analytics, and advanced monitoring technologies that can reshape firms/companies' maintenance plans, decrease idle time, and increase asset reliability and performance in complex systems.

12.7 CHALLENGES AND FUTURE DIRECTIONS

Whether it is the organizational limits, the inaccessible models, or the data restriction, implementing RCM for multistate and city systems will come with a lot of limitations and problems. To address these obstacles, in addition to creative approaches to therapy, the search for novel research chains is required to increase the results of therapy in the management of more complicated systems that include spaceflight. The following are some of the difficulties and restrictions related to RCM for multistate systems:

1. **Availability of Data:** One of the most important problems in the process of RCM implementation is the provision of a large volume of accurate and sufficient data for reliability analysis of multistate systems. Multiple-state systems, intertwined to a certain extent, have then one's own so dataset collecting and analyzing is complex and also hard to resource.
2. **Complexity of Model:** Interstate systems entail intricate interconnections and linkages between constituent entities, and thus, they are prone to increase in level of model intricacy. It is difficult to model the reliability of complex networks by using dependency techniques that require complex mathematical calculations and might need huge computing resources.

3. **Organizational Barriers:** Organizational heights like a miss of managerial assistance, inadequacy of resources, and reluctance to change can be the main reasons for resistance to materializing the RCM in companies. Ensuring expeditious organizational approval, drawing public participation, and communicating well are all crucial for dealing with the issues at hand.

12.8 FUTURE RESEARCH DIRECTIONS AND EMERGING TRENDS

1. **Enhancement of Data Analysis:** Research directed at the RCM methodology to overcome issues related to parameters' accessibility should lay great emphasis on applying advanced data analysis techniques such as machine learning and artificial intelligence. Such methodological techniques make possible an effective analysis of huge datasets and produce vital findings in support of multistate system operations as well as criticality analysis.
2. **Integration with Digital Twins:** Addressing the underlying causes of data accessibility in future studies is one of the main priorities of research on RCM methodology that would be dealing with machine learning and AI techniques implementation to the greatest extent. These methods can be applied for the analysis of massive datasets and also achieve beneficial information that can present sound conclusions on multistate systems and dependability analysis.
3. **Prognostics and Health Management (PHM):** Just-in-time information, along with real-time condition monitoring, allows the PHM to assess and manage a remaining useful life and component health. Following the PHM technologies, current research projects will be aimed at integrating them with the RCM tactics. The combination of such technologies will promote resource allocation necessary for complex systems and enable the authorities to make proactive decisions regarding maintenance activities.
4. **Human Factors and Organizational Culture:** Organizational culture and human factors as research areas are essential and beneficial for overcoming barriers to RCM adoption in organizations. Future studies might focus on creating an organizational environment that has evenly practically focusing on the safety and upkeep of staff and resources, as well as developing correspondent tools and templates for positive engagement with stakeholders and change management.
5. **Resilience Engineering:** Fostering system resilience by using a comprehensive methodology to improve system performance and reliability is what characterizes resilience engineering methods in multistate systems that are continuously growing more complex and interconnected.

Specifically, we can use resilience systems engineering concepts along with RCM-based methods to come up with better approaches for reliability in the face of unknown problems and unexpected disruption.

Briefly, the implementation of RCM for multistate systems is complicated by the fact that data accessibility is limited, models are complex, and organizational barriers exist. Up against these obstacles, new research ideas in the high level of data processing, digital replica integration, health supervision and predictive maintenance, human factor and workplace culture as well as robust engineering should be introduced. By way of accepting new strategies and considering a systematic approach, companies can be able to improve the effectiveness of RCM techniques used in managing complex systems for utmost reliability and efficiency.

12.9 CONCLUSION

Finally, the chapter on RCM for systems in various states has delivered summaries on how preventive maintenance programs are used to increase the availability of the system, its efficiency, and reliability. A number of deserving issues have been exposed in the course of the discussion which signifies the relevance of RCM and its play in technical and operational settings today. One of the major advantages of RCM is an organized approach to the efficient strategy of maintenance through determining the critical parts, defining possible failures, and devising personalized maintenance schedules as a result of risk analysis. Through RCM, industries can in a rational way allocate resources, reduce downtime, and hence, increase asset reliability by doing maintenance works based on their probability and relevance. Besides, for multistate systems, the combination of RCM with CBM and predictive analytics strengthens the reliability and performance assessment. Current asset status with possible future projections is suitable for replacing old maintenance methods with predictive maintenance through using data. Identifying maintenance jobs based on the anticipated state of the asset and its predisposition to fail. With RCM and smart monitoring technologies enabling the ability to proactively make maintenance decisions, condition monitoring becomes possible with the combination of digital twins, cloud-based analytics platforms, and the Internet of Things sensors. Via the implementation of complex data analytics, simulation tools, and on-time monitoring, these technologies enable enterprises to increase productivity as well as reduce downtime and make the optimal maintenance methods. On the other hand, for an RCM in multiple-state systems, there are some issues as well. Getting through RCM profitability and making it work includes dealing with challenges like models, data availability, and organizational impediments. To overcome these challenges businesses can be creative and come up with strategies like the integration of digital twins and advanced data analytics, top down reliability cultures. What is more, future developments in RCM approaches and the emergence of new research can make maintenance procedures more proactive through the implementation of unforeseen improvements. Data analytics, resilience engineering, PHM, human factors, and the usage of digital twins are the main areas where RCM might be improved, and further innovations may be created. To sum up, RCM is one of the essentials for planning which increases the availability, performance, and reliability of the compounds of a complex system. Even if the role of RCM is still important as the utilities have to adapt to the fast-changing techniques and operations, the techniques and principles of RCM will still be utilized. Organizations can fully exploit RCM and get practical benefits in the fields of operational excellence, efficiency, and reliability by means of implementing continuous improvement and with the help of creativity.

REFERENCES

Potdar, P., 2013. Reliability analysis of moulded case circuit breaker mechanism based on stress strength interference with degradation analysis. Int J Eng Sci Innov Technol IJESIT 2, 26–34.

Rane, S.B., Narvel, Y.A.M., 2016. Reliability assessment and improvement of air circuit breaker (ACB) mechanism by identifying and eliminating the root causes. Int J Syst Assur Eng Manag 7, 305–321. https://doi.org/10.1007/s13198-015-0405-z

Rane, S.B., Narvel, Y.A.M., Khatua, N., 2017. Development of mechanism for mounting secondary isolating contacts (SICs) in air circuit breakers (ACBs) with high operational reliability. Int J Syst Assur Eng Manag 8, 1816–1831. https://doi.org/10.1007/s13198-017-0678-5

Rane, S., Pai, R., Pai, A., Rane, S.B., 2021. 13 – Multiresponse maintenance modeling using desirability function and Taguchi methods, in: Pham, H., Ram, M. (Eds.), Safety and Reliability Modeling and Its Applications, Advances in Reliability Science. Elsevier, pp. 353–372. https://doi.org/10.1016/B978-0-12-823323-8.00013-1

Rane, S.B., Potdar, P.R., Rane, S., 2019. Accelerated life testing for reliability improvement: a case study on moulded case circuit breaker (MCCB) mechanism. Int J Syst Assur Eng Manage 10, 1668–1690. https://doi.org/10.1007/s13198-019-00914-6

Index

K

L

M

N

O

P

Q

R

S

T

U

W

For Product Safety Concerns and Information please contact our EU representative GPSR@taylorandfrancis.com Taylor & Francis Verlag GmbH, Kaufingerstraße 24, 80331 München, Germany

Batch number: 10397790

Printed by Printforce, the Netherlands